Text Mining with MATLAB®

Rafael E. Banchs

Text Mining
with MATLAB®

Second Edition

 Springer

Rafael E. Banchs
Mountain View, CA, USA

ISBN 978-3-030-87694-4   ISBN 978-3-030-87695-1   (eBook)
https://doi.org/10.1007/978-3-030-87695-1

This Springer imprint is published by the registered company Springer Nature Switzerland AG
The registered company address is: Gewerbestrasse 11, 6330 Cham, Switzerland

Preface

This book is the result of a multidisciplinary journey during the last few years of my career. As an Electrical Engineer, who did his dissertation on Electromagnetic Field Theory, I found myself surprised about writing a book on Text Mining! It was in 2004, when I first got involved in natural language processing research, at the TALP research centre of Univesitat Politècnica the Catalunya, in Barcelona. There, I participated in the European Project TC-STAR, which focused on the problem of speech-to-speech translation. Later, at Barcelona Media Innovation Centre, I got the opportunity to further explore other natural language applications and problems such as information retrieval and sentiment analysis. Finally, during my years at the Institute for Infocomm Research, in Singapore, I had the opportunity to work on knowledge representation, question answering and dialogue.

In my experience as Electrical Engineer, the MATLAB® programming platform has always been an excellent tool for conducting experimental research and proof of concepts, as well as for implementing prototypes and applications. However, in the natural language processing community, with the exception of a few machine learning practitioners that have entered in the community via text mining applications, there is not a well-established culture of using the MATLAB® platform. As a technical computing software that specializes in operating with matrices and vectors, MATLAB® offers and excellent framework for text mining and natural language processing research and development.

This book has been written with two objectives in mind. It aims at opening the doors of natural language research and applications to MATLAB® users from other disciplines, as well as introducing the new practitioners in the field to the many possibilities offered by the MATLAB® programming platform. The book has been conceived as an introductory book, which should be easy to follow and digest. All examples and figures presented in the book can be reproduced by following the same procedures described for each case.

Finally, I would like to thank all the persons that have encouraged and helped me to make this project come to life. Special thanks to the MATLAB® Book Program and the Springer Editorial teams for all the support provided. Thanks also to my colleagues who have helped reviewing the different chapters of the book. And special thanks to my wife, my children and my parents for their support and their patience!

Rafael E. Banchs
Mountain View, California, August 2021

Table of Contents

1 Introduction

The universality and ubiquity of the Internet in the current information society has changed human life in many different ways. One important element of this change is the possibility of accessing a virtually infinite amount of information in digital text format. Consequently, the text-oriented derivation of data mining, text mining, has been gaining attention as the available volume of text information grows at a rate that is by far higher than our human capacity to handle and process such a huge volume of information.

This book introduces some of the fundamental concepts of text mining from an experimental perspective. It presents and illustrates all practical issues and implementations by using MATLAB® technical computing software,[1] a highly specialized programming language for numerical computing. The main contents of the book are presented at an introductory level, which should be useful for those audiences without any previous experience on using the MATLAB® programming environment or without any previous knowledge about text mining applications and techniques.

This introductory chapter is organized as follows. First, in section 1.1, a brief discussion on text mining and the suitability of the MATLAB® product for text mining applications is presented. Next, in section 1.2, more detailed information is provided about what to expect from this book and how to use it. In section 1.3, a very quick introduction to the MATLAB® programming environment is given and, in section 1.4, the Text Analytics Toolbox™ is briefly introduced.[2]

1.1 About Text Mining and MATLAB®

Data mining, also referred to as knowledge discovery in data, can be defined as "the science of extracting useful knowledge from [...] huge data repositories".[3] In accordance with this, text mining refers to such a knowledge discovery process when the source data under consideration is text.

---

[1] MATLAB® is a registered trademark of The MathWorks, Inc.

[2] https://www.mathworks.com/products/text-analytics.html Accessed 6 July 2021

[3] This definition is taken from the Curriculum Proposal of the ACM Special Interest Group on Knowledge Discovery and Data Mining, https://www.kdd.org/curriculum/view/introduction Accessed 6 July 2021

R. E. Banchs, *Text Mining with MATLAB®*, https://doi.org/10.1007/978-3-030-87695-1_1

Strictly speaking, rather than specific areas of knowledge by themselves, data mining and text mining in general should be regarded as application-oriented interdisciplinary fields. In the particular case of text mining, it can be found to be closely related to disciplines such as natural language processing, computational linguistics and information retrieval, as well as to rely on important contributions from statistics, machine learning and artificial intelligence, in general. Because of its close relationship with and dependence on these related disciplines, precise definitions of the scope of text mining and its frontiers with these other disciplines cannot be easily depicted. In this sense, the notion of text mining only becomes clear for a given technique or application under the endeavor of discovering knowledge from a large collection of text data.

Nowadays, with the increase of computational power and the access to a virtually unlimited amount of information in digital text format, text mining is becoming a very important tool for both providing competitive services to users and extracting valuable knowledge for business intelligence and marketing research applications.

As there are currently several text-oriented computational tools for text mining and data mining in general, you might be wondering why using a highly specialized numerical computing language such as the MATLAB® technical computing software product for developing and implementing text mining applications. There are actually lots of reasons for recommending its use for text mining purposes such as, for instance:

- it is a high-level application-oriented language which is also relatively easy to learn and use,
- it offers a large number of algorithms and methods, which are already programmed in the form of functions and toolboxes,
- it allows for interfacing with other programming languages such as Fortran, C++ and Java,
- it facilitates the creation of user interfaces and the generation of very high-quality graphics and plots,
- it allows for debugging and deploying standalone applications, and
- it offers the Text Analytics Toolbox™, a specialized library for processing, analyzing, and modeling text data.

Nevertheless, apart from all these reasons, there is a conceptual and fundamental reason that makes the MATLAB® technical computing software an ideal tool for text mining purposes. Its name derives from MATrix LABoratory, as it was originally conceived as a programming language for manipulating and operating with matrices. On the other hand, as you will see in chapter 9, one of the most popular ways of modeling and operating with text data collections is by means of the vector space model, in which a complete collection of documents can be represented by means of a matrix, and most of the basic language processing operations can be conducted by means of matrix and vector operations. In this way, we can actually think about the MATLAB® software as the perfect programming environment for developing, implementing and deploying text mining applications and services.

According to this, the present book is an attempt to simultaneously introduce the unfamiliar reader to the basic concepts of text mining, as well as demonstrate the main advantages of using the MATLAB® technical computing software for implementing text mining applications.

1.2 About this Book

Before getting into technical matters, let us present in more detail what is and what is not this book about, as well as provide some basic but useful tips on how to use this book.

The book is structured in three main parts: *Fundamentals*, *Mathematical Models* and *Techniques and Applications*. The first part, *Fundamentals*, is devoted to introducing basic procedures and methods for manipulating and operating with text within the MATLAB® programming environment. It comprises chapters 2 to 6, in which text variables, regular expressions, string operations, file read/write operations and some basic concepts of computational linguistics are introduced. More specifically:

- Chapter 2 focuses on the different types of variables that can be used for manipulating text, as well as it introduces some basic MATLAB® and Text Analytics Toolbox™ built in functions for operating with strings.
- Chapter 3 is completely devoted to the specificities and use of regular expressions in the MATLAB® programming environment.
- Chapter 4 focuses on basic operations with strings, such as search, replacement, segmentation, concatenation, and some basic set operations that can be applied to string and character sets.
- Chapter 5 deals with reading and writing text files and describes some commonly used file formats. Also, in chapter 5, some basic functions for operating with directories and files are presented and described.
- Chapter 6 focuses on the structure of language. Some functionalities of the Text Analytics Toolbox™ that are useful to process and analyze text data at different language structural levels are introduced in this chapter.

The second part of the book, *Mathematical Models*, is devoted to motivating, introducing, and explaining the two main paradigms of mathematical models most commonly used for representing text data: the statistical approach and the geometrical approach. It comprises chapters 7 to 10. More specifically:

- Chapter 7 introduces the main concepts related to corpus statistics. First, it presents some fundamental properties of language such as the Zipf's law of frequencies and the burstiness phenomenon. Then, it introduces the notion of word co-occurrences and the incidence of word order information.
- Chapter 8 is devoted to the statistical modeling approach. It introduces the basic n-gram model and the fundamental concepts of discounting and model interpolation. Additionally, statistical bag-of-words and topic models are also presented and discussed.
- Chapter 9 focuses on the geometrical modeling approach. It starts by introducing the concept of term-document matrix and vector space model representations. Then, the geometrical modeling notions of distance, similarity and association scores are introduced.
- Chapter 10 is devoted to the specific problem of dimensionality reduction. It presents the ideas of vocabulary pruning and merging, as well as some fundamental linear and non-linear projection methods. Additionally, it introduces the notion of the embedded representations.

The third part of the book, *Techniques and Applications*, is devoted to some general problems in text mining and natural language processing applications. More specifically, the problems of document categorization, document search, content analysis, summarization, question answering, and conversational systems are addressed in chapters 11 to 15.

- Chapter 11 focuses on the problem of document categorization. It presents and illustrates basic techniques for unsupervised clustering and supervised classification. The case of supervised classification is addressed from both, vector space and statistical modeling approaches. Also, in this chapter, basic methods for extracting terminology that is relevant to a given document category are illustrated.
- Chapter 12 focuses on the problem of document search. More specifically, the binary search and the vector-based search approaches are described and illustrated. This chapter also introduces the basic metrics of precision and recall, as well as some other fundamental concepts of information retrieval, such as query expansion and relevance ranking. Finally, the problem of cross-language search is introduced.
- Chapter 13 deals with the problem of content analysis. Although this is indeed a very broad concept, this chapter focuses on two specific types of content analysis: polarity estimation and property extraction. In the first case, the problems of detecting polarity and estimating its intensity within the context of opinionated texts is presented and discussed. In the second case, the problem of extracting properties and other specific informational elements by means of text-pattern matching is introduced and illustrated.
- Chapter 14 focuses on the problems of keyword extraction and document summarization. With regards to keyword extraction, the concept of word

centrality is presented and two standard approaches to keyword extraction are introduced. In the case of summarization, the extractive summarization approach to text summarization is introduced and discussed.

- Chapter 15 focuses on the topics of question answering and dialogue systems. In the first case, the generic problem of question answering system is introduced along with the specific problems of question understanding and intent detection. With regards to dialogue, the basic operation of a dialogue manager system is described, briefly introducing the specific problems of dialogue state tracking and response selection.

The main audience this book was conceived for are persons with very little or no previous knowledge about text mining techniques and applications who are also not familiar with the MATLAB® programming environment and its Text Analytics Toolbox™. If you belong to this group, you should be able to take maximum benefit from, as well as enjoy, all the chapters in this book.

This book might also be useful for experienced text mining practitioners who are not familiar with the MATLAB® technical computing software and its Text Analytics Toolbox™. In this case, you should focus your attention on those chapters contained in the first and second parts of the book: *Fundamentals* and *Mathematical Models*.

Similarly, this book might also be useful for experienced MATLAB® users without any previous experience in text mining applications. In this case, you should focus your attention on those chapters contained in the second and third parts of the book, *Mathematical Models* and *Techniques and Applications*. You might also need to review chapters 3 and 6, which introduce regular expressions and the structure of language, respectively.

Otherwise, if you are both familiar with the MATLAB® programming environment and its Text Analytics Toolbox™ and, also, you have previous experience with text mining: this book is not for you! You should be already acquainted with most of the materials presented along the book.

In addition to the main technical sections, each chapter in the book also contains three additional sections: *Further Reading, Proposed Exercises* and *References*. All these sections provide complementary materials aimed at reinforcing the main concepts covered by the technical sections, as well as further exploring some related concepts. The chapters in the second and third parts of the book also include an additional section entitled *Short Projects*, which proposes more broad and challenging exercises related to the problems described within the chapter.

All examples illustrated in the book are fully reproducible from the MATLAB® command window. In this sense, you should be able to follow the explanations in each section of the book and reproduce the very same results presented therein in your own computer. Most of the required data files and functions you will need to reproduce the examples along the book are available from the companion website www.textmininglab.net. For some specific examples, in which you will need to get the data by yourself, the pointers to the corresponding sources are provided within the book and in the companion website.

The specific MATLAB® version that was used in the preparation of all examples in this book is 9.9.0.1467703 (R2020b) along with Text Analytics Toolbox™ version 1.6. You might expect small differences or inconsistencies in some examples if you are using different versions, especially if you are using older ones. For more information about possible clashes among different MATLAB® versions you must consult the corresponding release notes of the product and toolbox.

All the code presented in this book has been created with two specific objectives in mind: intelligibility and demonstrativeness. In this way, all example codes are meant to be understandable and able to demonstrate the different potentialities offered by both the MATLAB® product and its Text Analytics Toolbox™, but they are not meant to be efficient! Efficiency has not been considered as a major criterion for code development in this book. Nevertheless, some important efficiency issues are noted and left to you as exercises in the *Proposed Exercises* sections of the corresponding chapters.

The examples in each chapter are totally independent from the examples in the others, so you must be able to reproduce the examples in a given chapter without the need for executing any code from previous chapters. However, this is not the case for the examples within the same chapter, as in most of the cases the results of a given example are used as inputs for the subsequent ones. In this sense, you must be acquainted with the use of MATLAB® functions `save` and `load`, which will allow you to save your working session and restore it at a later time. A brief presentation of these functions is given in the following section, while a more detailed description is provided in section 5.1.

1.3 A (very) Brief Introduction to MATLAB®

MATLAB® stands for MATrix LABoratory. It is a highly specialized programming language for numerical computing, which has been specially designed for efficiently operating with matrices. It is an interpreted language, which means that you need the MATLAB® interpreter to execute MATLAB® code. However, it also provides specific tools for creating standalone applications. Here, we will restrict our use of the MATLAB® software as an interpreter.

The first thing you need to do is to launch the MATLAB® environment. It will open a window including different elements on it. The most important ones are the *command window* and the *workspace*. The *command window* allows you to execute MATLAB® commands one at a time. The *workspace* contains and displays all the variables that are currently accessible from the *command window*.

Once you have launched MATLAB® environment, you can try reproducing the following examples from the *command window*. First, let us create a matrix:

```
>> matrix = [1 2 3;4,5,6]                                    (1.1)
matrix =

     1     2     3
     4     5     6
```

From this example you can see that creating a matrix by assigning its entry values is indeed very simple. The list of values must be given into brackets. The semicolon is used for vertical concatenation and the comma, or alternatively the white space, is used for horizontal concatenation.

Indexing operations for accessing specific elements within the matrix are also very simple and intuitive:

```
>> matrix(2,2) % retrieves the element in row 2 and column 2    (1.2a)
ans =

     5

>> matrix(2,:) % retrieves all elements in row 2              (1.2b)
ans =

     4     5     6

>> matrix(:,2) % retrieves all elements in column 2          (1.2c)
ans =

     2

     5

>> matrix(2,1:2) % retrieves the elements in row 2 and columns 1 to 2   (1.2d)
ans =

     4     5
```

Notice from the examples in (1.2) how parentheses have been used for retrieving the matrix contents. Notice also that if the output of an operation is not explicitly assigned to a variable, it is assigned to a default variable called `ans`.

The content of the *workspace* can be displayed at any time by using the `whos` function:

```
>> whos                                                      (1.3)
  Name        Size          Bytes  Class      Attributes
  ans         1x2              16  double
  matrix      2x3              48  double
```

You can save all the variables in the workspace by means of the function `save`. This will create a binary file called `matlab.mat`. You can restore your work session by using the function `load` to read the file `matlab.mat` and upload your saved variables back into the *workspace*. Let us illustrate this in the following example:

```
>> save % saves the workspace into the file matlab.mat          (1.4a)
Saving to: C:\Users\Documents\MATLAB\matlab.mat

>> clear % erases all variables in the workspace                (1.4b)

>> whos % the workspace is empty!                               (1.4c)


>> load % restores the variables from matlab.mat into the workspace  (1.4d)
Loading from: C:\Users\Documents\MATLAB\matlab.mat

>> whos                                                         (1.4e)
  Name        Size            Bytes  Class      Attributes
  ans         1x2                16  double
  matrix      2x3                48  double
```

The MATLAB® programming environment also supports most of the commonly used statements in other programming languages, such as `for`, `while`, `if`, `then`, etc. For displaying a description on how to use them, you can use the `help` function. For instance, you can execute: `help for`, `help while`, `help if`, and so on.

Regarding the use of `for`, it is worth mentioning that in a wide variety of cases, and due to MATALB® matrix-oriented design, it can be avoided. Suppose, for instance, that you need to create a vector containing odd integers between *0* and *10*. While, conventionally, you would need to do something like this:

```
>> for k=0:4                                                   (1.5a)
      odd_number = 2*k+1;
      vector(k+1) = odd_number;
   end

>> vector                                                      (1.5b)
vector =
      1    3    5    7    9
```

within the MATLAB® environment the same vector can be created as follows:

```
>> newvector = 1:2:10                                          (1.6)
newvector =
      1    3    5    7    9
```

Two observations can be made with respect to the example in (1.5a). First, notice that the use of a semicolon at the end of a command line prevents from displaying the output of the operation in the *command window*. This is especially useful when dealing with large matrices and vectors. In general, unless you are really interested in looking at the result of a given operation, the common practice will be to end each command line with a semicolon.

The second observation is that, different from *C++* and other languages in which array indexes start with zero, MATLAB® array indexes must start with *1*.

Matrix operations are also very simple and intuitive. In the following example we create a *2×3* matrix by multiplying a column vector created from the first two elements of **newvector** times a row vector containing the three last elements of **newvector**:

```
>> newmatrix = newvector(1:2)'*newvector(end-2:end)                    (1.7)
newmatrix =
     5      7      9
    15     21     27
```

Notice from (1.7) how the apostrophe has been used for transposing the *1×2* row vector **newvector(1:2)** into a *2×1* column vector. Notice also how the **end** keyword has been used to refer to the last index of the vector.

Two types of mathematical operations are to be distinguished within the MATLAB® programming environment, namely matrix operations and array operations. In the case of addition and subtraction, they are totally equivalent, but in the cases of multiplication, division, and exponentiation they produce completely different results. While matrix operations refer to the conventional mathematical definition of matrix operations, array operations refer to operations that are carried out on an element-by-element basis. The following example illustrates such a difference for the specific case of multiplication:

```
>> matrix*newmatrix' % conventional matrix multiplication             (1.8a)
ans =
    46    138
   109    327

>> matrix.*newmatrix % element-by-element array multiplication        (1.8b)
ans =
     5     14     27
    60    105    162
```

As seen from (1.8a), the matrix multiplication implements the mathematical definition of such kind of operation, where the element *ij* in the resulting matrix is computed by adding up the corresponding products between elements in row *i* of the first matrix and elements in column *j* of the second matrix. For instance, the *46* in (1.8a) results from the following operation *1*5+2*7+3*9*. Notice how the multiplication of a *2×3* matrix (**matrix**) times a *3×2* matrix (**newmatrix'**) resulted in a *2×2* matrix.

On the other hand, as seen from (1.8b), the array multiplication implements an element-by-element multiplication, in which the element *ij* in the resulting matrix corresponds to the product of the element *ij* of the first matrix times the element *ij* of the second matrix. In this case, the dimensions of the resulting matrix are the same to those of the matrices being multiplied. Notice that the operator for array multiplication is given by `.*` and, in general, pre-appending a dot to an arithmetic operator invokes the corresponding array operation.

Recall that implementing any of the two operations in (1.8a) and (1.8b) in a conventional programming language requires two nested *for–end* loops, one for moving along the rows of the resulting matrix and the other for moving along the columns!

Although we will not be using it in this book, another very useful feature of the MATLAB® programming environment is that it allows for handling complex numbers in a very natural way too. For instance, let us create a complex-valued matrix with the real parts derived from `matrix` and the imaginary parts derived from `newmatrix`:

```
>> matrix+i*newmatrix                                                       (1.9)
ans =

   1.0000 + 5.0000i    2.0000 + 7.0000i    3.0000 + 9.0000i
   4.0000 +15.0000i    5.0000 +21.0000i    6.0000 +27.0000i
```

Operations with complex-valued matrices and vectors are carried out in exactly the same way operations with real-valued matrices and vectors are conducted.

Finally, before concluding this (very) brief introduction to the MATLAB® programming platform, let us discuss in more detail about the issue of writing and executing programs. In addition to the option of executing commands on a one-by-one fashion in the *command window*, it is also possible to write and save code in specific files denominated *m*-files.

An *m*-file is actually a text file containing MATLAB® code. The two main types of *m*-files are scripts and functions. Both types of *m*-files can be created in the MATLAB® text editor or, alternative, in any other text editor able to generate plain text files. Additionally, both types of *m*-files can be executed directly from the *command window*, as well as invoked from other *m*-files (either a script or a function).

They main difference between scripts and functions is that scripts are executed over the main *workspace*, the same you use when executing commands directly from the *command window*. This means that all variables created and used in the script are loaded into the *workspace*, as well as all previously existing variables in the *workspace* are accessible and can be used from the script. You can think of a script as a segment of code that you better write down into a text file and execute it all as a single command (the script file name), rather than having to input and execute the same code one command at a time.

On the other hand, a function is executed in a dedicated workspace that is created when the function is called and discarded when the function execution ends. According to this, functions do not have access to the variables in the main *work-*

space. So, they need to receive, as input variables during the function call, those *workspace* variables they must use. Equivalently, internal variables created during the function execution are not visible from the main *workspace*. So, variables of interest must be returned by the functions as output variables.

For declaring an *m*-file to be a function, the first line of the file must obey the following syntax:

```
function [out_1,out_2,...,out_N] = function_name(in_1,in_2,...,in_M)        (1.10)
```

where `out_n` refers to the output variables returned by the function, and `in_m` refers to the input variables received by the function. Both, the total number of output and input variables, `N` and `M`, can be any integer or zero. The *m*-file containing a given function `function_name` must be named as `function_name.m`

Although you can create your own functions, one of the main advantages of the MATLAB® technical computing software is that it has already available hundreds of functions for you to use. A simple example, for illustrating the use of a function here, can be the one for computing the inverse of a matrix:

```
>> inverse = inv([1 1;0 1])                                                 (1.11)
inverse =
     1    -1
     0     1
```

We will be using both scripts and functions along this book depending on what happens to be the most convenient thing to do in each specific situation.

One final advice regarding the issue of writing and using scripts and functions, which also applies to the creation and use of variables in general, is that you must be careful about not using existing script, function or variable names for your newly created scripts, functions or variables. This kind of omissions can result in mysterious bugs that are difficult to track and solve. For avoiding name collisions in the specific case of variables, the function `genvarname` can be used to generate valid variable names that are different from other existent variables.

1.4 The Text Analytics Toolbox™

The Text Analytics Toolbox™ comprises a collection of functions that are specific for processing, analyzing, and modeling text data, which can be used to support a multiplicity of different language-centered applications such as text classification, sentiment analysis, text summarization, document search, etc. Additionally, text data models and features can be used in combination with other functions and data sources to support a broader range of applications in areas such as business intelligence, product marketing and social analysis, among many others.

The functions available within the Text Analytics Toolbox™ are organized into ten different categories:

- Input. This category includes functions for reading and extracting text from text files in different formats such as PDF, HTML, MSWord, etc. Some of the functions in this category are described in chapter 5.
- Preprocessing. This category includes functions for manipulating text data, such as conversion of text representations and objects across different variable types, as well as applying basic language preprocesses to text variables, such as segmenting documents into sentences and sentences into words, and removing stop words, among others. Functions in this category are mainly covered in chapters 4 and 6, but also used in chapters 9 to 14.
- Data Collection. This category includes functions for manipulating collections of text data and constructing bag-of-word (BOW) models. Except for `editDistance`, which is introduced in chapter 4, most functions in this category are introduced in chapter 9, and also used in chapters 10 to 14.
- Dimensionality Reduction and Topic Models. This category includes functions for computing low-dimensional representations of text data collections, such as Latent Semantic Analysis (LSA) and Latent Dirichlet Allocation (LDA). Some functions in this category are covered in chapters 9 to 11.
- Word Embeddings. This category includes functions for computing continuous space representations of words, referred to as word embeddings, and functions for mapping words to embeddings and back. Most of the functions in this category are described in chapter 10.
- Word Encodings. This category includes functions for representing the vocabulary of a text data collection in the form of numeric indices, as well as mapping vocabulary words to indices and back. Functions in this category are introduced in the *Proposed Exercises* section in chapter 10.
- Display and Presentation. This category includes functions for displaying scatter plots of texts and word clouds. Some of the functions in this category are introduced in chapters 10 and 14.
- Similarity. This category includes two specific functions for computing similarity between documents. The `cosineSimilarity` function is introduced in chapter 11, the `bm25Similarity` function is described in chapter 12.
- Evaluation. This category includes specific functions for evaluating text outputs or responses against references. Functions within this category are presented in chapter 14.
- Summarization. This category includes functions to generate text summaries. Functions in this category are also introduced in chapter 14.

Not all the functions available within the Text Analytics Toolbox™ are covered in this book. However, a significant number of them are introduced along the book according to their relevance to the specific topics being discussed in the different chapters. Some of the non-covered functions are left to be explored and used as part of the exercises proposed at the end of the chapters.

To list all the functions and function categories contained in the Text Analytics Toolbox™, you can run the command `help textanalytics` in the *command window*. Additionally, to look at more detailed information about each of the functions in the toolbox you can either click on the highlighted name of the function or run the `help` command followed by the name of the function (e.g. `help tfidf`).

1.5 Further Reading

There are several introductory and advanced books related to text mining theory and applications. Some good examples include (Feldman and Sanger 2006), (Srivastava and Sahami 2009), and (Berry and Kogan 2010). Other good reference books in the related field of natural language processing include (Manning and Schütze 1999) and (Jurafsky and Martin 2000).

There are also several introductory books to MATLAB®. Some good references include (Palm III 2007), (Gilat 2008), (Pratap 2009), and (Etter 2010). However, the most comprehensive and updated guide to MATLAB® can be found in the corresponding online Product Documentation (The MathWorks 2020).

Similarly, the most comprehensive and updated source of information for the Text Analytics Toolbox™ can be found in the corresponding online Product Documentation (The MathWorks 2020).

1.6 References

Berry MW, Kogan J (2010) Text Mining: Applications and Theory. John Wiley & Sons

Etter DM (2010) Introduction to MATLAB (2nd Edition). Prentice Hall

Feldman R, Sanger J (2006) The Text Mining Handbook: Advanced Approaches in Analyzing Unstructured Data. Cambridge University Press

Gilat A (2008): MATLAB: An Introduction with Applications (3rd Edition). John Wiley & Sons

Jurafsky D, Martin JH (2000) Speech and Language Processing: An Introduction to Natural Language Processing, Computational Linguistics and Speech Recognition. Prentice-Hall

Manning CD, Schütze H (1999) Foundations of Statistical Natural Language Processing. The MIT Press

The MathWorks (2020) MATLAB Product Documentation, https://www.mathworks.com/help/matlab/index.html Accessed 6 July 2021

The MathWorks (2020) Text Analytics Toolbox Product Documentation, https://www.mathworks.com/help/textanalytics/index.html Accessed 6 July 2021

Palm III WJ (2007) A Concise Introduction to MATLAB. McGraw Hill Higher Education

Pratap R (2009) Getting Started with MATLAB: A Quick Introduction for Scientists and Engineers. Oxford University Press

Srivastava A, Sahami M (2009) Text Mining: Classification, Clustering, and Applications. Data Mining and Knowledge Discovery Series. Chapman and Hall/CRC

Part I: Fundamentals

"Fundamental brainwork,
is what makes the difference in all art."

Dante Gabriel Rossetti

2 Handling Text Data

This chapter introduces the main variable classes that are used in the MATLAB® programming environment for representing and handling text. First, in section 2.1, the basic variable type for representing text, which is the character array, is described. Then, in sections 2.2, 2.3 and 2.4, cell arrays, structures, and string arrays, which are the most commonly used variable classes for handling and operating with text, are described, respectively. Finally, in section 2.5, a brief overview is provided on the specific MATLAB® built-in functions for operating with text, as well as other useful functions worth to be known.

2.1 Character and Character Arrays

The simplest variable class for representing text in the MATLAB® programming environment is the character. Characters are used to represent symbols in some predefined encoding system. Depending on the encoding system being used, a character can be represented with one, two or more bytes. By default, when writing and reading text files from your system, the MATLAB® environment uses the default encoding of the operating system. However, in the case of MATLAB® data files, the *unicode* encoding system is used. This guarantees the portability of data files across systems. The specific encoding of a given text can be changed at any moment by using MATLAB® functions `native2unicode` and `unicode2native`. More details on the encoding scheme issue are given in Chapter 5, where we focus our attention in the problem of writing and reading text files.

A text string can be represented as an array (or matrix) of characters. For defining a variable as a character array, it is required that its text value is provided between apostrophes. Try the following example in the command line:

```
>> text_string = 'Hello, this is a string!'                    (2.1)
text_string =
   'Hello, this is a string!'
```

Such a command defines and initializes the variable `text_string` as a character array. By using MATLAB® command `whos` we can list all current variables in the workspace along with their corresponding sizes and classes:

```
>> whos                                                         (2.2)
  Name              Size          Bytes  Class    Attributes
  text_string       1x24             48  char
```

R. E. Banchs, *Text Mining with MATLAB®*, https://doi.org/10.1007/978-3-030-87695-1_2

You can get the numerical codes assigned to each character in the array by casting the variable `text_string` from character to integer as it is shown in the following example:

```
>> codes = cast(text_string,'int16')                           (2.3a)
codes =
  1x24 int16 row vector
  Columns 1 through 10
      72     101     108     108     111      44      32     116     104     105
  Columns 11 through 20
     115      32     105     115      32      97      32     115     116     114
  Columns 21 through 24
     105     110     103      33
>> whos                                                         (2.3b)
  Name                Size              Bytes    Class    Attributes
  codes               1x24                 48    int16
  text_string         1x24                 48    char
```

Similarly, you can recover the original variable `text_string` from variable `codes` by either casting it back from integer to character or, alternatively, by using the function `char`:

```
>> cast(codes,'char')                                          (2.4a)
ans =
    'Hello, this is a string!'
>> char(codes)                                                 (2.4b)
ans =
    'Hello, this is a string!'
```

In the same way any numerical variable is handled by default as a matrix, characters are also handled as matrices. Indeed, as already seen in example (2.2), a string of n characters is actually represented in the MATLAB® workspace as a $1 \times n$ matrix of characters. According to this, any list or array of texts can be represented by means of a matrix. Take a look at the following example:

```
>> list = ['This is the first string ';'This is the second string']  (2.5a)
list =
  2x25 char array
    'This is the first string '
    'This is the second string'
>> whos list                                                   (2.5b)
  Name      Size            Bytes    Class    Attributes
  list      2x25              100    char
```

The only problem with this particular way of representing string arrays is that all the strings in the array (rows in the matrix) are required to have the same number of characters. This is why we added a white space at the end of the first string in example (2.5a)!

If you try to reproduce the same example without including the trailing white space at the end of the first string, you will get the following error message:

```
>> list = ['This is the first string';'This is the second string']          (2.6)
Error using vertcat
Dimensions of arrays being concatenated are not consistent.
```

So, it is very important to remember that representing a text array by means of a character matrix will require padding with blanks all the text strings up to the length of the largest one. In this sense, and just with very specific exceptions, we do not recommend using character matrices for representing text data. Indeed, as we will see in the following sections, handling text by using other variable classes such a cell arrays, structures, and string arrays[1] is much more convenient.

Nevertheless, one of the main advantages of the character matrix representation discussed here is that any of the conventional MATLAB® matrix indexing strategies can be used with it. For instance, try reproducing the following sequence of examples in the command line:

```
>> dataset = ['name age';'mark    35';'beth    26';'peter   39']          (2.7a)
dataset =
  4x9 char array
    'name age'
    'mark    35'
    'beth    26'
    'peter   39'
>> names = dataset(2:end,1:end-4)                                          (2.7b)
names =
  3x5 char array
    'mark '
    'beth '
    'peter'
>> ages = dataset(2:end,end-1:end)                                         (2.7c)
ages =
  3x2 char array
    '35'
    '26'
    '39'
```

---

[1] The variable class *string array* was introduced in release R2016b.

```
>> people_in_their_30s = dataset(dataset(:,end-1)=='3',1:end-4)          (2.7d)
people_in_their_30s =
  2x5 char array
    'mark '
    'peter'
```

In the examples presented in (2.7), first a list of names and ages is defined and written into a variable called **dataset**, which is just a matrix (two-dimensional array) of characters. As seen from (2.7a), the defined array of characters looks like a table, the first row containing the headers for each column in the table: **'name'** and **'age'**, and the subsequent rows containing the corresponding data entries, one sample per row.

By just using matrix indexing operations, several different tasks can be accomplished. For instance, in (2.7b), the names in **dataset** were extracted by retaining columns *1* to *6* from rows two and onwards. Similarly, in (2.7c), the ages were extracted by retaining the last two columns from rows two and onwards. Finally, (2.7d) shows a more elaborated example, in which only the names for those people whose age's first digit was equal to *3* were extracted.

2.2 Handling Text with Cell Arrays

A more convenient way of representing text in the MATLAB® programming environment is by using a special class of variables denominated cell arrays. A cell array is a special class of structure that allows organizing variables into a matrix form regardless their size and type.

Consider the following example, in which a *2×2* cell array is defined:

```
>> cell_array = {'Hello World',eye(3);5+j,dataset}               (2.8)
cell_array =
  2x2 cell array
    {'Hello World'      }    {3x3 double}
    {[5.0000 + 1.0000i]}     {4x9 char  }
```

In this example, the variable **cell_array** consists of a *2×2* matrix of elements, where element *{1,1}* is a character array, element *{1,2}* is by itself a *3×3* numerical matrix, element *{2,1}* is a complex number, and element *{2,2}* is the same *4×9* matrix of characters created in (2.7a).

Different from the case of conventional arrays and matrices (whose elements are retrieved by indicating their indexes between parentheses), the elements within a cell array must be retrieved by using braces. So, the four elements of the *2×2* cell array defined in (2.8) can be retrieved as follows:

```
>> cell_array{1,1}                                          (2.9a)
ans =
    'Hello World'
>> cell_array{1,2}                                          (2.9b)
ans =
        1       0       0
        0       1       0
        0       0       1
>> cell_array{2,1}                                          (2.9c)
ans =
    5.0000 + 1.0000i
>> cell_array{2,2}                                          (2.9d)
ans =
  4x9 char array
    'name  age'
    'mark   35'
    'beth   26'
    'peter  39'
```

According to this, an alternative way to represent any set of text variables is by using cell arrays. In this kind of representation, each individual text value is represented by means of a character array, and a vector (or matrix) of character arrays can be constructed by means of a cell array.

Consider the following example, in which a list of character arrays is constructed by using the cell array variable class:

```
>> list = {'This is the 1st string';'This is the 2nd one';'And the 3rd'}  (2.10)
list =
  3x1 cell array
    {'This is the 1st string'}
    {'This is the 2nd one'    }
    {'And the 3rd'            }
```

Notice that with this kind of representation character arrays are not required to have the same number of characters. And, like in any cell array, each individual variable can be retrieved by using braces as follows:

```
>> list{1} % retrieves the first string in list             (2.11a)
ans =
    'This is the 1st string'
>> list{2} % retrieves the second string                    (2.11b)
ans =
    'This is the 2nd one'
```

```
>> list{3} % retrieves the third string                               (2.11c)
ans =
    'And the 3rd'
```

Additionally, each character or substring within the individual variable can be retrieved in the same way it is done in the case of character arrays:

```
>> % retrieves the first four characters in the first string          (2.12a)
>> list{1}(1:4)
ans =
    'This'

>> % retrieves the last three characters in the second string         (2.12b)
>> list{2}(end-2:end)
ans =
    'one'

>> % retrieves all non-white-space characters in the third string     (2.12c)
>> list{3}(not(list{3}==' '))
ans =
    'Andthe3rd'
```

It must be known that cell arrays also admit the use of parentheses for retrieving their contents. However, it is important to understand the difference between using parenthesis or braces for accessing cell array elements. When using parenthesis, the retrieved elements are the cells of the cell array (i.e. retrieved elements are also of the class cell array) while when using braces, the retrieved elements are the contents within the cells (i.e. whichever class is contained in each cell).

The following example better clarifies the difference between using parenthesis or braces for accessing cell array contents:

```
>> with_braces = list{1}                                              (2.13a)
with_braces =
    'This is the 1st string'

>> with_parenthesis = list(1)                                         (2.13b)
with_parenthesis =
  1x1 cell array
    {'This is the 1st string'}

>> whos list with_braces with_parenthesis                             (2.13c)
  Name               Size           Bytes  Class     Attributes
  list               3x1              416  cell
  with_braces        1x22              44  char
  with_parenthesis   1x1              148  cell
```

As can be seen from (2.13c), while the variable `with_braces` is a character array of size 1×22, the variable `with_parenthesis` is a cell array of size 1×1. In this latter case, we have retrieved cell *l* from cell array `list`, while in the first case, we have retrieved the character array within the cell. This difference might not seem important at this point, but it will definitively be very important later on.

2.3 Handling Text with Structures

Another alternative way of handling text is by using structures. As many other programming languages, the MATLAB® environment supports the use of structures. A structure is actually a set of variables that are indexed as fields of a common main variable. This kind of representation is especially useful for importing and exporting data from and to databases and markup language formats. Consider, for instance, the following example where a structure of *4* fields is created:

```
>> structure = struct('f1','Hello World','f2',eye(3),'f3',5+j);     (2.14a)

>> structure.f4.names = ['mark ';'beth ';'peter'];                  (2.14b)
>> structure.f4.ages = [35;26;39];
```

Notice from (2.14b) that the fourth field of `structure` is defined as a structure itself, which is composed of subfields `names` and `ages`. Structures, as well as cell arrays, admit nested constructs.

For accessing structure contents, each field in a structure can be retrieved by appending a dot to the name of the main variable followed by the name of the field:

```
>> structure.f1 % retrieves field f1                                (2.15a)
ans =
    'Hello World'

>> structure.f2 % retrieves field f2                                (2.15b)
ans =
     1     0     0
     0     1     0
     0     0     1

>> structure.f3 % retrieves field f3                                (2.15c)
ans =
   5.0000 + 1.0000i

>> structure.f4 % retrieves field f4                                (2.15d)
ans =
  struct with fields:
    names: [3x5 char]
     ages: [3x1 double]
```

In the particular case of nested structures, such as the one in `structure.f4`, each subfield element can be retrieved by concatenating with dots the names of the corresponding sequence of fields:

```
>> structure.f4.ages' % retrieves and transposes subfield ages of f4      (2.16a)
ans =
    35    26    39

>> structure.f4.names % retrieves subfield names of f4                     (2.16b)
ans =
  3x5 char array
    'mark '
    'beth '
    'peter'

>> % retrieves the second row of subfield names of f4                      (2.16c)
>> structure.f4.names(2,:)
ans =
    'beth '
```

Finally, very important and useful resources for converting structure-based text representations into cell-array-based text representations, and vice versa, are available through MATLAB® built-in functions `struct2cell`, `fieldnames` and `cell2struct`. In the remaining of this section, we illustrate the use of these three functions in detail.

Let us consider again the sample dataset defined in (2.7a), which contains information about the names and ages of three people. In (2.7a), this dataset was represented by using a two-dimensional array of characters. Now, let us consider a more appropriate representation by means of a structure array composed of the two fields `name` and `age`. In the following example, such a structure array is created and initialized with the corresponding three data entries:

```
>> data = struct('name',{'mark';'beth';'peter'},'age',{'35';'26';'39'})   (2.17)
data =
  3x1 struct array with fields:
    name
    age
```

As seen from (2.17), our sample dataset has been stored into a structure array of 3×1 elements (one element per data entry), and each element of the structure array is composed of two fields: `name` and `age`.

Now, let us consider the procedure for converting the structure-based representation of (2.17) into a cell-array-based representation. For this, the two functions `struct2cell` and `fieldnames` must be used. While the former maps the fields of the structure array into a cell array, the latter retrieves the names of the fields into another cell array. We proceed as follows:

```
>> datacell = struct2cell(data)                                    (2.18a)
datacell =
  2x3 cell array
    {'mark'}     {'beth'}     {'peter'}
    {'35'  }     {'26'  }     {'39'    }
>> datafields = fieldnames(data)                                   (2.18b)
datafields =
  2x1 cell array
    {'name'}
    {'age' }
```

Notice from (2.18a) how the resulting cell array `datacell` contains the same data entries as the structure array `data`. The mapping has been done such that elements and fields of the structure array correspond to columns and rows of the cell array, respectively, i.e. the 3×1 structure array has been mapped into a $2\times3\times1$ cell array. In general, `struct2cell` will map any $N\times M$ structure array of K fields into a $K\times N\times M$ cell array.

In addition to `struct2cell`, as seen from (2.18b), the function `fieldnames` should be used if we are also interested in retrieving the field names of the structure. In this case, the field names are retrieved into a one-dimensional cell array.

Let us now consider the problem of going back from the cell-array-based representation obtained in (2.18) to the original structure-based representation in (2.17). In this case, the function `cell2struct` must be used as it is illustrated in the following example:

```
>> databack = cell2struct(datacell,datafields,1)                  (2.19a)
databack =
  3x1 struct array with fields:
    name
    age

>> databack(1)                                                    (2.19b)
ans =
  struct with fields:
    name: 'mark'
     age: '35'

>> databack(2)                                                    (2.19c)
ans =
  struct with fields:
    name: 'beth'
     age: '26'

>> databack(3)                                                    (2.19d)
ans =
```

```
struct with fields:
  name: 'peter'
   age: '39'
```

As seen from (2.19a), function `cell2struct` requires three input parameters: the cell array to be mapped into the structure, a cell array of strings containing the names of the fields to be used in the structure definition, and the dimension along which the cell array is to be folded into the structure fields. To recover the same shape of the original structure array in (2.17), this last parameter must be *1*.

2.4 Handling Text with String Arrays

Starting with release R2016b, the MATLAB® programming environment supports the string array variable class. String arrays represent the most convenient way to store and manipulate sequences of characters. Different from character arrays, presented in (2.1), to define a text string as a variable of the class string array, the corresponding text values have to be provided within double quotes:

```
>> text_string = "Hello, this is a string!"                    (2.20a)
text_string =
    "Hello, this is a string!"

>> whos text_string                                            (2.20b)
  Name              Size            Bytes  Class      Attributes
  text_string       1x1              198   string
```

Notice that different from (2.1), where the variable `text_string` stored the text in the form of a character array of size *1x24*, in (2.20) the text is stored in the form of a string scalar, which is actually a string array of dimensions *1x1*.

String scalars can be converted into character arrays and vice versa by using the functions `char` and `string`, respectively:

```
>> char_array = char(text_string) % to character array          (2.21a)
char_array =
    'Hello, this is a string!'

>> string_scalar = string(char_array) % back to string scalar   (2.21b)
string_scalar =
    "Hello, this is a string!"

>> whos char_array string_scalar                                (2.21c)
  Name               Size            Bytes  Class      Attributes
  char_array         1x24               48  char
  string_scalar      1x1               198  string
```

Similar to cell arrays, as described in section 2.2, string arrays also allow for organizing text strings into matrix form:

```
>> strarray = ["mark","beth","peter";35,26,39]                    (2.22)
strarray =
  2x3 string array
    "mark"      "beth"      "peter"
    "35"        "26"        "39"
```

However, string arrays only admit variables of the string variable class within its elements. Notice from (2.22), for instance, how the numerical inputs in the second row of the matrix have been automatically converted into the string variable class in the resulting representation.

Also, similar to cell arrays, the elements in string arrays can be retrieved by using either parenthesis or braces. By using parenthesis, the corresponding string subarrays are retrieved. By using braces, on the other hand, the corresponding character arrays are retrieved instead.

```
>> strarray(:,2) % retrieves the string subarray in the second column    (2.23a)
ans =
  2x1 string array
    "beth"
    "26"

>> strarray{:,2} % retrieves the character arrays in the second column    (2.23b)
ans =
    'beth'
ans =
    '26'
```

Individual characters and character substrings can also be retrieved from string scalars within a string array:

```
>> strarray{1,2}(3:4)                                            (2.24a)
ans =
    'th'
>> strarray{1,3}(strarray{1,3}~='e')                             (2.24b)
ans =
    'ptr'
```

In both cases illustrated in example (2.24) we have used braces to retrieve the character arrays corresponding to specific string scalars within strarray. Then, we used parenthesis to retrieve substrings from the character arrays. In example (2.24a), we retrieve the third and fourth characters from the element in position {1,2} (i.e. *th* from *beth*); and in example (2.24b), we retrieved all characters different from *e* from the element in position {1,3} (i.e. *ptr* from *peter*).

Convenient and useful resources for converting string-array-based text representations into cell-array-based text representations, and vice versa, are available through MATLAB® built-in functions `cellstr` and `string`. This is illustrated in the following example:

```
>> cellarray = cellstr(strarray) % from string array to cell array        (2.25a)
cellarray =
  2x3 cell array
    {'mark'}    {'beth'}    {'peter'}
    {'35'  }    {'26'  }    {'39'  }

>> backtostrarray = string(cellarray) % back to string array              (2.25b)
backtostrarray =
  2x3 string array
    "mark"      "beth"      "peter"
    "35"        "26"        "39"

>> whos strarray cellarray backtostrarray                                 (2.25c)
  Name                Size          Bytes  Class     Attributes
  backtostrarray      2x3             420  string
  cellarray           2x3             664  cell
  strarray            2x3             420  string
```

Finally, let us briefly introduce the notion of *missing string*, which is the string equivalent to the not-a-number (*NaN*) representation in numeric arrays. Missing strings are different from empty strings (i.e. strings of zero length) and are specifically used to indicate missing values in a string array. The difference between empty and missing strings is illustrated in the following example:

```
>> strarray(2,4)=""                                                       (2.26)
strarray =
  2x4 string array
    "mark"      "beth"      "peter"     <missing>
    "35"        "26"        "39"        ""
```

As seen from the example, `strarray` from (2.22) has been updated by inserting the empty string `""` in position *(2,4)*. As a result, the original string array size of *2x3* has been extended to *2x4*, leaving a missing value at position *(1,4)*.

It is important to notice that missing strings are supported only in string arrays, and they are not reversible when converting string arrays into cell arrays or structures and then back into string arrays:

```
>> cellstr(strarray)                                                      (2.27a)
ans =
  2x4 cell array
    {'mark'}    {'beth'}    {'peter'}    {0×0 char}
    {'35'  }    {'26'  }    {'39'  }     {0×0 char}
```

```
>> string(cellstr(strarray))                                   (2.27b)
ans =
  2x4 string array
    "mark"      "beth"      "peter"      ""
    "35"        "26"        "39"         ""
```

As seen from the example, the missing string element in (2.26) is mapped into an empty string when the string array variable `strarray` is converted into a cell array in (2.27a). Notice how it remains an empty string when mapped back into a string array in (2.27b).

2.5 Some Useful Functions

After having seen the fundamental variable types that allow for handling text data, and before moving on into more advanced procedures and techniques, we devote this section to briefly introduce the most common MATLAB® built-in functions for handling and operating with character arrays and strings. We also introduce here some other general functions worth to be known. More than a comprehensive guide, this section is intended to be a quick reference.

Built-in functions for handling and operating with character arrays and strings are organized into six different categories:

- Conversion Functions. This category includes functions for creating different types of text variables and converting text representations among them. Most of the functions in this category are introduced in this chapter.
- Text Determination Functions. This category includes functions for testing whether a given variable belongs to specific classes of text variables. Some of these functions are left for self-studying as exercises in section 2.7.
- String Operations. This is the largest category of built-in functions for handling and operating with character arrays and strings. It includes functions for searching, removing, and replacing substrings, as well as functions for splitting, joining, and comparing strings, among others. Most of the functions in this category are introduced in chapters 3 and 4, while others are covered by some proposed exercises within those chapters.
- Character Set Conversion. This category includes functions for converting bytes into *unicode* and vice versa (`native2unicode` and `unicode2native`). These functions are left for self-studying as exercises in section 2.7.
- Conversion Between Text and Numbers. This category includes functions for converting numerical variables into formatted text representations and vice versa. Some of these functions are introduced in chapter 5.
- Base Number Conversion. This category includes functions for converting numerical variables and their formatted text representations among different

basis such as binary, decimal, hexadecimal, etc. Some of the functions in this category are left for self-studying as exercises in section 2.7.

Not all functions in these six different categories are covered in the book. Some of these functions were already introduced in this chapter, while others are studied and described in more detail in the following chapters and some others are left for self-studying in some of the exercises proposed in section 2.7.

To list all functions and function categories that are available for handling and operating with character arrays and strings, you can run `help strfun` in the *command window*; and to get more detailed information about each of the functions listed, you can either click on the highlighted function name or run the `help` command followed by its name (e.g. `help cellstr`).

Table 2.1 summarizes those functions within the six described categories that are described or used in any of the examples in this book along with the chapters in which they are introduced.

Table 2.1. Built-in functions for arrays and strings that are described or used in this book, their corresponding categories, and the chapters where they are introduced in

Category	Functions	Described in
Conversion	`char, string, cellstr`	Chapter 2
Conversion	`double, strings, blanks`	Chapter 2 (exercises)
Text Determination	`iscellsrt, ischar, isstring, isstrprop`	Chapter 2 (exercises)
String Operations	`deblank, lower, pad, reverse, upper`	Chapter 2 (exercises)
String Operations	`regexp, regexpi, regexprep`	Chapters 3 and 4
String Operations	`contains, count, erase, insertAfter, insertBefore, join, replace, replaceBetween, split, strcat, strcmp, strncmp, strncmpi, strfind, strip, strrep, strtok`	Chapter 4
String Operations	`endsWith, extractBetween, startsWith`	Chapter 4 (exercises)
String Operations	`strlength`	Chapter 7
String Operations	`compose`	Chapter 8
Character Set	`native2unicode, unicode2native`	Chapter 2
Text and Numbers	`sprintf`	Chapter 3 (exercises)
Text and Numbers	`str2num, sscanf`	Chapter 5
Base Conversion	`dec2base, dec2bin, dec2hex`	Chapter 2 (exercises)

In addition to the string related functions presented in Table 2.1, there are other MATLAB® functions of more general scope worth to be known too. These functions are summarized in Table 2.2 and described thereafter.

Table 2.2. Other non-string-specific functions worth to be known

Function	Description
help	Displays the help text contained within a function
lookfor	Searches for the specified keyword in all M-files
iskeyword	Check if the given string constitutes a MATLAB keyword
pause	Halts the execution and waits for a response from the user
keyboard	Halts the execution and invokes the keyboard
input	Prompts for the user to input either a number or a string
disp	Displays the values of the given variable
inputdlg	Dialogue box for receiving input values from the user
questdlg	Dialogue box for receiving an input selection from the user
msgbox	Message box for displaying a given string

Next, let us illustrate the use of the functions presented in Table 2.2. Consider, for instance, the following example in which we illustrate the use of `lookfor` and `help` for searching functions and looking at their descriptions:

```
>> lookfor blanks % searches for the string 'blanks' in all M-files    (2.28a)
blanks                        - Character vector of blanks
deblank                       - Remove trailing blanks

>> help blanks % looks at the description of function blanks           (2.28b)
 blanks Character vector of blanks
    blanks(n) returns a character vector of n blanks.
    Use with DISP, e.g.  DISP(['xxx' blanks(20) 'yyy']).
    DISP(blanks(n)') moves the cursor down n lines.
    See also newline, deblank, pad, strip, string, char
    Documentation for blanks
```

The function `iskeyword` allows for checking if a given string is a MATLAB® keyword. It is especially useful to check for the validity of variable names:

```
>> iskeyword('start') % valid variable name                           (2.29a)
ans =
  logical
    0

>> iskeyword('end') % not a valid name as it is a MATLAB keyword       (2.29b)
ans =
  logical
    1
```

The function `pause` allows for momentarily halting the execution either for some specified amount of time or until the user hits any key in the keyboard:

```
>> pause(5) % halts the execution for 5 seconds                    (2.30a)

>> pause % halts the execution until a key is hit                  (2.30b)
```

The function `keyboard` also halts the execution of the current program, but it gives access to the command window. The execution is resumed when the user enters the command `dbcont` and hits the *enter* key:

```
>> keyboard % halts the program and gives access to the command window  (2.31)
K>> a = 5+1;
K>> dbcont
```

The function `input` allows the user for entering either a numeric value or a string directly from the command window:

```
>> r = input('How old are you? ') % prompts for a numeric variable     (2.32a)
How old are you? 25
r =

    25

>> s = input('What is your name? ','s') % prompts for a text input      (2.32b)
What is your name? John
s =

    'John'
```

The function `disp` allows for displaying variables in the command window:

```
>> disp(dataset) % displays the variable 'dataset' defined in (2.7a)    (2.33)
name   age
mark   35
beth   26
peter  39
```

Finally, functions `inputdlg`, `questdlg` and `msgbox` allow for collecting inputs and displaying information by means of interactive dialogue and message boxes. In the case of `inputdlg`, the entered information is returned in a cell array; while in the case of `questdlg`, the selected option is returned in a character array.

The function `msgbox`, by default, does not halt the execution of the current program. If the execution is to be halted when displaying a message, the function `msgbox` should be used along with function `uiwait` and the option `'modal'` must be included in the function call.

The following examples illustrate the use of these last three functions. The resulting dialogue and message boxes are depicted in Figure 2.1.

```
>> q1 = 'What day is it?'; q2 = 'What time is it?';                (2.34a)
>> today = inputdlg({q1,q2});

>> q1 = 'What is your favorite fruit?';                           (2.34b)
```

```
>> fruit = questdlg(q1,'','apple','banana','mango','mango');

>> msg = 'This is the end of section 2.5.';                    (2.34c)
>> uiwait(msgbox(msg,'','modal'));
```

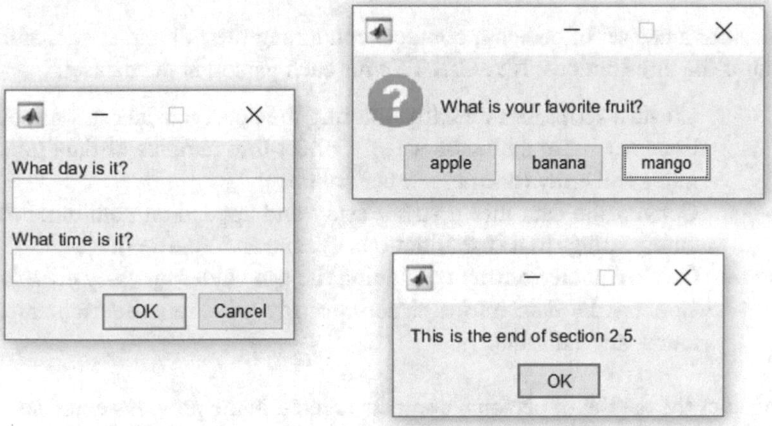

Fig. 2.1. Dialogue and message boxes resulting from examples in (2.34)

2.6 Further Reading

For more detailed information on the *unicode* character set you must refer to (The Unicode Consortium 2020). Some details about some other character encodings are also provided in section 5.1 when describing the function `fopen`.

For a more comprehensive description on string handling, as well as all string related functions described in this chapter you should refer to the MATLAB® online Product Documentation (The Mathworks 2020a). Similarly, more detailed information on dialogue boxes and other user interface functions is available from (The Mathworks 2020b).

2.7 Proposed Exercises

1. Create the *2×N* cell array `ascii_codes` and store in it the corresponding code values and symbols for *ASCII* codes in the range from *32* to *127*.

   – Store the numerical values of the codes (integers) in the first row of the cell array, i.e. `ascii_codes{1,:}`.

    – Store the corresponding symbols (characters) in the second row of the cell array, i.e. `ascii_codes{2,:}`.

    – Consider using `num2cell` to create a cell array from a numerical array and `char` for casting code's integer values into characters.

2. Consider a dataset of personal contacts containing the full name, age, and state (out of the three options: NY, CA, TX) for each person in the dataset.

    – Create a script for manually entering the data (consider using the functions `inputdlg` and `questdlg`). Collect few samples of data (about *5*) into a cell array (one person per column).

    – Convert the data into a string array and append an additional row of empty strings to it (use functions `string` and `strings`).

    – Create a numeric array containing the age values by using `double`.

    – Convert the data into a structure array with four fields: name, surname, age and state.

3. Consider the dataset of personal contacts created in the previous exercise.

    – Create a script for converting the structure into a cell array.

    – Retrieve the field names and save them into another cell array.

    – Sort the collection into alphabetical order according to the contact person's surnames (consider using the function `sort`).

    – Use function `disp` in combination with functions `pad` and `blanks` to display a header containing the four field names.

    – Just below the header line, generate a printout of the sorted dataset (notice that you will need to implement a loop for doing this).

4. Create a function to convert a list of words among character array, cell array and string array representations.

    – Use function `input` to manually enter the *10* words. Save the words into a cell array (one word per cell).

    – Create the function `convert` that converts its input (either char, cell, or string arrays) into an output (either char, cell, or string arrays).

    – The function should automatically identify whether the input variable is a character array, a cell array, or a string array (use `ischar`, `iscellstr`, and `isstring` to identify the class of the input variable).

    – Consider using `questdlg` to request the desired class of output variable and use either `char`, `cellstr`, or `string` to convert the input the requested class.

    – Modify the function to receive the desired class of output variable as an additional input parameter.

    – Modify the function to display an error message with `msgbox` when an input different from a character, cell, or string array is given.

5. Create a function **strops** that implements the basic string operations of lower-casing, uppercasing, and reversing a given text string.

    – The input must be a string scalar and the output must be a structure with the four fields original, lower, upper, and reverse, containing the corresponding copies of the input string.
    – Trailing blanks in the input string array, if any, must be removed.
    – Consider using functions **deblank**, **lower**, **upper**, and **reverse**.

6. Modify the function **strops** such as it also supports base conversions from decimal to binary, octal and hexadecimal for a given decimal number.

    – If the input is an integer, then the output must be a structure with the four fields bin, oct, dec and hex containing the corresponding numeric string representations.
    – Consider using functions **dec2base**, **dec2bin**, and **dec2hex**.

7. Create a function **blanktouscore** that receives any string and returns the same string with all white spaces " " replaced by underscores "_".

    – As a hint, consider using function **isstrprop** for identifying white-spaces and then replace them with the underscore character "_".[2]
    – Alternatively, you can either consider the procedure illustrated in (2.12c) and (2.24b) for identifying white-space characters and replace them with "_" or find the *ASCII* code for the white space (*32*) and re-place its occurrences with the code for the underscore (*95*).
    – Modify the function so that it can receive either a character array, a cell array, or a string array. It must return the same data type with all white spaces in each individual text string replaced by underscores.
    – The function should be able to automatically detect whether the input is a single string, a cell array, or a string array and proceed according-ly (consider using the text determination functions: **iscellsrt**, **ischar**, **isstring**).

2.8 References

The Unicode Consortium (2020) Unicode 6.0.0, http://www.unicode.org/versions/Unicode6.0.0/ Accessed 6 July 2021
The MathWorks (2020a) MATLAB Product Documentation: Strings, https://www.mathworks.com/help/matlab/ref/strings.html Accessed 6 July 2021
The MathWorks (2020b) MATLAB Product Documentation: Dialog, https://www.mathworks.com/help/matlab/ref/dialog.html Accessed 6 July 2021

---

[2] More details on character replacement will be given in section 4.2.

3 Regular Expressions

This chapter introduces the use of regular expressions in the MATLAB® programming environment. The main objectives of this chapter are either to provide experienced MATLAB® users not familiar with text processing applications with a basic guide and quick introduction to the topic of regular expressions, or to provide experienced NLP practitioners not familiar with MATLAB® with a reference guide for MATLAB® regular expressions and their related functions.

The chapter is structured as follows. First, in section 3.1, basic operators for matching characters are presented along with MATLAB® built in functions for matching regular expressions. Second, operators for matching sequences of characters are presented in section 3.2, followed by conditional operators, which are presented in section 3.3. Finally, section 3.4 is devoted to explaining the concept of tokens and their use in regular expression matching.

3.1 Basic Operators for Matching Characters

The basic MATLAB® built in function for matching regular expressions is the function `regexp`. This function allows for matching a regular expression in a given string of text. Its syntax is as follows:

```
[out1,out2,...] = regexp(string,expression,parameter1,parameter2,...)      (3.1)
```

where `expression` is the regular expression that is to be matched within the text string `string`, and `parameter#` is a qualifying argument that determines the kind of output `out#` that should be returned. `regexp` admits up to six different types of qualifying arguments that can be used individually or in combination: `'match'`, `'start'`, `'end'`, `'tokens'`, `'names'` and `'tokenExtents'`. It is important to notice that, in any case, the total number of parameters used in the function call must be equal to the total number of outputs that are specified.[1]

In this section, individual character matching is described. Table 3.1 presents the four basic operators for matching characters.

The best way for understanding how regular expressions work is by experimenting with them. The following group of examples, from (3.2) to (3.6), aims at

---

[1] Although a formal description of `regexp` is not provided here, the use of the function is fully illustrated with examples along the chapter. For a more detailed description of `regexp` you can refer to the corresponding reference page in the MATLAB® help browser.

R. E. Banchs, *Text Mining with MATLAB®*, https://doi.org/10.1007/978-3-030-87695-1_3

illustrating the use of the four basic operators presented in Table 3.1. Try reproducing these examples, as well as experimenting with some other similar examples of your own creation.

Table 3.1. The four basic operators for matching characters

Operator	Type	Description
.	any	Matches any character
[ ]	disjoint	Matches any occurrence of characters specified within the square brackets
[-]	range	Matches any character within the specified range `char1-char2`
[^]	negated	Matches any occurrence of characters not specified

Let us consider the following string (for reference purposes, in Figure 3.1, we
illustrate all characters within the proposed string along with their corresponding
indexes):

```
>> string = "This is SECTION 3.1";                                      (3.2)
```

```
String -> T   h   i   s       i   s       S   E   C   T   I   O   N       3   .   1
Indexes -> 01  02  03  04  05  06  07  08  09  10  11  12  13  14  15  16  17  18  19
```

Fig. 3.1. Characters and their corresponding indexes for sample string defined in (3.2)

In the following example, the *any character* operator is used for matching all
characters in the string, and all matches are returned by using the qualifying argument `'match'`:

```
>> match = regexp(string,'.','match')                                   (3.3)
match =
  1x19 string array
  Columns 1 through 11
    "T"    "h"    "i"    "s"    " "    "i"    "s"    " "    "S"    "E"    "C"
  Columns 12 through 19
    "T"    "I"    "O"    "N"    " "    "3"    "."    "1"
```

On the other hand, the *disjoint* operator must be used if we are interested in
matching a specific set of characters. For instance, we can only match vowels:

```
>> [match,index] = regexp(string,'[aeiouAEIOU]','match','start')        (3.4)
match =
  1x5 string array
    "i"    "i"    "E"    "I"    "O"
index =
     3     6    10    13    14
```

Notice how in this case the qualifying arguments `'match'` and `'start'` were used for returning both the matches and their corresponding starting indexes, respectively.

If the set of characters to be matched are sequential in terms of their encodings, the *range* operator can be used to define the set. For instance, let us match any digit and return the matches and their indexes:

```
>> [match,index] = regexp(string,'[0-9]','match','start')                    (3.5)
match =
  1x2 string array
    "3"    "1"
index =
    17    19
```

or, for example, we can consider matching any non alphanumeric character:

```
>> [match,index] = regexp(string,'[^0-9a-zA-Z]','match','start')             (3.6)
match =
  1x4 string array
    " "     " "     " "      "."
index =
     5     8    16    18
```

Another MATLAB® built-in function for matching regular expressions is the function `regexpi`. Different from `regexp`, this function is case insensitive, which means that it will match any lowercase and uppercase occurrence of a given character regardless the case used for such a character in the regular expression definition. The following examples repeat examples (3.4) and (3.6) presented above, but using the function `regexpi` instead of `regexp`.

```
>> match = regexpi(string,'[aeiou]','match')                                 (3.7a)
match =
  1x5 string array
    "i"     "i"     "E"     "I"     "O"

>> index = regexpi(string,'[^0-9A-Z]')                                       (3.7b)
index =
     5     8    16    18
```

Some commonly used regular expressions are also available as operators. This is the case of numeric digits, alphanumeric characters, and blanks. Table 3.2 presents these operators along with their corresponding regular expressions.

Table 3.2. Specific operators for matching digits, alphanumeric characters, and blanks

Operator	Regular Expression	Description
\d	[0-9]	Any character that is a numeric digit

\D	[^0-9]	Any character that is not a numeric digit
\w	[0-9a-z-A-Z]	Any character that is an alphanumeric character
\W	[^0-9-z-A-Z]	Any character that is not an alphanumeric character
\s	[\t\r\n\f\v]	Any character that is a blank/whitespace character
\S	[^ \t\r\n\f\v]	Any character that is not a blank/whitespace character

The following two examples illustrate how examples (3.5), on matching numeric digits, and (3.6), on matching non alphanumeric characters, can be carried out by using the corresponding operators presented in Table 3.2.

```
>> match = regexp(string,'\d','match')                              (3.8a)
match =
  1x2 string array
    "3"    "1"
>> match = regexp(string,'\W','match')                              (3.8b)
match =
  1x4 string array
    " "    " "    " "    "."
```

Finally, it is important to mention that for being able to match any special or reserved character within a regular expression, the character should be preceded by the symbol '\'. Look at, for instance, the example in (3.3). In such a case, the character '.' was interpreted as the *any character* operator and all the characters in the string were matched. So, in the case we are required to match the dots in a given string of text we should refer to the character itself in the following way '\.'. This is illustrated in the following example:

```
>> whereisthedot = regexp(string,'\.','start')                     (3.9)
whereisthedot =
    18
```

Up to this point, we have considered the problem of matching individual characters. In the following section, the problem of matching specific sequences of characters is considered.

3.2 Matching Sequences of Characters

For matching specific sequences of characters instead of individual characters, two basic types of operators can be used: quantifiers and logical operators. Quantifiers are a special type of operators that are used to specify the number of times a given expression must repeat itself for a match to occur. Logical operators, on the other hand, are used to group expressions and define alternative matches.

Table 3.3 presents the two basic quantifier operators along with other four commonly used quantifiers, which are special cases of the two basic ones. For the latter ones, corresponding equivalences in terms of the two basic quantifiers are also provided in the table.

Let us consider the following sample string:

```
>> string = "'sample 1','sample a','sample20','sample 321'"          (3.10)
string =
    "'sample 1','sample a','sample20','sample 321'"
```

Table 3.3. Most commonly used basic and special quantifier operators

Operator	Equivalence	Description
{n,}	−	*n* or more occurrences of the preceding regular expression
{n,k}	−	Between *n* and *k* occurrences of the preceding regular expression
{n}	{n,n}	Exactly *n* occurrences of the preceding regular expression
*	{0,}	*0* or more occurrences of the preceding regular expression
+	{1,}	*1* or more occurrences of the preceding regular expression
?	{0,1}	*0* or *1* occurrence of the preceding regular expression

Next, we provide three simple examples about using quantifiers for matching different sequences of alphanumeric characters and digits. The first one illustrates the matching of any consecutive sequence of alphanumeric characters within the proposed sample string:

```
>> regexp(string,'\w+','match')                                      (3.11)
ans =
  1x7 string array
    "sample"    "1"    "sample"    "a"    "sample20"    "sample"    "321"
```

In the second example, we illustrate the matching of any sequence of alphanumeric characters containing a white space:

```
>> regexp(string,'\w+\s\w+','match')                                 (3.12)
ans =
  1x3 string array
    "sample 1"    "sample a"    "sample 321"
```

Finally, in the third example, we illustrate how to match a sequence of alphanumeric characters followed by a sequence of digits with or without a white space in between:

```
>> regexp(string,'\w+\s?\d+','match')                                (3.13)
ans =
  1x3 string array
    "sample 1"    "sample20"    "sample 321"
```

When used as presented in Table 3.3, just as illustrated above, the quantifier operators behave as *greedy* quantifiers. This means that the operators will consider the largest possible string that satisfies the matching condition to produce a match. However, in some cases, one could be interested in matching the shortest strings that satisfy the given matching condition. In this sense, quantifiers can be turned into *lazy* quantifiers by appending a question mark symbol to them.

The following two examples illustrate the difference between using a *greedy* or a *lazy* quantifier. The first one looks for the largest possible match (*greedy*):

```
>> regexp(string,"'.+'",'match')                                    (3.14)
ans =

   "'sample 1','sample a','sample20','sample 321'"
```

while the second one looks for the shortest possible matches (*lazy*):

```
>> regexp(string,"'.+?'",'match')                                   (3.15)
ans =

  1x4 string array

   "'sample 1'"      "'sample a'"      "'sample20'"      "'sample 321'"
```

Another important group of operators are the logical operators. In the remaining of this section, we will introduce the basic logical operators, and next section will be specifically devoted to conditional operators. Two different kinds of basic logical operators can be distinguished: expression grouping and alternative matching. The syntax of these operators is described in Table 3.4 and some illustrative examples are provided below.

Table 3.4. Basic logical operators

Operator	Type	Description
`(rexp)`	grouping	Groups expression `rexp` and captures a token[2] when a match occurs
`(?:rexp)`	grouping	Groups regular expression `rexp` without capturing a token[2]
`rexpa|rexpb`	alternative	Matches either regular expression `rexpa` or `rexpb` (with that specific precedence order)

Let us consider the sample string:

```
>> string = "This processor's code is: i-351!"                     (3.16)
string =

   "This processor's code is: i-351!"
```

---

[2] More details about tokens will be provided in section 3.4, but it worth mentioning here that apart from the token capturing issue, both operators `(rexp)` and `(?:rexp)` behave exactly in the same way.

The following example is intended to show how to match *vowel-consonant* sequences:

```
>> regexp(string,'[aeiou][^aeiou\s\W\d]{2}','match')                    (3.17)
ans =

    "ess"
```

Notice that, in this case, as no grouping operator was used, the quantifier {2} only applies to the second disjoint operator, which matches consonants. According to this, only patterns of the form *vowel-consonant-consonant* are matched. On the other hand, if we group together both disjoint operators, patterns of the form *vowel-consonant-vowel-consonant* are matched:

```
>> regexp(string,'(?:[aeiou][^aeiou\s\W\d]){2}','match')                (3.18)
ans =

    "oces"
```

Next, let us illustrate the use of the alternative matching operator. In the following example, occurrences of either three substrings: *is*, *i* or *s* are matched, however, precedence is given to the first substring *is*.

```
>> regexp(string,'(?:is)|[is]','match')                                 (3.19)
ans =

  1x6 string array
    "is"      "s"      "s"      "s"      "is"      "i"
```

On the contrary, if we give precedence to *i* and *s* over *is*, we will get the following result:

```
>> regexp(string,'[is]|(?:is)','match')                                 (3.20)
ans =

  1x8 string array
    "i"      "s"      "s"      "s"      "s"      "i"      "s"      "i"
```

It is important to notice, as illustrated in the last two examples (3.19) and (3.20), that the *alternative matching* operator is not invariant with respect to the order of its arguments. Indeed, this operator gives precedence to the first argument, and a possible match for the second argument is evaluated only if a failure occurs for the first one. Notice that for the regular expression used in example (3.20), the substring *is* will never be matched!

3.3 Conditional Matching

A more sophisticated class of logical operators allows for conditional matching. These operators, which are presented in Table 3.5, can be categorized into two groups: anchors and look-arounds. While anchor operators allow for matching regular expressions only when they occur at specific locations such as the start or the end of a word or a string; look-around operators allow for matching regular expressions only when a test or probe pattern is also matched.

Table 3.5. Operators for performing conditional matching

Operator	Type	Description
`\<rexp`	anchor	Matches `rexp` if it occurs at the start of a word
`rexp\>`	anchor	Matches `rexp` if it occurs at the end of a word
`^rexp`	anchor	Matches `rexp` if it occurs at the start of the string
`rexp$`	anchor	Matches `rexp` if it occurs at the end of the string
`rexp(?=test)`	look-ahead	Matches `rexp` if it is followed by expression `test`
`rexp(?!test)`	look-ahead	Matches `rexp` if it is not followed by expression `test`
`(?<=test)rexp`	look-behind	Matches `rexp` if it is preceded by expression `test`
`(?<!test)rexp`	look-behind	Matches `rexp` if it is not preceded by expression `test`

Let us consider the following string (for which some relevant indexes are illustrated in Figure 3.2):

```
>> % sample string for illustrating the use of operators in Table 3.5      (3.21)
>> s = "attention on how to match an expression at any section of the chat";
```

Fig. 3.2. Characters and some relevant indexes for sample string defined in (3.21)

Next examples illustrate the use of anchor operators for matching substring `'at'` at:

```
>> regexp(s,'\<at') % the start of a word                                  (3.22a)
ans =
     1    41
```

```
>> regexp(s,'at\>') % the end of a word                                    (3.22b)
ans =
    41    65
```

```
>> regexp(s,'\<at|at\>') % either the start or the end of a word           (3.22c)
ans =
     1    41    65
```

```
>> regexp(s,'\<at\>') % both the start and the end of a word³              (3.22d)
ans =
   41
>> regexp(s,'^at') % the start of the string                               (3.22e)
ans =
   1
>> regexp(s,'at$') % and, the end of the string                           (3.22f)
ans =
   65
```

Similarly, next examples illustrate the use of look-around operators for matching substring `'at'` only if:

```
>> regexp(s,'at(?=\w)') % followed by an alphanumeric character           (3.23a)
ans =
   1    22
>> regexp(s,'at(?!\w)') % not followed by an alphanumeric character       (3.23b)
ans =
   41    65
>> regexp(s,'(?<=\w)at') % preceded by an alphanumeric character          (3.23c)
ans =
   22    65
>> regexp(s,'(?<!\w)at') % not preceded by an alphanumeric character      (3.23d)
ans =
   1    41
```

3.4 Working with Tokens

In the context of regular expressions, a token is defined as any string that can be captured by using the grouping operator within a given regular expression. Capturing tokens within a given expression allows for the following operations: first, to capture and memorize partial matches for re-matching them in subsequent sections of the same string; and second, to capture and memorize specific partial matches for recovering them after an overall match is produced.

Let us see, for instance, how to re-match a partial match by using tokens. Consider the following string which constitutes a typical sample section of some XML-formatted text:

---

[3] Notice how, in this particular case, the start-of-a-word and the end-of-a-word anchor operators can be combined into the same regular expression in order to force an exact word match.

```
>> s = "<name>Peter</name><age>44</age><city>Boston</city>";                    (3.24)
```

We can use tokens for matching units of the type *<attribute>value</attribute>* as follows:

```
>> matches = regexp(s,'<(\w+)>(.*?)</\1>','match')                              (3.25)
matches =
  1x3 string array
    "<name>Peter</name>"      "<age>44</age>"      "<city>Boston</city>"
```

In this example, the patterns provided inside the two grouped expressions `(\w+)` and `(.*?)` define two different tokens within the overall regular expression `<(\w+)>(.*?)</\1>`, where `\1` is used for matching an exact repetition of the partial match captured in the first token `(\w+)`. In general, as many tokens as defined in a given regular expression can be re-matched again in the same regular expression, in which case `\n` must be used for referring to any re-match of the nth token.

Now, let us illustrate how captured tokens can be recovered. This may be done by using the qualifier argument `'tokens'` in the `regexp` function call, which prompts the function to return all captured tokens for each match:

```
>> tokens = regexp(s,'<(\w+)>(.*?)</\1>','tokens')                              (3.26)
tokens =
  1x3 cell array
    {1×2 string}    {1×2 string}    {1×2 string}
```

Notice that, different from the case illustrated in example (3.25), when `regexp` is prompted to return the captured tokens, the function returns a cell array of string arrays instead of single string array. This is because more than one token can be defined in a regular expression; then, a string array of tokens is returned for each match occurrence and, consequently, captured tokens for all matches are arranged in a cell array of string arrays.

Therefore, in order to display the contents of the output variable `tokens` in (3.26), we must explicitly list the string arrays in it:

```
>> tokens{:}                                                                    (3.27)
ans =
  1x2 string array
    "name"     "Peter"
ans =
  1x2 string array
    "age"     "44"
ans =
  1x2 string array
    "city"     "Boston"
```

Differently from (3.26), captured tokens might be alternatively recovered in the form of a structure. However, for this to be possible, a name must be assigned to each token. This can be done by using the following syntax with the grouping operator `(?<name>rexp)`; and, in correspondence, named tokens can be further referenced within the regular expression by using `\k<name>` instead of `\n`.

Once a name has been assigned to each token, the qualifier argument `'names'` should be used for retrieving captured tokens in a structure array:

```
>> expression = '<(?<attribute>\w+)>(?<value>.*?)</\k<attribute>>';      (3.28a)

>> structure = regexp(s,expression,'names')                              (3.28b)
structure =
  1x3 struct array with fields:
    attribute
    value
```

where the fields of the resulting structure array are given by the names assigned to the tokens, and each element in the array corresponds to one successful match:

```
>> structure(1)                                                          (3.29a)
ans =
  struct with fields:
    attribute: "name"
        value: "Peter"

>> structure(2)                                                          (3.29b)
ans =
  struct with fields:
    attribute: "age"
        value: "44"

>> structure(3)                                                          (3.29c)
ans =
  struct with fields:
    attribute: "city"
        value: "Boston"
```

Table 3.6 summarizes the basic operators for working with tokens that have been discussed here.

Table 3.6. Operators and arguments for working with tokens

Operator	Type	Description
`(rexp)`	grouping	Captures a token when `rexp` is matched
`\n`	reference	Matches the nth captured token
`(?<name>rexp)`	grouping	Captures and names a token when `rexp` is matched
`\k<name>`	reference	Matches the captured token named as `name`

`'tokens'`	argument	Returns captured tokens in a cell array of string arrays
`'names'`	argument	Returns captured tokens in a structure array (only works with named tokens)

3.5 Further Reading

For a more detailed description of the `regexp` function you can refer to the specific MATLAB® online documentation page (The MathWorks 2020).

The mathematical foundations of regular expressions are to be found on automata theory. A good introduction to this topic is available on (Hopcroft and Ullman 1979), and a more advanced reference on regular expressions can be found in (Friedl 1997).

3.6 Proposed Exercises

1. Create a regular expression for matching all words contained into a text string.

    – You should consider a "word" as any sequence of alphanumeric characters containing neither white spaces nor punctuation marks.

2. Create a regular expression for matching any number occurrence within a text string.

    – You should take into consideration all possible formats such as: *123, 123.25, 1,233.24, –0.4, +12*, etc.

3. Consider typical *html* sequences for inserting:

    – Hyperlinks: `<a href="link">text</a>`
    – Images: `<img width="x" height="y" src="image_location" />`
    – Meta-tags: `<meta name="mt_name" content="mt_content" />`
    – Comments: `<!-- body_of_the_comment -->`
    – Create four regular expressions, one for matching each of the above sequence types, and capture into tokens the relevant parameters for each case.
    – Create a single regular expression such that it matches sequences corresponding to both hyperlinks and comments, but not sequences corresponding to images and meta-tags.

- Create a single regular expression such that it matches sequences corresponding to any of the four sequences described above, but not sequences of the general form `<text>`.

4. Consider the following segment of *html* code that has been extracted from a news webpage:

```
<ul>
    <li class="on"><a href="/" id="index">Main Page</a></li>
    <li><a href="/international/" >International</a></li>
    <li><a href="/national/" >National</a></li>
    <li><a href="/sports/" >Sports</a></li>
    <li><a href="/economy/" >Economy</a></li>
    <li><a href="/technology/" >Technology</a></li>
    <li><a href="/culture/" >Culture</a></li>
    <li><a href="/society/" >Society</a></li>
    <li><a href="/blogs/" >Blogs</a></li>
    <li><a href="/participation/" >Give us your opinion</a></li>
</ul>
```

- Create a regular expression for capturing both the link and its corresponding text into two tokens named *link* and *title*, respectively, for each element in the list and return all captured tokens in a structure array format.
- By using MATLAB® functions `disp` and `sprintf`, generate a report listing all pairs of titles and links matched in the previous step. (For a quick overview on `sprintf` you can refer to sections 4.3 and 5.1).

5. Consider the following string `"table elephant door engine car"`, which contains a list of five words. Create and test specific regular expressions for matching words satisfying each of the following conditions:

- Words containing at least an `'a'`.
- Words not containing any `'a'`.
- Words containing at least an `'e'`.
- Words not containing any `'e'`.
- Words containing `'a'`, `'e'`, or both.
- Words containing `'a'` but not `'e'`.
- Words containing `'e'` but not `'a'`.
- Words containing either `'a'` or `'e'`, but not both.
- Words containing `'a'` and `'e'`.
- Words containing neither `'a'` nor `'e'`.

6. Consider the problem of number normalization, in which we are interested in identifying numbers given in alphabetic form (non-digit form) from *one* to *nine hundred ninety nine* and provide their corresponding digit representation.

- Create a function that uses the following syntax:

  `[index,numbers,digits] = extractmumbers(text_string),`
- where `text_string` is the input text to be analyzed, `index` is an integer array pointing to each detected number position in `text_string`, `numbers` is a string array containing the identified numbers (one per cell), and `digits` is also a string array containing the corresponding text digit form.
- For instance, the input `'a total of seven people, out of the seventy five or so taking part in the riot, were arrested'`, will produce the outputs: `index = [12,37]`, `numbers = ["seven", "seventy five"]`, and `digits = ["7","75"]`.

7. Consider the problem of contraction expansion. In this case we are interested in creating a function for identifying English contractions and proposing their corresponding full forms.

    - Similar to the previous exercise, the proposed function should use the following syntax:

      `[index,contract,fullf] = extractcontract(text_string),`
    - where `text_string` is the input text to be analyzed, `index` is an integer array pointing to contraction positions in `text_string`, `contract` is a string array containing the identified contractions, and `fullf` is also a string array containing the proposed full forms.
    - Notice that this is actually a very complex problem as some of the cases may be ambiguous. Consider for instance the case *he's*, which can refer to either *he is* or *he has*. Additionally, in the case *Peter's car*, *'s* should not be considered a contraction! For a list of standard contractions in English, you can check the following link http://grammar.about.com/od/words/a/EnglishContractions.htm Accessed 6 July 2021.

8. Another interesting but hard problem is identifying dates. Try to create a function for identifying a date reference in a given segment of text and retrieving the corresponding date parameters, if any.

    - Consider the syntax: `[isdt,param] = extractdate(text_string),`
    - where `text_string` is the input text to be analyzed, `isdt` is a flag indicating whether a date reference has been found or not, and `param` is a structure containing the corresponding date parameters.
    - The proposed fields and subfields for the `param` structure are:
    - `day.dayofweek` which contains the corresponding day in case it is given: `'Monday'`, `'Tuesday'`, `'Wednesday'`, etc.
    - `day.dayofmonth` which contains the corresponding number of the day in case it is given: `'28th'`, `'fifth'`, `'eleven'`, etc.
    - `day.reference` which contains other day references, if any, such as `'today'`, `'yesterday'`, `'tomorrow'`, etc.

- month.monthofyear which contains the corresponding month's name in case it is given: 'January', 'February', 'March', etc.
- year.yearofcalendar which contains the corresponding year in case it is given: '1985', '2010', '87', etc.
- Consider the following suggestions: assume that just one date reference, if any, is to be detected in the input string of text, and consider dividing the problem into simpler sub-problems by creating specific functions for identifying references to days, months, etc.

9. Similar to the previous exercise, try to create a function for identifying a time reference in a given segment of text and retrieving the corresponding time parameters, if any.

- Consider the syntax: [istm,param] = extracttime(text_string),
- where text_string is the input text to be analyzed, istm is a flag indicating whether a time reference has been found or not, and param is a structure containing the following fields:
- hour which contains the hour's digits: '1', '7', '18', etc.
- minute which contains minute's digits: '15', '43', etc.
- seconds which contains second's digits: '15', '43', etc.
- momentofday which contains the corresponding moment of the day, in case it is explicitly given or can be inferred from the existing information: 'morning', 'noon', 'a.m.', 'p.m.', etc.
- timeformat which should be either '12-hour' or '24-hour' depending on the hour format used in field param.hour
- For simplicity, assume that just a single time reference, if any, is to be detected in the input string of text.

3.7 References

Friedl JEF (1997) Mastering Regular Expressions. O'Reilly & Associates, Sebastopol, CA
Hopcroft JE, Ullman JD (1979) Introduction to Automata Theory, Languages and Computation. Addison-Wesley, Reading, MA
The MathWorks (2020) MATLAB Product Documentation: regexp, https://www.mathworks.com/help/matlab/ref/regexp.html Accessed 6 July 2021

4 Basic Operations with Strings

This chapter presents some basic procedures for operating with strings and the MATLAB® built-in functions for performing the most commonly required string operations. The chapter is structured as follows, first, in section 4.1, basic functions for searching and comparing strings are presented along with the concept of edit distance. Then, in section 4.2, basic functions and procedures for string replacement and insertion are described. In section 4.3, the specific problems of string segmentation and concatenation are addressed. Last, in section 4.4, set operations with strings are presented.

4.1 Searching and Comparing

Most common procedures in text mining applications involve computing statistics, which, in short, is based on counting occurrences of specific substrings in a given set of data. Regular expressions, as already described in the previous chapter, constitute a basic tool for implementing string-based searching and comparing functionalities. However, in addition to `regexp`, the MATLAB® environment offers other specific functions for searching, comparing, and counting. In this section we describe and illustrate the use of these functions, as well as we introduce the fundamental concept of edit distance.

First, let us consider the string search function `strfind`. Its syntax is:

```
output = strfind(content,pattern);
```
(4.1)

where `content` is the text in which the string `pattern` is to be found. It is important to notice that different from `regexp`, where the pattern input variable is a regular expression, in this case, the pattern input variable must be either a string containing the literal text to be matched or a variable of the class `pattern`. As output, this function returns the starting indexes of the occurrences found, if any. In case no occurrences are found, the function returns an empty variable.

Let us consider, for instance, the following example:

```
>> text_string = "This is SECTION 4.1";
```
(4.2)

Now, suppose we want to match occurrences of the character sequence `'is'`. We can use `strfind` to find the occurrences of `'is'` within `text_string`, and the overall number of occurrences can be computed by using the function `length`. This is illustrated in the following example:

© The Author(s), under exclusive license to Springer Nature Switzerland AG 2021
R. E. Banchs, *Text Mining with MATLAB®*, https://doi.org/10.1007/978-3-030-87695-1_4

```
>> indexes = strfind(text_string,'is')                    (4.3a)
indexes =
     3     6

>> nmatch = length(indexes)                               (4.3b)
nmatch =
     2
```

As mentioned above, variables of the class **pattern**[1] can also be used with **strfind**. In the next example we use the **digitsPattern** to find numbers:

```
>> patt = digitsPattern                                   (4.4a)
patt =
  pattern
  Matching:
    digitsPattern

>> indexes = strfind(text_string,patt)                    (4.4b)
indexes =
    17    19

>> nmatch = length(indexes)                               (4.4c)
nmatch =
     2
```

Notice that the described procedure is also valid when no matches are found:

```
>> indexes = strfind(text_string,'example')               (4.5a)
indexes =
    []

>> nmatch = length(indexes)                               (4.5b)
nmatch =
     0
```

The function **strfind** also admits string arrays and cell arrays of character vectors as its first input variable **content**. In such a case, the output is a cell array containing the indexes of the corresponding pattern matches:

```
>> content = ["this is line one","and line two","and three"]   (4.6a)
content =
  1x3 string array
    "this is line one"    "and line two"    "and three"

>> indexes = strfind(content,'ne')                        (4.6b)
indexes =
  1x3 cell array
```

---

[1] For a more comprehensive description of the class **pattern**, its methods, and functions, you can refer to the corresponding documentation by executing the command **help pattern**.

```
    {1×2 double}      {[7]}      {0×0 double}
>> indexes{1}                                                            (4.6c)
ans =
    11    15
>> indexes{2}                                                            (4.6d)
ans =
    7
>> indexes{3}                                                            (4.6e)
ans =
    []
```

If we are not interested in finding the specific locations at which a given pattern occurs within a string, but we just want to check whether the patter occurs or how many times it occurs, we can use the functions `contains` and `count`.

```
>> contains(content,'ne') % checks for pattern occurrences              (4.7a)
ans =
  1x3 logical array
   1   1   0
>> count(content,'ne') % counts pattern occurrences                     (4.7b)
ans =
    2    1    0
```

As seen from (4.7) the function `contains` returns a logical array of the same size of the input string array indicating whether the pattern is contained in the corresponding strings: `'1'` when the string contains the pattern, and `'0'` when it does not. The function `count`, on the other hand, returns a numeric array indicating how many times the pattern occurs in the corresponding strings.

If we are interested in looking for occurrences of a single character within a given character verctor, an alternative and faster procedure can be implemented by means of MATLAB® `eq` (equal) relational operator. However, it is important to know that this procedure will work only with character vectors.

For instance, if we want to search for the occurrences of character `'T'` within the string in example (4.2), we can proceed as follows:

```
>> occurrences = eq(char(text_string),'T')                              (4.8a)
occurrences =
  1x19 logical array
   1   0   0   0   0   0   0   0   0   0   0   1   0   0   0   0   0   0   0
>> % uses an alternative syntax for the equal relational operator       (4.8b)
>> occurrences = char(text_string) == 'T'
occurrences =
```

```
1x19 logical array
  1   0   0   0   0   0   0   0   0   0   0   0   1   0   0   0   0   0   0   0
```

Notice, however, that the matches are returned in the form of a logical array. Then, the resulting number of occurrences can be computed by simply adding up the values in the logical array by means of MATLAB® function **sum**:

```
>> nmatch = sum(occurrences)
nmatch =
    2
```
(4.9)

and indexes to matches can be recovered by means of MATLAB® function **find** as follows:

```
>> indexes = find(occurrences)
indexes =
    1    12
```
(4.10)

In general, much more complex search procedures at the character level can be directly implemented by means of relational and logical operators. For instance, consider the following example in which capital letters are searched for within the string defined in example (4.2):

```
>> disp(text_string)
This is SECTION 4.1
```
(4.11a)

```
>> indexes = find( char(text_string)>='A' & char(text_string)<='Z' )
indexes =
    1    9    10    11    12    13    14    15
```
(4.11b)

More examples like the one illustrated in (4.11) are discussed in more detail in section 4.2, where string replacement and insertion operations are introduced and described.

Another basic MATLAB® function to be considered in this section is the string comparison function **strcmp**. Its syntax is as follows:

```
output = strcmp(content1,content2);
```
(4.12)

where **content1** and **content2** are two strings, or character vectors, to be compared. As output, this function returns a logical variable, or a logical array, which indicates the result of the comparisons: '1' for a successful comparison, and '0' for an unsuccessful one.

The following example illustrates the use of the function **strcmp** when both input variables **content1** and **content2** are strings:

```
>> content1 = "comparable"; content2 = "comparison";
```
(4.13a)

```
>> samestrings = strcmp(content1,content1)
samestrings =
```
(4.13b)

```
  logical
   1
```

```
>> samestrings = strcmp(content1,content2)                        (4.13c)
samestrings =
  logical
   0
```

An alternative way for string comparison can be implemented by means of the equal relational operator. However, notice that, in this case, the behavior when comparing strings is different from when comparing character vectors:

```
>> samestrings = content1 == content2 % compares strings         (4.14a)
samestrings =
  logical
   0
```

```
>> samechars = char(content1) == char(content2) % compares characters   (4.14b)
samechars =
  1x10 logical array
   1   1   1   1   1   1   0   0   0   0
```

```
>> samechars = char(content1) == 'compare'                       (4.14c)
Matrix dimensions must agree.
```

Notice from (4.14b) that when the equal relational operator is used for comparing character vectors, the comparison is actually carried out at the character level instead of at the string level. Notice also from (4.14c) that comparing character vectors in this way requires both character vectors to be of the same length.

Another possible use of **strcmp** involves the comparison of a string array or cell array of character vectors with a single string. In this case, **strcmp** will return a logical array corresponding to the individual comparisons between each element of the array and the single string. Consider the following example:

```
>> content = ["txt","pdf","txt","jpeg","pdf","txt"];             (4.15a)
```

```
>> istxt = strcmp(content,'txt')                                 (4.15b)
istxt =
  1x6 logical array
   1   0   1   0   0   1
```

where, similarly to examples (4.9) and (4.10), MATLAB® functions **sum** and **find** can be used to compute the total number of occurrences and to retrieve the corresponding indexes, respectively.

A third and last possible use of **strcmp** involves the comparison between two string arrays or cell arrays of character vectors. In this case, both arrays must have

the same dimensions, and the output will be a logical array indicating the results
of the comparisons between each corresponding pair of elements:

```
>> output = strcmp(content,["pdf","pdf","pdf","jpeg","jpeg","jpeg"])        (4.16)
output =
  1x6 logical array
   0   1   0   1   0   0
```

It is important to mention that both `strcmp` comparison modalities illustrated in
(4.15) and (4.16) can be implemented by means of the equal relational operator
too. However, this is only possible with string arrays:

```
>> content == "txt"                                                         (4.17a)
ans =
  1x6 logical array
   1   0   1   0   0   1
```

```
>> content == ["pdf","pdf","pdf","jpeg","jpeg","jpeg"]                       (4.17b)
ans =
  1x6 logical array
   0   1   0   1   0   0
```

Three convenient variants of the `strcmp` function are also available: `strncmp`,
`strcmpi`, and `strncmpi`, which allow for partially matching the first n characters,
ignoring case, and both, respectively. All different variants of `strcmp` are summarized in Table 4.1.

Table 4.1. The four different variants of function `strcmp`

Function	Compares	Case Sensitiveness
`strcmp(content,pattern)`	full strings	case sensitive
`strcmpi(content,pattern)`	full strings	case insensitive
`strncmp(content,pattern,n)`	first n characters	case sensitive
`strncmpi(content,pattern,n)`	first n characters	case insensitive

Up to this point, the comparison between two strings has been categorically assessed in a binary sense: either the strings are equal, or they are not. An alternative
powerful and useful concept in text mining, and natural language processing in
general, is the notion of distance between two strings. According to this idea, the
comparison between two strings can result into a numerical value within a discrete
or continuous scale indicating a degree of similarity which ranges from totally different (maximum distance) to completely equal (minimum distance).

This notion can be implemented by means of several different metrics. For instance, the number of matches at the character level can be used as a metric for
string similarity, see example (4.14b). Here, we focus our attention on the most
commonly used definition of string similarity: the edit distance. The edit distance

between two strings, also known as Levenshtein distance,[2] is defined as the minimum number of character editions that are required to transform one string into another. Three different types of character editions can be considered: insertions, deletions and replacements.

Consider, for instance, strings `s1='days'` and `s2='tray'`. These two strings have in common the substring `'ay'`, so we can say they are not totally different. Following the definition of edit distance we just gave above, we can see that we would need one replacement `'d'->'t'`, one insertion `'r'` and one deletion `'s'` in order to transform `s1` into `s2`. According to this, the edit distance between the strings `'days'` and `'tray'` is *3*.

Notice that the same result can be obtained by applying the following sequence of operations: one insertion `'t'`, one replacement `'d'->'r'` and one deletion `'s'`. In general, we can transform `s1` into `s2` by many different edit sequences, but the edit distance accounts only for those that use the minimum number of operations.

The function `editDistance`, available within the Text Analytics Toolbox™, allows for calculating the edit distance between two strings:

```
>> s1 = "days"; s2 = "tray";                                          (4.18a)

>> editDistance(s1,s2)                                                 (4.18b)
ans =
     3

>> editDistance(s2,s1)                                                 (4.18c)
ans =
     3
```

Notice from (4.18) that the edit distance is a symmetric operator. This means that if we compute the minimum number of edit operations required for transforming a string `s2` into `s1` the result must be the same as the minimum number of edit operations required for transforming `s1` into `s2`. This can be easily verified, for our previous example, by noticing that we can reconstruct `s1` from `s2` by applying exactly the same replacements we used to transform `s1` into `s2` and changing the insertions by deletions and vice versa, so the minimum number of edit operations required remains the same.

The basic operation of the edit distance algorithm, which is illustrated in Figure 4.1 for the specific case of `s1='days'` and `s2='tray'`, can be described as follows:

- First, a matrix of size `(length(s1)+1)×(length(s2)+1)` is initialized such as the elements in its first column and first row represent the edit distances between the *null* string (a string with no characters) and each possible substring of `s1` and `s2`, respectively.

---

[2] Proposed by the Russian mathematician Vladimir Levenshtein in 1965.

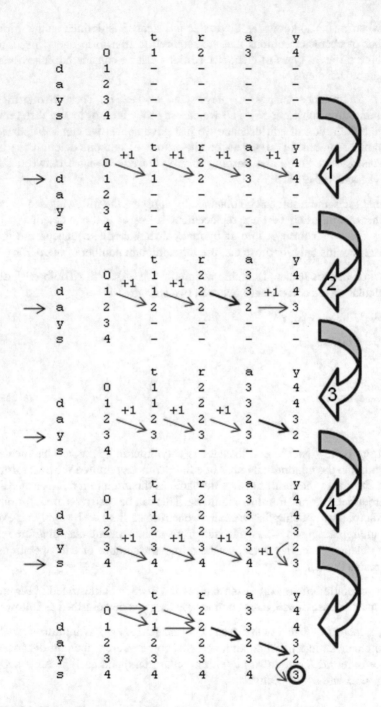

Fig. 4.1. Illustrative example of the edit distance computation between *days* **and** *tray*

- Second, a recursive procedure, implemented by means of two nested loops (the outer running over the matrix rows, and the inner running over the matrix columns), computes the edit distance increments resulting when each new pair of characters from s1 and s2 is taken into account. If the two characters are the same, the edit distance count is maintained; otherwise, the minimum of the three adjacent distances already computed is incremented by one and used as the new edit distance count.
- Finally, after all matrix cell values have been computed, the total edit distance between s1 and s2 corresponds to the element in the last column and row of the matrix.

As can be verified from the last matrix representation at the bottom of Figure 4.1, two different sequences of edit operations can produce the minimum edit distance of 3. Notice also that each of the entries in the matrix provides the actual edit distance between the corresponding substrings of s1 and s2.

4.2 Replacement and Insertion

Up to this point, we have seen some functions and procedures for matching patterns, searching, and comparing strings. These procedures are very useful for both identifying and extracting specific segments of text from a larger body or collection of text contents. However, sometimes it is also useful to be able to modify the identified segments of text by replacing or inserting substrings at specific locations of the text under consideration. In this sense, this section introduces some basic functions for replacing and inserting substrings within a given string.

First of all, let us consider the simple case of single character replacement, in which we are interested in replacing the occurrences of specific characters by a replacement character. In this specific case, the most effective and simple way to proceed is by making use of MATLAB® relational and logical operators. However, be aware this works for character vectors only.

Consider, for instance, the following examples. In the first one, all occurrences of any digit are replaced by the hash character '#'. In the second one, all occurrences of any non-alphanumeric character are replaced by the star character '*'.

```
>> txt = 'Today''s 2nd winner is ticket number 638-775-321 !';      (4.19a)

>> digit = (txt>='0')&(txt<='9');                                    (4.19b)
>> s = txt; s(digit) = '#'
s =
    'Today's #nd winner is ticket number ###-###-### !'

>> alpha = ((txt>='A')&(txt<='Z')) | ((txt>='a')&(txt<='z'));        (4.19c)
```

```
>> s = txt; s(not(digit|alpha)) = '*'
s =
   'Today*s*2nd*winner*is*ticket*number*638*775*321**'
```

As can be seen from the examples, the illustrated procedure performs an in-place computation, where all characters within the "input" string s corresponding to digits, in example (4.19b), and to non-alphanumeric characters, in example (4.19c), are overwritten with replacement characters '#' and '*', respectively.

It is worth mentioning that this procedure can also be used for deleting characters. This can be achieved by replacing characters matches with the null string ''. However, the same trick is not allowed for inserting characters. Consider the following cases:

```
>> s = txt; s(not(digit|alpha)) = ''                                    (4.20a)
s =
   'Todays2ndwinneristicketnumber638775321'
```

```
>> s = txt; s(not(digit|alpha)) = '<*>'                                 (4.20b)
Unable to perform assignment because the left and right sides have a different
number of elements.
```

Although efficient and simple, the single character replacement method just described is indeed very limited. It only allows for character-by-character replacement strategies and cannot be directly used for insertions or replacements involving substrings. In this sense, the MATLAB® programming environment offers other functions, which are much more convenient than the single character replacement method, for conducting more advanced string replacement and insertion operations. In the remaining of this section, we will discuss two of these functions in more detail, while others will be left for self-study in section 4.6.

First, let us consider the function regexprep. This function constitutes an extension to the regular expression function regexp that was already described in detail in section 3.1. It allows for performing substring substitutions on patterns that have been matched by means of regular expressions. The syntax of regexprep is as follows:

```
output = regexprep(string,expression,replacement);                      (4.21)
```

where replacement is the substring to be used for replacing all occurring instances of expression within string. As output, this function returns an updated version of string in which all replacements, if any, have been performed.

Consider the following example in which the function regexprep is used to replace plain numbers and percentages with the tag <NUM>:

```
>> txtstr = "In 1994, 20% of investments produced 80% of revenues";     (4.22a)
```

```
>> regexprep(txtstr,'\d+\%*','<NUM>')                                    (4.22b)
```

```
ans =
    "In <NUM>, <NUM> of investments produced <NUM> of revenues"
```

Alternatively, `regexprep` also allows for using those tokens extracted during the pattern matching process as replacement substrings. Consider, for instance, the following example in which the *html* specifications for an image's width and height are swapped:

```
>> charv = '<img width="250" height="300" src="image_location" />';      (4.23a)
>> expression = '(<img.+?width=)(.+?)(".+?height=)(.+?)(".+?/>)';

>> regexprep(string(charv),expression,'$1$4$3$2$5')                      (4.23b)
ans =
    "<img width="300" height="250" src="image_location" />"
```

In general, `regexprep` allows for substring replacements under any of the regular expression matching conditions allowed by `regexp`. For a more detailed description on the usage of `regexp` you must refer to chapter 3.

Another MATLAB® function for substring replacement is `replace`. Similar to `regexprep` it allows for replacing all occurrences of a given substring with a replacement substring, but different from it, `replace` does not operates with regular expressions. Instead, similar to `strfind`, the pattern input variable in `replace` must be either a string containing the literal text to be matched or a variable of the class `pattern`, see example (4.4). The syntax of `replace` is as follows:

```
output = replace(string,pattern,replacement);                            (4.24)
```

where `replacement` is the substring to be used for replacing all occurring instances of `pattern` within `string`. As `output`, an updated version of `string` in which all replacements have been performed is returned.

The following examples illustrates the use of `replace`. In particular, example (4.25a) reproduces example (4.22):

```
>> replace(txtstr,(digitsPattern+'%')|digitsPattern,'<NUM>')             (4.25a)
ans =
    "In <NUM>, <NUM> of investments produced <NUM> of revenues"

>> replace(txtstr,'produced','generated')                                (4.25b)
ans =
    "In 1994, 20% of investments generated 80% of revenues"
```

Additionally, `replace` admits string arrays and cell arrays of character vectors as inputs too. This is illustrated in the following examples:

```
>> txtstr = ["next Monday","next Tuesday","last Friday"];                (4.26a)

>> replace(txtstr,'next','last')                                         (4.26b)
ans =
```

```
 1x3 string array
    "last Monday"     "last Tuesday"     "last Friday"
>> patt = ["Monday","Tuesday","Friday"];                                (4.26c)
>> rplc = ["week","month","year"];
>> replace(txtstr,patt,rplc)
ans =
 1x3 string array
    "next week"      "next month"     "last year"
```

Other functions that perform more specific replacement and insertion operations on strings include **erase**, **strip**, **insertAfter**, **insertBefore**, **replaceBe-
tween**, etc. With the exception of **erase** and **strip**, which are introduced in the
following section, some of these functions are explored in more detail in some of
the exercises proposed in section 4.6.

4.3 Segmentation and Concatenation

Let us now consider the specific problem of segmenting a text string into multiple
substrings. A common application of this problem involves separating alphanumeric from non-alphanumeric characters in a given body of text, which is fundamental to more sophisticated problems involving string segmentation, such as tokenization, for example. One of the MATLAB® functions for splitting text strings
is **split**, which is illustrated in the following example:

```
>> txtstr = " He'd said: ""I just don't care!!!"""                       (4.27a)
txtstr =
   " He'd said: "I just don't care!!!""

>> split(txtstr)'                                                       (4.27b)
ans =
 1x7 string array
    ""      "He'd"     "said:"    ""I"     "just"     "don't"      "care!!!""
```

As seen from (4.27b), the function **split** breaks down a given input string into
substrings by splitting the original string at whitespace locations. The output variable is a string array of the corresponding size.

The function **split** also allows for specifying the string splitting locations:

```
>> % defining split locations with a cell array of characters           (4.28a)
>> split(txtstr,{':','!','"',' '})'
ans =
  1x13 string array
  Columns 1 through 9
```

```
    ""      "He'd"     "said"     ""      ""      "I"      "just"     "don't"     "care"
  Columns 10 through 13
    ""        ""        ""        ""
```

```
>> % defining split locations with a pattern                                  (4.28b)
>> split(txtstr,alphanumericBoundary)'
ans =
  1x17 string array
  Columns 1 through 9
    " "      "He"      "'"      "d"      " "      "said"      ": ""      "I"      " "
  Columns 10 through 17
    "just"      " "      "don"      "'"      "t"      " "      "care"      "!!!""
```

More elaborated procedures attempting to better split an input string into words can use **split** in combination with other string functions. Consider, for instance, the following example in which **erase** is used to remove non-word characters, the function **strip** is used to remove preceding and trailing whitespaces, and **split** is finally used to break the input string into words:

```
>> patt = characterListPattern(':,.;"?!'); % defines the pattern            (4.29a)
>> temp = erase(txtstr,patt) % removes non-word characters
temp =
    " He'd said I just don't care"
```

```
>> temp = strip(temp) % removes preceding and trailing whitespaces          (4.29b)
temp =
    "He'd said I just don't care"
```

```
>> split(temp)' % splits the input string into words                        (4.29c)
ans =
  1x6 string array
    "He'd"      "said"      "I"      "just"      "don't"      "care"
```

An alternative procedure for extracting tokens from a given input string can be implemented by means of MATLAB® function **strtok**. This function allows for extracting tokens delimited by a given set of characters in a sequential manner. The syntax of **strtok** is as follows:

```
[token,remainder] = strtok(string,delimiters);                              (4.30)
```

where **token** is the first token in **string**, which is delimited by any of the characters provided in **delimiters**, and **remainder** contains the remaining portion of **string**. Any leading delimiter character in **string** is ignored, so **strtok** will always return the first valid token within **string**. In case the variable **delimiters** is omitted, the whitespace is used as a default delimiter character.

If we apply this function to the same input string defined in (4.27a) and specify the same set of delimiter characters as in (4.29a) plus the whitespace character, we will get the same first token as in (4.29c):

```
>> [token,remainder] = strtok(txtstr,':.,;"?! ')                        (4.31)
token =
    "He'd"
remainder =
    " said: "I just don't care!!!""
```

If we are interested in extracting all tokens from the input string, a recursive procedure must be implemented. In the following example, such a recursive procedure is illustrated, and the extracted tokens are saved into a string array:

```
>> newstring = txtstr; index = 0;                                       (4.32)
>> while 1
        [temp,newstring] = strtok(newstring,':.,;"?! ');
        if temp=="", break; end;
        index = index+1;
        tokens(index) = temp;
    end;
>> tokens
tokens =
  1x6 string array
    "He'd"     "said"     "I"     "just"     "don't"     "care"
```

An alternative and less obvious way to perform token extraction without a recursive implementation like the one in (4.32) involves the use of MATLAB® function **eval** along with some replacement and insertion operations. Consider, for instance, the following example:

```
>> % removes non-word characters, preceding and trailing whitespaces    (4.33a)
>> patt = characterListPattern(':,.;"?!');
>> temp = strip(erase(txtstr,patt));

>> % replaces all whitespaces between words with the sequence ","        (4.33b)
>> temp = replace(temp,' ','","')
temp =
    "He'd","said","I","just","don't","care"

>> % executes the command tokens=["..."] to create the string array     (4.33c)
>> eval(['tokens=["',char(temp),'"]'])
tokens =
  1x6 string array
    "He'd"     "said"     "I"     "just"     "don't"     "care"
```

The procedure described in (4.33) consists of three steps. First, any consecutive sequence of non-word characters and preceding and trailing whitespaces are removed (4.33a). Then, all remaining whitespaces between consecutive words are replaced with the sequence `","` (4.33b). At this point, the auxiliary variable `temp` contains a string of the form: `token_1","token_2",...,"token_n`, which follows the basic syntax for defining string arrays. Finally, the command for creating the string array `tokens=["token_1","token_2",...,"token_n"]` is constructed and evaluated by using MATLAB® function `eval` (4.33c).

The procedures illustrated in (4.29), (4.32) and (4.33) constitute naïve approximations to the complex problem of word tokenization. Those can be certainly improved for producing more desirable segmentations in which, for example, contractions are separated from their corresponding words (i.e. `He 'd` and `don 't` instead of `He'd` and `don't`), abbreviations maintain the dots (i.e. `Mr.` instead of `Mr`), numerical expressions are not segmented (i.e. `1,230.50` instead of `1 230 50`), and so on.[3]

In the same way we are interested in segmenting a given string into substrings and tokenized forms, we are also interested in going back from these representations to the corresponding original strings. However, before describing this procedure in detail, let us present some MATLAB® functions for string concatenation and formatting.

String concatenation can be done in several different ways. The simplest way is by using the `join` function, which is illustrated in the following example:

```
>> data = ["This is"," an example ","on string concatenation"]          (4.34a)
data =

  1x3 string array

    "This is"      " an example "      "on string concaten…"

>> join(data) % joins strings with whitespaces                          (4.34b)
ans =

    "This is  an example  on string concatenation"

>> join(data,'') % joins strings with the given token                   (4.34c)
ans =

    "This is an example on string concatenation"

>> join(data,{' :','-> '}) % joins strings with different tokens        (4.34d)
ans =

    "This is : an example -> on string concatenation"
```

As seen from (4.34b), `join` concatenates strings with whitespaces by default. However, delimiters can also be specified: either a common one for all strings (4.34c) or specific ones for each pair of strings (3.34d).

---

[3] More about word tokenization is explored in exercise 4.6-7 in section 4.6. This problem is also revisited in chapter 6, where the function `tokenizedDocument`, available as part of the Text Analytics Toolbox™, is introduced and discussed in detail.

Two alternative functions for text concatenation are `strcat` (horizontal concatenation) and `strvcat` (vertical concatenation). However, these functions operate with character vectors instead of strings, producing as output a one-dimensional character vector or a two-dimensional character array, respectively. The following example illustrates the operation of these two functions:

```
>> horcat = strcat(data{:})                                          (4.35a)
horcat =
    'This is an exampleon string concatenation'

>> vercat = strvcat(data{:})                                         (4.35b)
vercat =
  3x23 char array
    'This is               '
    ' an example           '
    'on string concatenation'

>> whos horcat vercat                                                (4.35c)
  Name          Size           Bytes  Class    Attributes
  horcat        1x41              82   char
  vercat        3x23             138   char
```

Some important observations can be derived from example (4.35). First, notice the way the input string array `data` has been indexed `data{:}`, i.e. using braces. This is actually retrieving individual elements from the string array, so the provided inputs to both functions `strcat` and `strvcat` are actually character vectors rather than strings.

Another important observation is that, in the first case, when `strcat` is used, trailing whitespaces are removed from the input character vectors. Notice from (4.35a) how the words `example` and `on` have been merged together into a single token `exampleon`. If preserving trailing whitespaces is a requirement, a different concatenation strategy must be considered. In the second case, on the other hand, elements of the input string array are mapped into a two-dimensional character array. Notice that, in this case, each element is padded with whitespaces so each row in the array has the same number of characters.

Alternative procedures for horizontal and vertical concatenation of strings can be implemented by means of bracket concatenation and the function `char`, respectively. Consider the following examples:

```
>> bracket_horcat = [data{:}] % horizontal concatenation            (4.36a)
bracket_horcat =
    'This is an example on string concatenation'

>> char_vercat = char(data{:}) % vertical concatenation             (4.36b)
char_vercat =
  3x23 char array
```

```
'This is                    '
' an example                '
'on string concatenation'
```

Notice how, different from `strcat`, the use of bracket-based concatenation preserves trailing whitespaces. In the case of vertical concatenation with `char`, the result is totally equivalent to the one produced by `strvcat`.

Although it is not always possible to go back from a horizontal concatenated output to its original input string array (unless some indexing variable or segmentation rule is explicitly created or provided); it is certainly possible to map back a vertical concatenation to its original string array, where each row in the two-dimensional character array will be placed into a separated element of the resulting string array. This can be easily done by using the function `string`, as illustrated in the following example:

```
>> newdata = string(vercat)                                        (4.37a)
newdata =
  3x1 string array
    "This is                    "
    " an example                "
    "on string concatenation"
```

```
>> newdata = strip(newdata)'                                       (4.37b)
newdata =
  1x3 string array
    "This is"    "an example"    "on string concaten…"
```

Notice that the transposition operator ' has been used in (4.37b) to obtain the original 1×3 cell array, as function `string` always generates by default an $n \times 1$ string array when the input is a two-dimensional character array. Notice also from (4.37b) that `strip` has been used to remove all preceding and trailing whitespaces from the string array elements, as function `string` preserves them.

An additional alternative procedure for concatenating text can be implemented by using the string-formatting function `sprintf`, which (along with `fprintf`, described in detail in section 5.1) constitute the MATLAB® implementation of the ANSI C standard for text formatting. The basic syntax of `sprintf` is as follows:

```
formatted_text = sprintf(format,input,...)                         (4.38)
```

where `format` is a format-specification string containing the desired conversion characters following C language syntax and definitions, and `input` is a data variable to be formatted into either a string or a character vector `formatted_text`. In general, the function admits as many input data variables as specified by the format string `format`. Here, we will restrict our description of `sprintf` to our specific purpose of string concatenation. For a more detailed description of this function, you can refer to chapter 5 (where `fprintf` is described in detail), or consult the function's self-contained description by executing `help sprintf` in the MATLAB® command window.

Concatenating the elements of the string array `data`, previously defined in (4.34a), by means of `sprintf` only requires using the formatting conversion specification for strings `%s`. In this way, this concatenation can be performed as follows:

```
>> txtstr = sprintf("%s",data) % returns a string                          (4.39a)
txtstr =
    "This is an example on string concatenation"
>> charv = sprintf('%s',data) % returns a character vector                  (4.39b)
charv =
    'This is an example on string concatenation'
>> whos txtstr charv                                                        (4.39c)
  Name          Size            Bytes  Class      Attributes
  charv         1x42               84  char
  txtstr        1x1               230  string
```

The main advantage of using `sprintf` for string concatenation is that additional text formatting can be applied to the data. Consider the following example:

```
>> sprintf("-> %s%s\n    %s!",data)                                        (4.40)
ans =
    "-> This is an example
        on string concatenation!"
```

A last and very convenient form of string concatenation is the use of the addition (+) operator, but be aware that this will work only with strings:

```
>> data(1) + data(2) + data(3)                                             (4.41)
ans =
    "This is an example on string concatenation"
```

Table 4.2 summarizes the concatenation options described in this section along with their most important characteristics.

Table 4.2. String concatenation alternatives and main characteristics

Option	Type	Output	What is important to know
`join`	horizontal	string / char	Preserves trailing whitespaces and inserts delimiters
`+`	horizontal	string only	Preserves trailing whitespaces
`strcat`	horizontal	char only	Does not preserve trailing whitespaces
`strvcat`	vertical	char only	Pads rows with white spaces to fit largest substring
`[...]`	horizontal	char only	Preserves trailing whitespaces
`char`	vertical	char only	Pads rows with white spaces to fit largest substring
`sprintf`	horizontal	string / char	Preserves trailing whitespaces and allows for formatting

Finally, let us reconsider the problem of going back from a tokenized representation, as the one previously illustrated in (4.29c), (4.32) and (4.33c), into a single string. Following either (4.34) or (4.39), this conversion can be easily done by using `join` or `sprintf` as follows:

```
>> join(tokens)                                                    (4.42a)
ans =
    "He'd said I just don't care"
>> strip(sprintf("%s ",tokens))                                    (4.42b)
ans =
    "He'd said I just don't care"
```

If the original string from (4.27a) is to be reconstructed, the information about the corresponding delimiters, as well as preceding and trailing characters, has to be provided too:

```
>> join([" ",tokens,"!!!"""],{'',' ',': "',' ',' ',' ',''})        (4.43a)
ans =
    " He'd said: "I just don't care!!!""
>> sprintf(" %s %s: ""%s %s %s %s!!!""",tokens)                     (4.43b)
ans =
    " He'd said: "I just don't care!!!""
```

4.4 Set Operations

Another important type of operations that can be performed with text is set operations. Set operations with text contents can be conducted at both the character level (where strings are considered to be sets and characters constitute the set's elements) and the token level (where string arrays are the sets and individual strings constitute the set's elements). In the first case, character vectors are used for text representation; and, in the second case, string arrays are used. Table 4.3 summarizes the six basic functions the MATLAB® programming environment provides for performing set operations. In this section, the basic operation of these six functions is illustrated.

Table 4.3. Basic functions for set operations

Function	Inputs	Symmetric	Main description
unique	1	–	Lists the elements in the input set (without repetitions)
union	2	yes*	Lists the elements contained in at least one of the two input sets
intersect	2	yes	Lists the elements that are contained in both input sets

setxor	2	yes	Lists the elements that are in either input sets, but not in both
setdiff	2	no	Lists the elements that are contained in set *1* but not in set *2*
ismember	2	no	A binary index to elements contained in set *1* but not in set *2*

* Although the union set is the same regardless the order in which the two input sets are provided, the indexes to elements in the original sets are different, see (4.49).

Let us consider the following two strings for which both, single string and token-per-cell, representations are provided:

```
>> s1 = "is this a test ?";                                              (4.44a)
>> c1 = split(s1)';

>> s2 = "what a test !";                                                 (4.44b)
>> c2 = split(s2)';
```

Set operations can be applied at the character level by using character vector representations of **s1** and **s2** as input sets. Consider, for instance, the following example illustrating the use of **unique** for extracting the sets of characters composing strings **s1** and **s2**:

```
>> unique(char(s1))                                                      (4.45a)
ans =

    ' ?aehist'

>> unique(char(s2))                                                      (4.45b)
ans =

    ' !aehstw'
```

In addition to the set of unique elements composing a given string, **unique** can also return indexes relating both the original set and its corresponding set of unique elements:

```
>> % gets the list of unique elements along with the indexes            (4.46a)
>> [uniqueset,fwdidx,bwdidx] = unique(char(s1));

>> % maps the original string into the set of unique elements           (4.46b)
>> s1{1}(fwdidx)
ans =

    ' ?aehist'

>> % maps back the set of unique elements into the original string      (4.46c)
>> uniqueset(bwdidx)
ans =

    'is this a test ?'
```

As seen from example (4.46), the first index **fwdidx** points to the specific character locations within string **s1** corresponding to the elements in **uniqueset**, and the second index **bwdidx** allows for reconstructing the original string **s1** from the set of elements contained in **uniqueset**.

Consider now the following three examples, where the use of functions `union`, `intersect` and `setxor` is illustrated:

```
>> union(char(s1),char(s2))                                           (4.47a)
ans =
    ' !?aehistw'
>> intersect(char(s1),char(s2))                                       (4.47b)
ans =
    ' aehst'
>> setxor(char(s1),char(s2))                                          (4.47c)
ans =
    '!?iw'
```

Similar to `unique`, these functions can also return indexes which, in this case, point to the corresponding elements in each of the input sets:

```
>> [un,id1,id2] = union(char(s1),char(s2));                           (4.48a)
>> id1'
ans =
      3     16      9     12      5      1      2      4
>> id2'
ans =
     13      1
>> sort([s1{1}(id1),s2{1}(id2)])
ans =
    ' !?aehistw'
>> [in,id1,id2] = intersect(char(s1),char(s2));                       (4.48b)
>> id1'
ans =
      3      9     12      5      2      4
>> id2'
ans =
      5      3      9      2     10      4
>> [s1{1}(id1),s2{1}(id2)]
ans =
    ' aehst aehst'
>> [xor,id1,id2] = setxor(char(s1),char(s2));                         (4.48c)
>> id1'
ans =
     16      1
>> id2'
ans =
     13      1
```

```
>> sort([s1{1}(id1),s2{1}(id2)])
ans =
    '!?iw'
```

These three functions (`union`, `intersect` and `setxor`) are symmetric functions in the sense than changing the order of the two input sets produces the same output set. However, for the case of the `union` function, it can be seen that the indexes pointing to the elements in the original sets depend on the order in which the input sets are provided. Consider, for instance, the following example:

```
>> [un,id2,id1] = union(char(s2),char(s1));                              (4.49)
>> id1'
ans =
    16     1
>> id2'
ans =
     5    13     3     9     2    10     4     1
>> sort([s1{1}(id1),s2{1}(id2)])
ans =
    ' !?aehistw'
```

Notice from (4.48a) and (4.49) how the same union set `' !?aehistw'` was obtained in both cases. However, while in the former case, the set was obtained by adding to the unique set of elements in `s1` those elements in `s2` that are not contained in `s1`; in the latter case, the set was obtained by adding to the unique set of elements in `s2` those elements in `s1` that are not contained in `s2`.

To better visualize the relations among the four described functions (`unique`, `union`, `intersect` and `setxor`), Figure 4.2 presents a diagram illustrating the different sets generated by each function. In the figure, the whitespace has been replaced by a dash for clarity purposes.

As seen from the picture, unless one set is entirely contained within the other, the set generated by function `setxor` is composed by two parts: the elements of `s2` that are not contained in `s1`, and the elements of `s1` that are not contained in `s2`. These are precisely the sets defined by `id2` and `id1` in (4.48a) and (4.49), respectively; which are also the sets that the last two functions in Table 4.3 (`setdiff` and `ismember`) allow for finding.

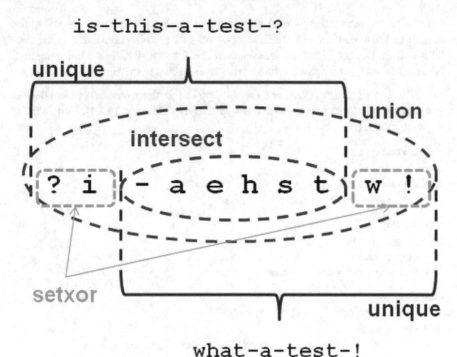

Fig. 4.2. Illustrative diagram of the different sets generated by the four functions `unique`, `union`, `intersection` **and** `setxor`

More specifically, `setdiff` and `ismember` allow for finding the sets of non-common elements between the two input sets and their corresponding locations:

```
>> setdiff(char(s1),char(s2))                              (4.50a)
ans =
    '?i'
>> ismember(char(s1),char(s2))                             (4.50b)
ans =
  1x16 logical array
   0   1   1   1   1   0   1   1   1   1   1   1   1   1   1   0
>> setdiff(char(s2),char(s1))                              (4.50c)
ans =
    '!w'
>> ismember(char(s2),char(s1))                             (4.50d)
ans =
  1x13 logical array
   0   1   1   1   1   1   1   1   1   1   1   1   0
```

Actually, as seen from (4.50b) and (4.50d), `ismember` returns a binary index pointing to those elements in the first input set that are also contained in the second input set. So, the elements contained in the first set, which are not contained in the second one, correspond to those with zero entries in the binary index.

Finally, the same set operations can be applied at the token level by using as input sets the string array representations `c1` and `c2` defined in (4.44). Consider the following examples:

```
>> unique(c1)                                              (4.51a)
ans =
  1x5 string array
    "?"     "a"     "is"     "test"     "this"

>> unique(c2)                                              (4.51b)
ans =
  1x4 string array
    "!"     "a"     "test"     "what"

>> union(c1,c2)                                            (4.51c)
ans =
  1x7 string array
    "!"     "?"     "a"     "is"     "test"     "this"     "what"

>> intersect(c1,c2)                                        (4.51d)
ans =
  1x2 string array
    "a"     "test"

>> setxor(c1,c2)                                           (4.51e)
ans =
  1x5 string array
    "!"     "?"     "is"     "this"     "what"

>> setdiff(c1,c2)                                          (4.51f)
ans =
  1x3 string array
    "?"     "is"     "this"

>> setdiff(c2,c1)                                          (4.51g)
ans =
  1x2 string array
    "!"     "what"

>> ismember(c1,c2)                                         (4.51h)
ans =
  1x5 logical array
   0   0   1   1   0
```

```
>> ismember(c2,c1)                                                    (4.51i)
ans =
  1x4 logical array
   0   1   1   0
```

4.5 Further Reading

For a more detailed description of the functions presented in this chapter you can refer to the MATLAB® online Product Documentation (The MathWorks 2020).

The concept of edit distance described in section 4.1 was first introduced by Levenshtein (1966). A more detailed discussion on the algorithm is available in (Wagner and Fischer 1974). A nice introduction to set theory, set operations and logic can be found in (Stoll 1979).

4.6 Proposed Exercises

1. Create a function for either lowercasing or uppercasing a given string.
   - The function must have the following syntax `recase(input,type)`, where `input` is the string to be re-cased and `type` is a parameter that indicates the type of re-casing to be applied: `'lc'` for lowercasing, `'uc'` for uppercasing and `'inv'` for inverting the case.
   - Use MATLAB® functions `upper` and `lower` for `'lc'` and `'uc'`.
   - For `'inv'`, consider developing your own uppercasing and lowercasing mechanism following the strategy presented in (4.19c) for identifying uppercase and lowercase characters. Notice that the difference between *ASCII* codes of lowercase and uppercase is *32*.

2. Consider the following script, which implements the edit distance algorithm illustrated in Figure 4.1:

```
mtx = (1:length(s1)+1)'*(1:length(s2)+1)-1; % initialization
for n=1:length(s1) % external loop over string s1 (matrix rows)
    for k=1:length(s2) % internal loop over string s2 (matrix columns)
        temp = mtx(n,k); % current position's edit distance is loaded
        if not(s1(n) == s2(k)) % if characters do not match
        % increments by one the minimum of the three adjacent distances
            temp = min([temp,mtx(n,k+1),mtx(n+1,k)])+1;
        end;
```

```
        mtx(n+1,k+1) = temp; % updates the new edit distance
    end;
end;
edist = mtx(end,end) % the edit distance is the last element
```

- Use the script to create the function `charedist(s1,s2,type)` for computing the edit distance between two strings.
- The input variables `s1` and `s2` are the two strings to be compared.
- The parameter `type` indicates the type of result to be returned by the function: `'dist'` for the edit distance between both strings and `'mtx'` for the complete matrix containing the incremental edit distances.
- What kind of practical applications do you think function `charedist` can be useful for?

3. Consider the function `charedist` created in the previous exercise:

- Modify the function to create the new function `wordedist` for computing the edit distance between two strings at the word (or token) level.
- Again, the input variables `s1` and `s2` are two strings to be compared (you can assume words (or tokens) are separated by whitespaces).
- And `type` indicates the type of result to be returned by the function: `'dist'` for the edit distance between both strings and `'mtx'` for the complete matrix containing the incremental edit distances.
- What kind of practical applications do you think function `wordedist` can be useful for?

4. Create a function for tokenizing text by splitting a string into substrings, each of which contains a sequence of either alphanumeric (words and numbers) or non-alphanumeric (punctuation) characters:

- Use the following syntax: `simple_tokenizer(input,type)`.
- Where `input` is the text to be tokenized and `type` a parameter that indicates whether non-alphanumeric characters are to be removed (`'remove'`) or retained (`'retain'`).
- Consider using functions like `insertAfter`, `insertBefore`, `replace`, `erase`, etc. along with some `patterns`.

5. Consider the same segment of *html* code from exercise 3.6-4:

```
<ul>
<li class="on"><a href="/" id="index">Main Page</a></li>
<li><a href="/international/" >International</a></li>
<li><a href="/national/" >National</a></li>
<li><a href="/sports/" >Sports</a></li>
<li><a href="/economy/" >Economy</a></li>
<li><a href="/technology/" >Technology</a></li>
```

```
<li><a href="/culture/" >Culture</a></li>
<li><a href="/society/" >Society</a></li>
<li><a href="/blogs/" >Blogs</a></li>
<li><a href="/participation/" >Give us your opinion</a></li>
</ul>
```

- Write a script to extract all links along with their corresponding texts by using functions like `replace, split, extract, extractBetween, startsWith, endsWith`, etc.
- Save the extracted links and texts in a *nx2* string array, which containing the links in its first column and the texts in the second column.
- Print a report showing the links and their corresponding titles.

6. Consider a stream of system logs that are collected into a string array, one log entry per string array element:

```
"warning: 6477747464 {webservice: time out after 5 secs}"
"error: 6477748976 {webservice: system down}"
"error: 6477773423 {data base: non responsive}"
"warning: 6477823349 {backend process: memory at 90%}"
"status update: 6477829765 {webservice: service restarted}"
"error: 6478319565 {backend process: out of memory}"
```

- Write a script that reads and parses the logs one by one and builds a report table containing only error information.
- The table must contain three columns: *timestamp*, *source* and *problem*.
- Store the table in a *nx3* string array, including the column headers in the first row.
- Again, consider using functions like `replace, split, extract, extractBetween, startsWith, endsWith`, etc
- Print a report showing the table, including the column headers.

7. Consider the problem of vocabulary extraction: given a text input, we are interested in extracting the unique set of different tokens (both alphanumeric and non-alphanumeric sequences of characters) conforming such text.

- Create a function for performing vocabulary extraction assuming that the input text is given either a string or character vector format.
- The proposed syntax is: `[vocab,index] = vocabextract(input)`.
- Where `input` is the input text, `vocab` is a string array containing the vocabulary (one token per element), and `index` is an integer array containing the corresponding locations of each vocabulary token into the original input, i.e. `input` can be reconstructed from `vocab(index)`.
- Consider using the function `unique` as described in (4.48).

8. Create the function `vocabmerge` for merging two given vocabularies `vocab1` and `vocab2` (as generated by function `vocabextract` in the previous exercise) into one single vocabulary set `vocab`.

- The function must also be able to generate and return converted indexes `idx1` and `idx2` that allow for reconstructing the original texts that originated `vocab1` and `vocab2`.
- For instance, the string that produced vocabulary `vocab1`, should be recovered from `vocab(idx1)`. Similarly, the string that produced vocabulary `vocab2`, should be recovered from `vocab(idx2)`.
- The proposed syntax is as follows:
  `[vocab,idx1,idx2] = vocabmerge(vocab1,index1,vocab2,index2)`
- You should consider using both functions `union` and `intersect`.

4.7 References

Levenshtein V (1966) Binary codes capable of correcting deletions, insertions, and reversals. Soviet Physics Doklady 10: 707-716

The MathWorks (2020) MATLAB Product Documentation, https://www.mathworks.com/help/matlab/characters-and-strings.html Accessed 6 July 2021

Stoll RR (1979) Set Theory and Logic. Dover Publications

Wagner RA, Fischer MJ (1974) The string-to-string correction problem. Journal of the ACM 21(1): 168-173

5 Reading and Writing Files

This chapter introduces and describes the main functions for reading and writing files. First, in section 5.1, the most commonly used file formats are presented along with specific functions for handling them. These include the MATLAB® proprietary binary MAT-file format, as well as more conventional formatted and unformatted text files which are commonly referred to as *plain text* files. Then, in section 5.2, functions for reading and writing some other commonly used file formats such as CSV, Row-Column-Value and HTML, among others, are described. Finally, in section 5.3, some useful tools for working with datasets and document collections are presented.

5.1 Basic File Formats

The easiest way of writing and reading files in the MATLAB® environment is by means of the functions **save** and **load**. While the former allows for saving all or some of the current workspace variables into a file, the latter allows for reading variables previously stored in a file and loading them into the workspace.

By default, the two functions **save** and **load**, use the MATLAB® proprietary format, in which variables names and their corresponding values are written to binary files with the extension **.mat**. In this sense, notice that MAT-files are not text files, but they can contain text data in any of the specific representations that were described in previous chapters: character arrays, cell arrays of character vectors, string arrays and/or structures.

Let us now consider the most basic usage of **save** and **load**. Using them without any arguments simply saves into and restores from a file named **matlab.mat** a copy of the workspace:

```
>> clear % deletes all variables in the workspace                    (5.1a)

>> a = [1 2; 3 4]; % creates some variables                          (5.1b)
>> b = {'Hello','World','!'}; c.f1 = "Hello"; c.f2 = "World!";

>> save % saves the workspace into default file matlab.mat           (5.1c)
Saving to: C:\Users\Documents\MATLAB\matlab.mat

>> clear % deletes all variables in the workspace                    (5.1d)

>> load % restores the workspace from default file matlab.mat        (5.1e)
Loading from: C:\Users\Documents\MATLAB\matlab.mat
```

© The Author(s), under exclusive license to Springer Nature Switzerland AG 2021
R. E. Banchs, *Text Mining with MATLAB®*, https://doi.org/10.1007/978-3-030-87695-1_5

```
>> whos                                                                          (5.1f)

   Name        Size              Bytes  Class      Attributes

   a           2x2                  32  double

   b           1x3                 334  cell

   c           1x1                 636  struct
```

This option is especially useful for saving and restoring a work session directly from the MATLAB® command window.

Alternatively, either the complete workspace or some specific variables can be saved to and loaded from a specific target file by listing the variables to be saved:

```
>> % save all variables into the file allworkspace.mat                           (5.2a)
>> save('allworkspace');

>> % save only variables 'b' and 'c' into the file somevariables.mat             (5.2b)
>> save('somevariables','b','c');

>> ls                                                                            (5.2c)
.                   allworkspace.mat    somevariables.mat
..                  matlab.mat

>> clear % deletes all variables in the workspace                                (5.2d)

>> % loads only variable 'a' from the file allworkspace.mat                      (5.2e)
>> load('allworkspace','a');

>> % loads variables b and c from the file somevariables.mat                     (5.2f)
>> load('somevariables');
```

It is important to know that both **save** and **load** admit command syntax too. So, the following examples are totally equivalent to (5.2a), (5.2b), (5.2e) and (5.2f).

```
>> save allworkspace                                                             (5.3a)

>> save somevariables b c                                                        (5.3b)

>> load allworkspace a                                                           (5.3c)

>> load somevariables                                                            (5.3d)
```

In addition to the basic operations just described, **save** and **load** also offer other useful functionalities that can be specified with additional input parameters. Table 5.1 summarizes these additional functionalities, along with their specific parameters. More detailed descriptions, as well as some illustrative examples are provided thereafter.

The first option in Table 5.1, -**append**, is specific to function **save**. It allows for adding new variables to an existing MAT-file without the need to rewrite all previously saved variables. In the following example, for instance, the -**append** op-

tion is used to add variable a from example (5.1b) into the already existing file
`somevariables.mat`:

```
>> save somevariables a -append                                              (5.4a)

>> whos -file somevariables                                                  (5.4b)
   Name       Size             Bytes  Class      Attributes
   a          2x2                 32  double
   b          1x3                334  cell
   c          1x1                636  struct
```

Table 5.1. Advanced functionalities of file functions save and load

Options	Additional Parameters	Description
-append	–	Appends variables to an existing MAT-file
-regexp	regular expression	Saves variables whose names match regular expression
-struct	structure and field names	Saves specified fields of structure as single variables
-mat | -ascii	-tabs -double (only ascii)	Specifies the file format
-v7.3 | -v7.0 ...	–	Specifies supported MATLAB version

In the case of option -regexp, only those variables whose names match the
given regular expression will be saved or loaded. Consider for instance the following example:

```
>> clear; load somevariables -regexp '[^a]'                                  (5.5a)

>> whos                                                                       (5.5b)
   Name       Size             Bytes  Class      Attributes
   b          1x3                334  cell
   c          1x1                636  struct
```

In the case of option -struct, which is specific to function save, the specified
fields of a structure are saved as independent variables. In both the function and
command syntax, structure's fields must be listed immediately after the structure's
name. In case no fields are specified, all fields are saved.

```
>> save structfile -struct c f1 f2                                           (5.6a)

>> whos -file structfile                                                     (5.6b)
   Name       Size             Bytes  Class      Attributes
   f1         -                  150  string
   f2         -                  150  string
```

Opposite to saving structure's fields as single variables in a MAT-file, independent variables within a MAT-file can be loaded into a structure variable as follows:

```
>> newstructure = load('structfile')                                    (5.7)
newstructure =
  struct with fields:
    f1: "Hello"
    f2: "World!"
```

Notice that, in this case, only the function syntax for `load` is admitted. The syntax in (5.7) also allows for specifying some variables as additional input parameters, in such case, only the specified variables are loaded into the structure.

The next option in Table 5.1 is the file format option. This allows for changing the default MAT-file format `-mat` for a text format `-ascii`. However, it is important to recall that such text format does not allow any arbitrary kind of text content. It is actually a very specific and restricted data format, which only allows for saving a matrix (two-dimensional array) of numerical values into a text file.

Consider, for instance, the following example in which the numerical matrix `a` from example (5.1b) is written to a text file:

```
>> clear; load allworkspace                                             (5.8a)
```

```
>> save matrix.txt a -ascii                                             (5.8b)
```

The resulting file `matrix.txt` is a text file in the sense that it can be opened, visualized, and modified with any text editor or text processor, as it is illustrated in Figure 5.1.

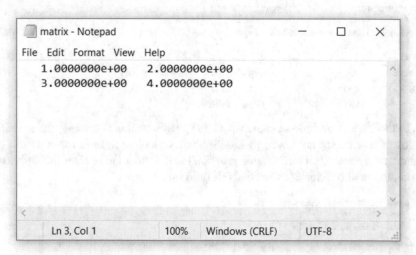

Fig. 5.1. View of file `matrix.txt` created by using `save` with the option `-ascii`

Alternatively, any text file like the one in Figure 5.1 (strictly containing a matrix of numerical values!) can be read and stored into a variable by using the function `load`. As seen from the following example, the option `-ascii` can be omitted as `load` is able to infer that it is actually a text file. In any case, including this option in the function call will make `load` to consider the file as a text file regardless its extension.

```
>> load matrix.txt                                                              (5.9a)

>> whos                                                                          (5.9b)
   Name         Size           Bytes  Class     Attributes
   a            2x2               32  double
   b            1x3              334  cell
   c            1x1              636  struct
   matrix       2x2               32  double
```

Notice, however, that in this case `load` is unable to restore the original name of the variable as this information is not included in the text file. By default, the variable is given the same name of the file without the file extension.

The content of the file read in (5.9a) can be saved to a specific variable by using the same syntax as in (5.7). However, different from (5.7), in this case a two-dimensional array is created instead of a structure:

```
>> a = load('matrix.txt')                                                       (5.10)
a =

     1      2
     3      4
```

Finally, in the last row of Table 5.1, the version option is described. This option allows for specifying the supported MATLAB® version and its corresponding features. Newer versions are always compatible and support older ones, but the opposite is not necessarily true. An important thing to know about option `-v7.3` is that it supports data variables larger than 2GB. This seems to be more a problem for image and audio processing applications; however, when handling huge text data collections, this limit can be reached, and the activation of this option is required if it is not the default option for MAT-files in your preference settings.

In addition to saving and reading variables from files, most of the data we will be interested in saving and reading from files are textual in nature. Different from the very restrictive format allowed by `save` and `load` through the `-ascii` option, we are interested in reading and writing any form of formatted and unformatted text contents. In the remainder of this section, we will focus our attention on reviewing some standard functions implemented in the MATALB® programming environment for reading and writing text files. More specifically, the functions `fopen`, `fclose`, `fprintf` and `fscanf` are presented and described in detail.[1]

---

[1] Descriptions presented here are not comprehensive. They are actually focused on the specific requirements that we will encounter when working with text files along the different applications and exercises in the book. For a more comprehensive description on these functions, you should refer to the appropriate MATLAB® technical documentation.

The basic procedure for working with files by using these functions is as follows: first, the file must be opened, and a file identifier must be obtained by using **fopen**. Then, either writing or reading operations are performed to the file by using **fprintf** or **fscanf**, respectively. The specific file to operate with is identified by means of its file identifier. Finally, the file must be closed to allow other programs to access it; **fclose** must be used for closing the file.

The typical syntax for **fopen** is:

```
fid = fopen(filename,permission,format,encoding)
```
(5.11)

where the output and input variables can be described as follows:

- **fid** is the file identifier, in case the file was opened successfully; otherwise, if the file could not be opened, a value of **-1** is returned in **fid**.

- **filename** is the full name of the file to be opened (including the extension), which can either contain an absolute or relative path to the actual file location.

- **permission** is a string that specifies the indented operation for which the file has been opened. Some of the possible values for this parameter are:
  - **'rt'** as a text file for reading.
  - **'wt'** as a text file for writing (existing contents are overwritten!).
  - **'at'** as a text file for writing, but in which new contents are to be appended at the end of the file.

- **format** specifies the machine format to be used. Some options are:
  - **'native'** uses the local machine format. It is used by default if the format is not specified.
  - **'ieee-le'** uses little-endian IEEE floating point format.
  - **'ieee-be'** uses big-endian IEEE floating point format.

- **encoding** specifies the character encoding scheme to be used. If the encoding is not specified, MATLAB® default encoding is used. Some commonly used encodings are:
  - **'US-ASCII'**, which is the most commonly used code set for English,
  - **'UTF-8'**, which is intended to be a universally standard character set,
  - **'ISO-8859-1'** or **'Latin1'**, for Western European languages,
  - **'GB2312'** for simplified Chinese,
  - **'Big5'**, for traditional Chinese,
  - **'ISO-8859-6'**, for Arabic, and
  - **'Shift_JIS'**, for Japanese.

On the other hand, the syntax for **fclose** is given by:

```
status = fclose(fid)
```
(5.12)

where `fid` is the file identifier of the file to be closed (which was previously generated with `fopen`), and `status` reports the result of the file-closing action: `0` if successful or `-1` otherwise.

Let us now consider the write-to-file function `fprintf`. Similar to `sprintf`, which was briefly described in section 4.3, `fprintf` allows for string editing and formatting, but different from `sprintf` it writes the resulting string into a specified file. The syntax of `fprintf` is as follows:

```
fprintf(fid,format,input,...)                                    (5.13)
```

where `fid` is the file identifier of the file in which the formatted string is to be written to, `format` is a format-specification string containing the desired conversion characters following C language syntax and definitions, and `input` is a data variable to be formatted into the text string. In general, `fprintf` admits as many input data variables as specified by the format string `format`.

The format-specification string describes how input variables are formatted in the output string. It can include any combination of the following three kinds of data: plain text, escape characters and conversion characters and its operators.

Plain text is printed into the output string just as it is included in the format-specification:[2]

```
>> fprintf('this is an example on printing plain text')          (5.14)
this is an example on printing plain text
```

Escape characters, on the other hand, are used to print reserved and control characters. They include escape sequences for reserved characters such as `%%` percent, `\\` backslash, and `''` apostrophe; and escape sequences for control characters such as `\b` backspace, `\f` form-feed, `\n` newline, `\r` return, `\t` horizontal tab, `\v` vertical tab, etc.

```
>> fprintf('this is an example\non ''escape characters''\n')     (5.15)
this is an example
on 'escape characters'
```

Finally, conversion characters and its operators are used for embedding the input variables into the output string with some specific formatting characteristics. Conversion characters (C) admit five different types of operators: identifier (I), flags (F), field width (W), precision (P) and subtype (S). Operators are optional but, when used, they must appear according to the following syntax:

```
%IFW.PSC                                                         (5.16)
```

Table 5.2 summarizes commonly used conversion characters and operators.

---

[2] For convenience, examples are illustrated while omitting the file identifier input variable `fid`. As seen from the examples, if `fid` is omitted, the result is displayed on the screen.

Table 5.2. Most commonly used conversion characters and operators

Type	Symbol	Description
conversion	d	signed integers (decimal)
conversion	u o x	unsigned integers (decimal, octal and hexadecimal)
conversion	f e	floating point numbers (fixed-point and exponential notation)
conversion	c s	characters (single character and character array)
identifier	n$	indicates the argument to be formatted (n must be an integer)
flags	- + blank 0	left justify, print sign, insert blanks, pad with zeros, respectively
field width	m	minimum number of characters to be printed (must be an integer)
precision	n	number of decimal digits to be printed (must be an integer, $n < m$)
subtype	h l t b	16 bits, 64 bits unsigned, single precision, double precision float

Let us consider now some illustrative examples about the use of the conversion characters and operators depicted in Table 5.2.

```
>> % uses integer conversion characters d, o and x                    (5.17a)
>> fprintf('decimal: %1$d  octal: %1$o  hexadecimal: %1$x\n',110)
decimal: 110  octal: 156  hexadecimal: 6e

>> % uses floating point conversion characters f and e               (5.17b)
>> fprintf('fixed point: %1$f  exponential: %1$e\n',110)
fixed point: 110.000000  exponential: 1.100000e+02

>> % uses character conversion characters c and s                    (5.17c)
>> fprintf('character: %1$c  string: %1$s\n',110)
character: n  string: n
```

Examples (5.17a), (5.17b) and (5.17c) show integer, floating point, and character representations, respectively, for the input value *110*. Notice how in all cases, the identifier operator 1$ has been used to refer to the first input argument. Notice also, that in the particular case of character conversion, the numerical input value has been automatically converted into a character type. In general, input variables of type character should be given as follows:

```
>> fprintf('character: %1$c  string: %1$s\n',char(110))            (5.18)
character: n  string: n
```

Consider now the following example, which illustrates the use of some flags, as well as the field width and precision operators:

```
>> fprintf(' %+10.4f %+10.4f\n %010.4f %010.4f\n',pi,-1,2,-pi)      (5.19)
   +3.1416    -1.0000
00002.0000 -0003.1416
```

Notice from (5.19) the effects produced by the different operators. In all four cases, a field width of *10* and a precision of *4* have been specified. In those cases, using the o flag (second row), the printed values have been padded with zeros until completing the specified field size. Notice also from the second row, that the minus sign is printed for those negative-valued inputs regardless of the presence of the sign flag +. Finally, notice how in this case the identifier operator n\$ has been omitted, so correspondences between format descriptors and input arguments are assumed to follow the same relative order they appear in.

For more examples on format specifications, see exercise 5.5-1 in section 5.5 at the end of this chapter.

Let us now finish this section by considering the read-from-file function `fscanf`. The syntax of the function is as follows:

```
[data,count] = fscanf(fid,format,size)
```
$\hfill$(5.20)

where `fid` is the file identifier of the file to be read, `format` is a format-specification string indicating the format of the data values to be read, `size` specifies the dimensions of the output variable `data` and `count` returns the total number of read elements. The input parameter `size` admits the following three possible values: `inf`, data elements are read until the end of the file is reached; n, the first n elements are read from the file and are arranged into a column vector in `data`, and `[n,m]`, the first n times m elements are read and arranged column-wise into a $n \times m$ matrix in `data`.

Consider, for instance, the following example in which the elements contained in the text file created in example (5.8b) are read and loaded into a numeric vector by using `fscanf`:

```
>> % opens the file and gets the file identifier                      (5.21a)
>> fid = fopen('matrix.txt','rt','native','US-ASCII');

>> [data,count] = fscanf(fid,'%e',[1,4]) % reads four data elements   (5.21b)
data =
     1     2     3     4
count =
     4

>> fclose(fid); % closes the file                                     (5.21c)
```

Alternatively, the file can be read, and the data loaded into a character array instead of a numeric array:

```
>> fid = fopen('matrix.txt','rt','native','US-ASCII');                (5.22a)

>> [data,count] = fscanf(fid,'%c',inf) % reads characters from file   (5.22b)
data =
    '  1.0000000e+00   2.0000000e+00
       3.0000000e+00   4.0000000e+00

    '
```

```
count =
   66
```

```
>> fclose(fid);                                                                    (5.22c)
```

Notice that, in (5.22b), we have used `%c` instead of `%s` as format-specification string. This is actually the appropriate way to do it as using `%s` will not recover the whitespaces and newline characters! (You can try it by yourself.)

Similar to `fprintf`, `fscanf` also has a string version: `sscanf`. This function reads formatted data in the same way `fscanf` does it, but instead of reading from a file, it reads from a string variable.

Finally, we must mention some other useful functions that are complementary to the ones described in this section. More specifically, these functions are: `ftell`, `fseek`, `frewind`, and `fgetl`. They are covered in exercises 5.5-2 and 5.5-3 in section 5.5 at the end of this chapter.

5.2 Other Useful Formats

In this section we will present some other useful and commonly used formats for text documents, as well as the corresponding MATLAB® functions to be used for reading and writing files according to those formats. More specifically, we focus our attention on CSV, Row-Column-Value, XLS, XML and HTML file formats.

CSV (Comma Separated Value) files are typically used to represent data that has been originally formatted into a tabular structure. The format of a CSV file is actually very simple: each line of the file contains a row of the table, and the different column values within each row are separated by commas. Consider the file `sample_file.csv`, which is illustrated in Figure 5.2.

As seen from the figure, the file contains the entries of a table composed of four rows and three columns. In the first line of the file, headers of the three columns are specified. In the subsequent three lines of the file, the entries corresponding to three data samples are provided.

Notice that using quotation marks " for complementing field delimitation is not mandatory, unless the contents in the field could produce ambiguities in the format interpretation. According to this, in the example presented in Figure 5.2, quotation marks are only necessary for the fields in the third columns of the second, third and fourth rows; otherwise, the commas within those fields would be erroneously interpreted as field delimiters. In general, the format specification for CSV files determines that a field must be provided within quotation marks when the field itself contains commas, quotation marks or line breaks.

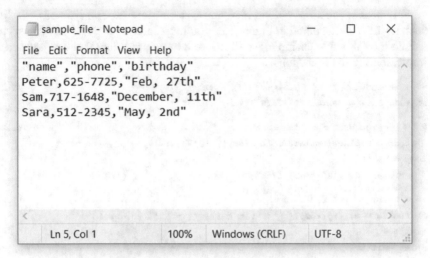

Fig. 5.2. Example of a CSV (Comma Separated Value) file

There are two MATLAB® functions for reading and saving a very specific type of CSV files: `csvwrite` and `csvread`. These functions allow for writing and reading a matrix of numerical values into a CSV formatted file. In this sense, these functions are similar to `save` and `load` when the `-ascii` option is used, but different from them, the numerical values within each file line are separated by commas. Consider for instance the following example:

```
>> mtx = rand(3,4); % creates a random-valued matrix of size 3x4          (5.23a)

>> csvwrite('mtx.csv',mtx); % saves the matrix into a csv file            (5.23b)

>> type mtx.csv                                                           (5.23c)
0.81472,0.91338,0.2785,0.96489
0.90579,0.63236,0.54688,0.15761
0.12699,0.09754,0.95751,0.97059

>> newmtx = csvread('mtx.csv') % loads the matrix into a new variable     (5.23d)
newmtx =
    0.8147    0.9134    0.2785    0.9649
    0.9058    0.6324    0.5469    0.1576
    0.1270    0.0975    0.9575    0.9706
```

Updated versions of `csvwrite` and `csvread` are implemented in `dlmwrite` and `dlmread`, which allow for specifying, as an additional input parameter, the delimiter character to be used for either writing or reading the file, respectively. However, similar to `csvwrite` and `csvread`, they are also restricted to numerical data. Alternatively, you might be also interested in looking at function `importdata`.

For specific functions able to read and write a more general type of CSV files, such as the one illustrated in Figure 5.2, you must refer to exercise 5.5-4 in section 5.5. In the following example, we illustrate a basic procedure that can be used for implementing such functions:

```
>> % reproduces the 2nd line of file sample_file.csv                    (5.24a)
>> file_line = "Peter,625-7725,""Feb, 27th""";
```

```
>> % preprocesses the line contents                                     (5.24b)
>> temp = replace(replace(file_line,',,',', ,'),',,',', ,') + ","
temp =
    "Peter,625-7725,"Feb, 27th","
```

```
>> % extracts the fields                                                (5.24c)
>> fields = regexp(temp,' *"(.*?)" *,|([^"].*?),','tokens')
fields =
  1x3 cell array
    {["Peter"]}      {["625-7725"]}      {["Feb, 27th"]}
```

The basic procedure illustrated in (5.24) operates as follows. First, it is assumed that each line of the CSV formatted file is read one at a time (5.24a). Second, the line contents are preprocessed to guarantee matching conditions for field extraction (5.24b). Two preprocesses are applied: a white space is inserted between any consecutive pair of commas (this will avoid missing empty fields), and a comma is appended at the end of the line (this will allow for matching the last field). Finally, the fields are extracted from the line contents by matching a regular expression (5.24c). Two patterns preceding a comma are attempted to be matched. The first one ` *"(.*?)" *,`, which is given priority over the second one, is intended to match any field appearing within quotation marks; the second one `([^"].*?),`, matches any sequence of characters not containing quotation marks.[3]

Let us now consider another important format, which is commonly known as the Sparse-Matrix format. Alternatively, it can be also referred to as row-column-value, i-j-value, or x-y-value. These last three connotations make reference to the 3-tuple structure of the file, in which each line of the file contains three elements: a row index (or x coordinate), a column index (or y coordinate), and a variable or attribute value. Notice that, strictly speaking, x-y-value is not the same as row-column or i-j-value, as coordinate values can be real numbers, but indexes must necessarily be positive integers.

On the other hand, the name of Sparse-Matrix format derives from the fact that this kind of format is typically used for representing data which is sparse in nature. As seen in chapter 7, in many text mining applications (as well as many other data mining applications, in general) we have to deal with large matrices for which

---

[3] Notice that the procedure described in (5.24) could fail if certain combinations of quotation marks and commas occur inside of a field!

most of their elements are equal to zero. So, instead of saving N times M elements (where N and M can be easily millions or hundreds of thousands), we just need to save the K non-zero elements of the matrix (where generally $K << N \times M$).

Figure 5.3 shows the contents of `sample_file.rcv`, which is in row-column-value format.

```
sample_file - Notepad                          —    □    ×

File  Edit  Format  View  Help
1    2    250
6    3    -110
7    1    152
7    4    049
9    6    000



        Ln 6, Col 1           100%    Windows (CRLF)    UTF-8
```

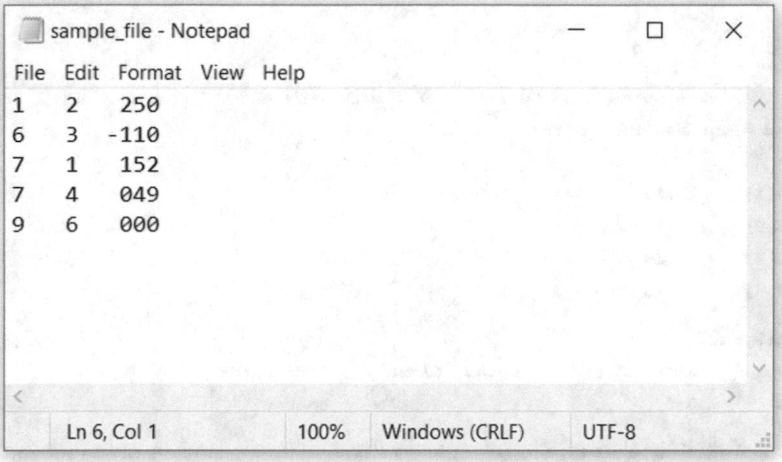

Fig. 5.3. Example of a Row-Column-Value file (also known as Sparse-Matrix file)

Notice from Figure 5.3 that, more than a file format, the Sparse-Matrix format is actually a data format specification. Indeed, the same matrix of data in Figure 5.3 can be perfectly stored in any other file format, such as CSV for example, while still fulfilling the Sparse-Matrix format specification. Accordingly, each row of the file illustrated in the figure represents a specific entry in the data matrix. The first two elements correspond to the row and column indexes, and the third element corresponds to the specific value of the matrix entry. Only those matrix entries with values different from zero are included, while all zero-valued matrix entries are omitted.

Notice, however, that the last row of the file represents a matrix entry which value is zero. The objective of specifying such an entry is to provide information about the size of the data matrix. In the particular case illustrated in the figure, the file is describing a 9×6 sparse matrix which only contains four non-zero elements.

A file like the one illustrated in Figure 5.3 can be read by using the function `load` as it was previously illustrated in examples (5.9) and (5.10). Afterwards, the transformation from the row-column-value format into an actual sparse matrix variable must be done. More specifically, this transformation can be performed by using function `spconvert`, which will generate a MATLAB® sparse-matrix variable. Consider, for instance the following example:

```
>> % row-column-value data sample from figure 5.3                    (5.25a)
>> rcvdata = [1,2,250;6,3,-110;7,1,152;7,4,49;9,6,0]
rcvdata =
      1     2    250
      6     3   -110
      7     1    152
      7     4     49
      9     6      0

>> % creates a sparse matrix variable from row-column-value data     (5.25b)
>> mtx = spconvert(rcvdata)
mtx =
    (7,1)       152
    (1,2)       250
    (6,3)      -110
    (7,4)        49

>> whos mtx                                                          (5.25c)
   Name       Size              Bytes  Class      Attributes
   mtx        9x6                 136  double     sparse
```

This function also admits complex-valued data. In such a case, the Sparse-Matrix data representation is required to contain four columns in total: the row and column indexes, and the real and imaginary parts of the complex values:

```
>> % appends an imaginary part to rcvdata from (5.25a)               (5.26a)
>> cdata = [rcvdata,[100;220;-150;100;0]];

>> % creates a complex sparse matrix variable                       (5.26b)
>> cmtx = spconvert(cdata)
cmtx =
  1.0e+002 *
    (7,1)       1.5200 - 1.5000i
    (1,2)       2.5000 + 1.0000i
    (6,3)      -1.1000 + 2.2000i
    (7,4)       0.4900 + 1.0000i

>> whos cmtx                                                         (5.26c)
   Name       Size              Bytes  Class      Attributes
   cmtx       9x6                 176  double     sparse, complex
```

MATLAB® matrix variables can be converted back and forth between sparse and full form representations by using functions full and sparse, respectively.

```
>> fmtx = full(mtx)                                                 (5.27a)
fmtx =
      0   250     0     0     0     0
      0     0     0     0     0     0
```

```
     0      0      0      0      0      0
     0      0      0      0      0      0
     0      0      0      0      0      0
     0      0   -110      0      0      0
   152      0      0     49      0      0
     0      0      0      0      0      0
     0      0      0      0      0      0
```

```
>> smtx = sparse(fmtx)                                           (5.27b)
smtx =
   (7,1)       152
   (1,2)       250
   (6,3)      -110
   (7,4)        49
```

```
>> whos fmtx smtx                                                (5.27c)
  Name        Size            Bytes  Class      Attributes
  fmtx        9x6               432  double
  smtx        9x6               120  double     sparse
```

Notice from (5.27c) the different storage requirements of the two variable types. As we will see in some of the following chapters, for sparse matrices with very large sizes, this difference can become very important.

Finally, consider the problem of getting back the row-column-value representation in (5.25a) from either the sparse or full representations in (5.27a) and (5.27b). The following procedure, which we illustrate for the sparse case, should work in any of the two cases:

```
>> % gets row, column, and value for all non-zero entries        (5.28a)
>> [r,c,v] = find(smtx);
```

```
>> % concatenates row, column, and value vectors                 (5.28b)
>> back2rcv = [r,c,v];
```

```
>> % appends and additional entry for making explicit the matrix size   (5.28c)
>> back2rcv = [back2rcv;size(smtx),0]
back2rcv =
     7     1   152
     1     2   250
     6     3  -110
     7     4    49
     9     6     0
```

Another file format commonly used in the data and text mining communities is the so-called feature-value format. It is actually a variant of the Sparse-Matrix format, in which each line of the file provides information corresponding to one row (data sample) of the sparse matrix. Additionally, it allows for including class or target data information, which makes it very convenient for supervised learning applications. Figure 5.4 shows an example of this file format.

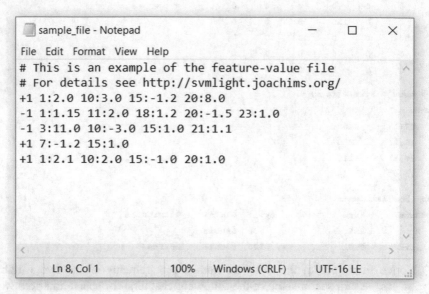

Fig. 5.4. Example of the Feature-Value format commonly used in data mining applications

As seen from the figure, the format admits comment lines at the beginning of the file. Comment lines must start with the # character. After the comment lines, data samples (matrix rows) are included one per line. Data lines must be provided according to the following format: first, the class or target value for the given data sample is included; afterwards, the sequence of the corresponding feature-value pairs is provided. The class or target value, represents either the category to which the data sample belongs to (classification framework), or the objective value associated to the data sample (regression/estimation framework). The feature-value pairs, on the other hand, correspond to the column and value parts of this particular row-column-value format. As seen from the figure, each feature and its corresponding value are separated by a colon, while feature-value pairs are separated among them by whitespaces.

Also notice that in this format, row indexes are assumed to follow the same sequence in which data lines are provided; although, in general, when using this kind of format, row ordering is assumed to be not important.[4]

There are not specific MATLAB® functions for writing and reading feature-value files like the one illustrated in Figure 5.4. However, creating these functions

---

[4] For a more detailed description of the feature-value data format described here, you must refer to https://www.cs.cornell.edu/people/tj/svm_light/ Accessed 6 July 2021.

is a relatively simple task. This problem is addressed and fully covered in exercise 5.5-6 in section 5.5.

Let us now consider the XLS file format, which is the format used in Microsoft's Excel® Spreadsheets. There are three basic functions available to work with this specific file format: `xlsfinfo`, `xlsread`, and `xlswrite`. Full functionality of these three functions is subject to the MATLAB® platform running the Excel ActiveX® Server. If the server is not available, the functions will operate in the `'basic'` input mode, which is explained further below.

Let us first consider the function `xlswrite`, which must be used according to the following syntax:

```
[status,message] = xlswrite(file,data,sheet,range);
```
                                                                                            (5.29)

where `file` is the name of the file to be created, `data` can be either a numeric array, string array or cell array containing the data to be written, and `sheet` and `range` are optional input parameters that determine the specific worksheet in which the data is to be written and the specific row-column range to be written, respectively. More specifically, the worksheet can be specified by using its name or its index number, and the range must be provided in the standard Excel® notation: initial-column-row:final-column-row (where columns are specified with letters and rows are specified with numbers). The output variable `status` returns the resulting status of the operation: `'1'` successful, `'0'` failure; and `message` returns the corresponding error message and code in case the operation fails.

Consider, for instance, the following example in which one of our micro datasets from chapter 2 is written into an Excel® Workbook. The resulting XLS file `sample_file.xls` is shown in figure 5.5.

```
>> data = {'name','age';'mark','35';'beth','26';'peter','39'};          (5.30a)

>> [status,message] = xlswrite('sample_file.xls',data)                  (5.30b)
status =
  logical
   1
message =
  struct with fields:
      message: ''
    identifier: ''
```

Additional data can be written to another sheet of the same workbook by specifying it with the `sheet` parameter. The following example writes a random-valued matrix into the second sheet of the recently created file `sample_file.xls`.

```
>> newdata = rand(3,5);                                                 (5.31a)

>> [status,message] = xlswrite('sample_file.xls',newdata,2);           (5.31b)
```

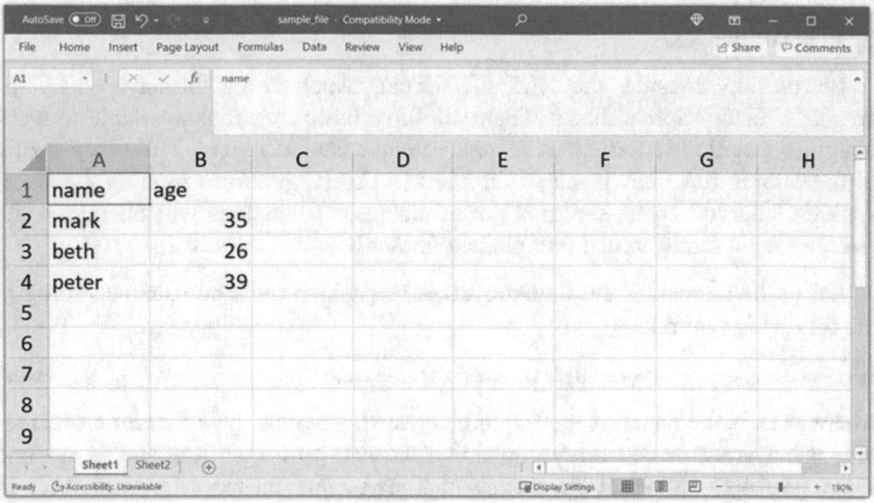

Fig. 5.5. View of the Excel® Workbook created with `xlswrite` in example (5.30)

Also, specific entries can be overwritten by using the **range** parameter. Consider the following example, in which cells in the second worksheet within the range *C2:E3* (rows from *2* to *3* and columns from *C* to *E*) are set to *1*.

```
>> [status,message] = xlswrite('sample_file.xls',ones(2,3),2,'C2:E3');     (5.32)
```

Let us now consider the syntax for the function `xlsfinfo`, which can be used to obtain relevant information related to a given file:

```
[isexcel,worksheets,format] = xlsfinfo(file);                              (5.33)
```

where **file** is the name of the file, **isexcel** is either a string indicating the file is a valid Excel® Spreadsheet or an empty variable, **worksheets** contains the names of the worksheets in the file, and **format** indicates the specific Excel® format used in the file.

In the following example we illustrate how to get the corresponding information for file **sample_file.xls** that was created in (5.30) and updated in (5.31) and (5.32):

```
>> [isexcel,worksheets,format] = xlsfinfo('sample_file.xls')              (5.34)
isexcel =
    'Microsoft Excel Spreadsheet'
worksheets =
  1x2 cell array
    {'Sheet1'}    {'Sheet2'}
format =
    'xlExcel8'
```

Finally, let us consider the reading function `xlsread`. Its syntax is as follows:

```
[num,txt,rawdata] = xlsread(file,sheet,range);
```
(5.35)

where the input parameters `file`, `sheet`, and `range` are the same as in `xlswrite`. In the case of the output parameters, `rawdata` is a cell array containing all the cells that have been read from the file, and `num` and `txt` are a numeric and a cell array, respectively, containing the numeric and textual portions of `rawdata`.

Consider, for instance, the following example in which name and age values from the table in `Sheet1` of `sample_file.xls` are read:

```
>> [num,txt,rawdata] = xlsread('sample_file.xls',1,'A2:B4')
num =
    35
    26
    39
txt =
  3x1 cell array
    {'mark' }
    {'beth' }
    {'peter'}
rawdata =
  3x2 cell array
    {'mark' }    {[35]}
    {'beth' }    {[26]}
    {'peter'}    {[39]}
```
(5.36)

The `'basic'` input mode (which is the same used when the Excel ActiveX® Server cannot be launched) can be invoked by including the string `'basic'` as an additional input parameter.

```
>> [num,txt,rawdata] = xlsread('sample_file.xls',1,'A2:B4','basic');
Warning: Range cannot be used in 'basic' mode. The entire sheet will be loaded
> In xlsread (line 208)
```
(5.37a)

```
>> rawdata
rawdata =
  4x2 cell array
    {'name' }    {'age'}
    {'mark' }    {[ 35]}
    {'beth' }    {[ 26]}
    {'peter'}    {[ 39]}
```
(5.37b)

Notice that in the `'basic'` input mode the specified range information is completely ignored, and the complete worksheet is imported.

Another file format of interest is XML (Extensible Markup Language), which is a protocol for encoding and formatting documents. It was conceived as a machine-readable format to be used on the Internet for documents and data structures, and it is fully compatible with the SGML (Standard Generalized Markup Language) specifications. Indeed, "*XML has been designed for ease of implementation and for interoperability with both SGML and HTML.*"[5]

In this section, our description of the XML format is presented from a very naïve point of view.[6] We consider it as means for structuring document contents by using a set of tags for defining elements, and we show how this kind of structured documents looks like. However, XML is much more than this. Valid XML documents must conform to some predefined grammars, commonly known as schema languages, which provide a set of rules and constraints that a given XML file must satisfy in order to be valid. Examples of these are the Document Type Definition (DTD) and XML Schema Definition (XSD) files.

Additionally, some Application Programming Interfaces for processing and parsing XML files are already available. One example is the Document Objet Model (DOM) interface, which is the one supported through MATLAB® functions `xmlread` and `xmlwrite`. The study of these two functions is left to you as an exercise, see 5.5-7 in section 5.5. In any case, a comprehensive description of XML specifications, schemas, and programming interfaces is beyond the scope of this book. You can refer to the original W3C specifications[7] or to any other supporting material to get more precise information on the XML subject.

For the effects of our discussion here, consider the XML file `sample_file.xml`, which is illustrated in Figure 5.6. As seen from the figure, an XML file is primarily composed of structural elements which conform to the following basic syntax:

```
<tag attribute="value"> content </tag>
```
                                                                                    (5.38)

Also, from the figure, the following characteristics of the XML format can be verified:

- Most elements start with an opening tag `<tag>` and end with its corresponding closing tag `</tag>`.
- Other elements consist of a single tag (such as the declaration element in the first line of the file, or the line break tag `<br/>` not shown here). Such elements are referred to as empty-element tags.

---

[5] W3C, Extensible Markup Language (XML) 1.0 (Fifth Edition), http://www.w3.org/TR/xml/ Accessed 6 July 2021

[6] After this section discussion, we will be only able to read and write XML files for which we explicitly know the structure and the expected types of tags and descriptors they contain. Very limited read and write procedures by using `fscanf` and `fprintf` will be presented here.

[7] See http://www.w3.org/TR/xml/ Accessed 6 July 2021

- The elements can be nested in the sense that one element's content can include other elements in it.
- Tags can contain none, one or more attributes, which define properties of the element. Specific values are to be assigned to attributes.

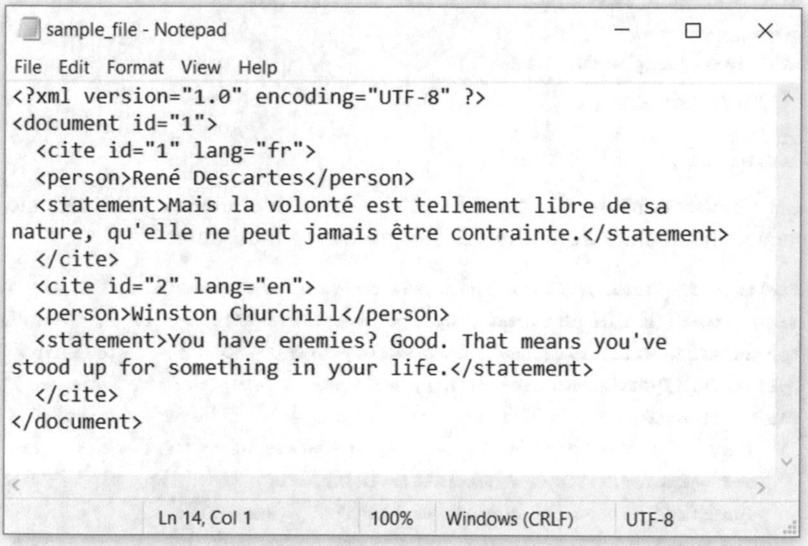

```
sample_file - Notepad                                    —    □    ×
File  Edit  Format  View  Help
<?xml version="1.0" encoding="UTF-8" ?>
<document id="1">
  <cite id="1" lang="fr">
  <person>René Descartes</person>
  <statement>Mais la volonté est tellement libre de sa
nature, qu'elle ne peut jamais être contrainte.</statement>
  </cite>
  <cite id="2" lang="en">
  <person>Winston Churchill</person>
  <statement>You have enemies? Good. That means you've
stood up for something in your life.</statement>
  </cite>
</document>

        Ln 14, Col 1        100%   Windows (CRLF)    UTF-8
```

Fig. 5.6. View of a simple XML file example

In the file presented in Figure 5.6, four different elements are used: *document*, *cite*, *person* and *statement*. In this simple case, the structure of the file can be easily inferred by visual inspection. *Documents* are composed of a collection of *cites*, and each *cite* is composed of a *person* and a *statement*. Both *documents* and *cites* have an attribute called *id*, which is an integer value for indexing purposes. Additionally, *cites* have a second attribute called *lang*, which specifies the language of the text provided in the corresponding *statement*.

With our knowledge about working with plain text files, from section 5.1, as well as our knowledge about handling strings and matching patterns, from previous chapters, we can build customized functions for reading and writing XML files as the one presented in Figure 5.6. We can do this by deriving structure information from a visual inspection of the file (without the need of any additional knowledge about the XML format or specification). However, these functions would be specific for files similar to the one in Figure 5.6 and will not be useful for reading or writing any other type of XML files.

Let us start by creating a data structure with the contents of the XML file under consideration:

```
>> language = {'fr','en'};                                          (5.39a)
```

```
>> name = {'René Descartes','Winston Churchill'};
>> data = {'Mais la volonté est tellement libre de sa nature, qu''elle ne peut
jamais être contrainte.','You have enemies? Good. That means you''ve stood up
for something in your life.'};

>> cite = struct('lang',language,'person',name,'statement',data)      (5.39b)
cite =
  1x2 struct array with fields:
    lang
    person
    statement
```

Let us now consider the problem of writing to file `sample_file.xml` the data contained in structure `cite`. Consider the following procedure:

```
>> fields = fieldnames(cite); % gets the names of the structure fields   (5.40)
>> fid = fopen('sample_file.xml','wt','native','UTF-8'); % creates the file
>> fprintf(fid,'<?xml version="1.0" encoding="UTF-8" ?>\n'); % first line
>> fprintf(fid,'<document id="1">\n'); % writes opening tag for document 1
>> for k=1:length(cite) % cite loop: writes one cite element per iteration
        % gets and writes cite's attribute and value within the opening tag
        atr = fields{1}; val = getfield(cite(k),atr);
        fprintf(fid,'  <cite id="%d" %s="%s">\n',k,atr,val);
        for n = 2:3 % cite elements loop: gets and writes person and statement
            tag = fields{n}; txt = getfield(cite(k),fields{n});
            fprintf(fid,'  <%1$s>%2$s</%1$s>\n',tag,txt);
        end;
        fprintf(fid,'  </cite>\n'); % writes closing tag for cite
   end;
>> fprintf(fid,'</document>\n'); % writes closing tag for document 1
>> fclose(fid); % closes the file
```

The procedure depicted in 5.40 consists of the following steps: initialization, two nested loops, and closing. First, in the initialization step, the names of the fields in the data structure are saved into the variable `fields`, the file is created and opened with writing permission, and the declaration line and the opening tag for element *document* are written. In the nested loops stage, the contents of *document* are written. The outer loop is responsible for opening and closing each *cite* and the inner loop is responsible for writing the corresponding *person* and *statement* elements. In the outer loop, the attribute name *lang* and its value are retrieved from `fields` and `cite`, respectively. Similarly, in the inner loop, the tags *person* and *statement* and their corresponding text contents are also retrieved from `fields` and `cite`, respectively. Finally, in the closing step, the closing tag for *document* is written and the file is closed.

Let us now address the problem of reading file `sample_file.xml` and load it into structure `cite`:

```
>> % reads the file and saves it into a string variable              (5.41a)
>> fid = fopen('sample_file.xml','rt','native','UTF-8');
>> datastring = fscanf(fid,'%c',inf);
>> fclose(fid);

>> % gets opening tag and content for each cite element in the file    (5.41b)
>> data = regexp(datastring,'(<cite.*?>)(.*?)</cite>','tokens');
>> % cite loop: extracts attributes and elements for each cite
>> for k=1:1:length(data)
       temp = regexp(data{k}{1},'id="(.*?)"','tokens'); % gets the id
       index = str2num(temp{1}{1}); % converts the id into a numeric value
       lang = regexp(data{k}{1},'lang="(.*?)"','tokens'); % gets the language
       % gets the person and statement elements
       pers = regexp(data{k}{2},'<person>(.*?)</person>','tokens');
       stat = regexp(data{k}{2},'<statement>(.*?)</statement>','tokens');
       % assigns values to corresponding fields and elements in structure cite
       cite(index) = struct('lang',lang{1}{1},'person',pers{1}{1},...
       'statement',stat{1}{1});
   end;
```

The procedure presented in (5.41) considers two steps. In the first step (5.41a), the contents of the XML file are copied to a string variable. In the second step (5.41b), the individual data elements are extracted from it. First, the opening tag and the content for each element *cite* in the file are extracted. Then, a loop is used to extract the four items of interest (*id*, *lang*, *person* and *statement*) for each *cite*. At the end of each iteration, the structure array `cite` is updated with the extracted information.

Although the procedures described in (5.40) and (5.41) illustrate an *ad hoc* approach to writing and reading XML files, the data structure defined in (5.39) is probably not the most appropriate one for the XML file presented in Figure 5.6. See exercise 5.5-8, for an alternative data structure and its corresponding writing and reading procedures.

Notice also that the procedure in (5.41) requires the previous knowledge of the tags and attributes present in the file, as well as the file structure. A more general procedure can be conceived in which the information related to the tags, attributes and structure is inferred from the data itself. Such a procedure would require two passes over the data: the first for extracting tag and attribute information and inferring the file structure, and the second for extracting the items.

Finally, the last file format we discuss in this section is the HTML (Hypertext Markup Language)[8] format, which is well known as it is the standard markup language used to create documents for web browsers. Figure 5.7 presents a simple example of a text file in HTML format.

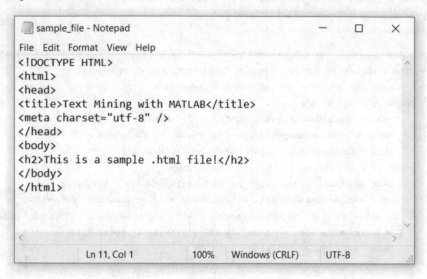

Fig. 5.7. View of a simple HTML file example

As seen from the figure, the HTML file is also composed of structural elements that are similar to the ones used in XML files. In the case of the file in Figure 5.7, we can see it starts with a DOCTYPE declaration in the first line, followed by a document with its content provided between the opening and closing tags *<html>* and *</html>*. The document is composed of two sections *head* and *body*, which are delimited by their corresponding opening and closing tags. Those sections are also composed of different types of elements, always following a syntax similar to the one described in (5.38).

In the following example we create the sample *.html* file from Figure 5.7:

```
>> % creates the sample .html file from Figure 5.7                    (5.42)
>> htmldata = ["<!DOCTYPE HTML>","<html>","<head>", ...
             "<title>Text Mining with MATLAB</title>",...
             "<meta charset=""utf-8"" />","</head>",...
             "<body>","<h2>This is a sample .html file</h2>",...
             "</body>","</html>"];
>> fid = fopen('sample_file.html','wt','native','UTF-8');
```

---

[8] For a comprehensive description about the HTML standard, you can refer to https://html.spec. whatwg.org/ Accessed 6 July 2021

```
>> for k=1:length(htmldata)
       fprintf(fid,'%s\n',htmldata(k));
   end
>> fclose(fid);
```

Now, let us illustrate how the contents in `sample_file.html` can be parsed by using the specific HTML parsing functions `htmlTree`, `findElement`, `getAttribute` and `extractHTMLText`.

```
>> htmldata = evalc('type(''sample_file.html'')'); % reads the file        (5.43a)
```

```
>> tree = htmlTree(htmldata) % parses the html content                      (5.43b)
tree =
  htmlTree:
   <HTML>
       <HEAD>
           <TITLE>Text Mining with MATLAB</TITLE>
           <META charset="utf-8"/>
       </HEAD>
       <BODY>
           <H2>This is a sample .html file</H2>
       </BODY>
   </HTML>
```

```
>> findElement(tree,'title') % extracts the 'TITLE' element                 (5.43c)
ans =
  htmlTree:
   <TITLE>Text Mining with MATLAB</TITLE>
```

```
>> findElement(tree,'body') % extracts the 'BODY' element                   (5.43d)
ans =
  htmlTree:
   <BODY>
       <H2>This is a sample .html file</H2>
   </BODY>
```

```
>> % extracts the 'CHARSET' attribute of the 'META' element                 (5.43e)
>> getAttribute(findElement(tree,'meta'),'charset')
ans =
    "utf-8"
```

```
>> strip(extractHTMLText(htmldata)) % extracts all text elements            (5.43f)
ans =
    'This is a sample .html file'
```

As seen from (5.43b), function `htmlTree` is used to create an HTML parse object from the raw text in the file. Then, in (5.43c) and (5.43d), specific elements are extracted by means of function `findElement` and, in (5.43e), an attribute of an element is retrieved by using `getAttribute`. Finally, function `extractHTMLText` is used in (5.43f) to extract text elements only.

An alternative way of directly extracting text elements from HTML files, as well as other different file formats such as PDF, MS Word® and plain text, is by using function `extractFileText`. This is illustrated in the following example for the specific case of the `sample_file.html` created in (5.42).

```
>> % extracts text elements directly from a formatted file          (5.44)
>> strip(extractFileText('sample_file.html'))
ans =
    "This is a sample .html file"
```

More specific examples on how to deal with the file formats described in this section are presented and discussed in some of the proposed exercises in section 5.5, at the end of this chapter.

5.3 Handling Files and Directories

This section focuses on some operational issues for working with files and collections of files. More specifically, we will describe specific functions for listing, selecting, copying, renaming, and deleting files within a given directory, as well as functions for managing and visualizing the contents of files. Four different categories of functions are described: operating-system commands, operating-system calls, user interfaces and document collection management.

The first group consists of functions that perform or reproduce basic operating system shell commands. This group includes functions for conducting operations at the directory and the file level. Table 5.3 provides a brief description of the functions in this group. Most of these functions admit both command and function syntaxes. When used as commands, they just list the result of the operation, if the operation implies an explicit result. On the other hand, when used as functions, some of them can return different types of variables such as error messages and codes, status indicators and, in some specific cases, structure arrays with parsed information resulting from the operation.

Table 5.3. Functions in the operating-system command category and their descriptions

Function	Level	Description
pwd	directory	returns the current directory
dir	directory	lists the files contained in the specified directory
ls	directory	lists the files contained in the specified directory
what	directory	lists MATLAB specific files in the specified directory
cd	directory	changes the current working directory

`mkdir`	directory	creates a new directory
`rmdir`	directory	removes the specified directory
`movefile`	directory /file	moves the specified file or directory to another file or directory
`copyfile`	directory /file	copies the specified file or directory to another file or directory
`fileattrib`	directory /file	sets or gets the properties of the specified file or directory
`delete`	file	either deletes or moves to the recycle bin the specified files
`recycle`	file	sets or resets the option "delete" for the `delete` function
`type`	file	displays the contents within a specified ascii file

As you probably are already familiar with some of the equivalent operating-system commands, we will not devote much time in describing the functions in Table 5.3. However, we encourage you to check the corresponding in-function descriptions by using the `help` command in order to get more specific details on how to use these functions. In this section, we will be only presenting examples of four of them: `what` and `type` just next, and `dir` and `ls` when discussing user interfaces.

Consider the following example in which the function `what` is used to list MATLAB® related files contained on the current directory:

```
>> what                                                        (5.45)
MATLAB Code files in the current folder C:\Users\Documents\MATLAB
chapter05_scripts
MAT-files in the current folder C:\Users\Documents\MATLAB
allworkspace    matlab          somevariables   structfile
```

Alternatively, this output can be saved into a structure as follows:

```
>> matlabfiles = what                                          (5.46a)
matlabfiles =
  struct with fields:
        path: 'C:\Users\Documents\MATLAB'
           m: {'chapter05_scripts.m'}
       mlapp: {0×1 cell}
         mlx: {0×1 cell}
         mat: {4x1 cell}
         mex: {0x1 cell}
         mdl: {0x1 cell}
         slx: {0×1 cell}
         sfx: {0×1 cell}
           p: {0x1 cell}
     classes: {0x1 cell}
    packages: {0x1 cell}

>> matlabfiles.m                                               (5.46b)
ans =
```

```
1×1 cell array
  {'chapter05_scripts.m'}
```

The next example shows the use of function **type** to list the contents of a file:

```
>> type matrix.txt                                                              (5.47)
  1.0000000e+00    2.0000000e+00
  3.0000000e+00    4.0000000e+00
```

However, notice that when **type** is used as a function, output variables are not generated:

```
>> output = type('matrix.txt')                                                  (5.48)
Error using type
Too many output arguments.
```

So, if you want to capture the output listed by **type**, you must invoke the command via **evalc** (**eval** with capture)[9], just as shown below:

```
>> output = evalc('type(''matrix.txt'')')                                       (5.49)
output =
  '
          1.0000000e+00    2.0000000e+00
          3.0000000e+00    4.0000000e+00
  '
```

Most of the functions in Table 5.3 somehow make use of operating-system commands to perform and complete the specific tasks they do. However, it is also possible to directly execute commands and/or programs from the operating system by means of operating-system calls. This second group of functions and operators are presented in Table 5.4.

Table 5.4. Functions and operators in the operating-system call category

Function/Operator	System	Description
dos	Windows®/MS-DOS®	Calls operating system
unix	Unix®	Calls operating system
system	Indifferent	Calls operating system
!	Indifferent	Calls operating system
&	Indifferent	Background execution

The corresponding syntaxes for the functions and operators presented in Table 5.4 are as follows:

```
[statuts,result] = dos('command');                                             (5.50a)
```

---

[9] Notice the procedure described here was previously introduced in example (5.43a).

```
[statuts,result] = unix('command');                           (5.50b)

[statuts,result] = system('command');                         (5.50c)

!command                                                       (5.50d)

!command &                                                     (5.50e)
```

For those functions in (5.50a), (5.50b) and (5.50c) the output variable `status` returns a status indicator for the attempted operation (as, for example, success `'0'` or failure `'-1'`), and `result` returns whatever the operating-system command returned to the standard output.

Consider, for instance, launching from your MATALB® command window a text editor for opening the file `matrix.txt` from (5.49), which was originally created in (5.8b). In the case of a Unix® operating system, you can enter the following command `[st,rs]=system('emacs matrix.txt');` and in the case of Windows® you can use `[st,rs]=system('<path>\notepad.exe matrix.txt')`.[10]

Let us present the third group of functions discussed in this section, which is related to user interfaces. The MATLAB® programming environment offers actually a large number of user-interface functions for allowing interactivity between the user and the scripts being executed. Nevertheless, here we only focus our attention on those that are related to the specific problem of working with files and directories: `listdlg`, `uigetdir`, `uigetfile`, `uiputfile`, and `uiimport`.

Let us first discuss function `listdlg`, which creates a dialogue box that allows the user to select strings from a list. This function, among many other possible applications, can be useful for displaying and selecting files from a directory. Consider, for instance, the following two alternative ways for reading a specific group of files names and saving them into a cell array:

```
>> files = dir('sample_file.*'); filelist = {files.name}';      (5.51a)

>> filelist = cellstr(ls('sample_file.*'));                     (5.51b)
```

Then, `listdlg` can be used as follows for generating the dialogue box that is displayed in figure 5.8.

```
>> [select,stat] = listdlg('ListString',filelist,'ListSize',[160,100]);   (5.52)
```

The dialogue box in Figure 5.8 allows for either a single selection or multiple selections. The returned value `stat` will be either `'1'` or `'0'` depending on whether the user has made a valid selection or has press the cancel button, respectively. In the case a valid selection has been made, `select` will contain a numeric array with the indexes of the selected items from the list. Otherwise, it will be an empty

---

[10] For verifying the exact location of the `notepad.exe` application in your system, you must check its property panel.

array. For getting more information on additional input parameters for `listdlg` you can use the `help` command.

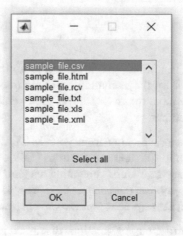

Fig. 5.8. Example of dialogue box for string selection created with `listdlg`

Let us now briefly introduce functions `uigetdir`, `uigetfile`, and `uiputfile`. These functions open specific browsing tools for directory selection (`uigetdir`), input file selection (`uigetfile`), and output file selection (`uiputfile`). The syntaxes for these functions are as follows:

```
dirname = uigetdir(startingpath,title);                                          (5.53a)
```

```
[filename,pathname,select] = uigetfile(filter,title);                            (5.53b)
```

```
[filename,pathname,select] = uiputfile(filter,title);                            (5.53c)
```

In (5.53a), `startingpath` refers to the initial directory the browsing tool will be showing when it is started and `dirname` returns the full path and name of the selected directory. In case no selection is made, `dirname` is set to zero.

In (5.53b) and (5.53c), `filter` specifies the type of files to be listed. It can be a single string, such as: `'*.mat'`, `'*.*'`, etc; or an $N \times 2$ cell array containing both the file types and their corresponding descriptions to be presented in the selection's menu. You can try using the following value for `filter`:

```
>> filter = {'sample_file.*','Files created in Chapter 5';...                    (5.54)
            '*.doc;*.xls','Microsoft Application''s files';...
            '*.txt','Plain text files';'*.*','All files'};
```

Regarding the output variables in (5.53b) and (5.53c), `filename` and `pathname` returns the specified file name and path, respectively; and `select` contains the index to the `filter` selection made by the user. In case no selection or specification is made, or the cancel button is pressed, all three output variables are set to zero.

In all three cases in (5.53), the input variable `title` is a string specifying the title to be given to the browsing tool's window. It will appear in the top left of the window, next to the MATLAB® icon.

Let us introduce now one last interactive function: `uiimport`. This function allows for importing data from a file in an interactive manner. When called, it opens an import wizard that can be used to interactively specify the details of the import process, after which `uiimport` will load the newly created variables directly into the workspace. It also provides the option to generate the necessary code to read the data from the file according to the specification defined in the import wizard.

Let us consider the following example, in which `uiimport` is used to import the random data from example (5.23):

```
>> uiimport('mtx.csv')
```
(5.55)

The display resulting from (5.55) is presented in figure 5.9.

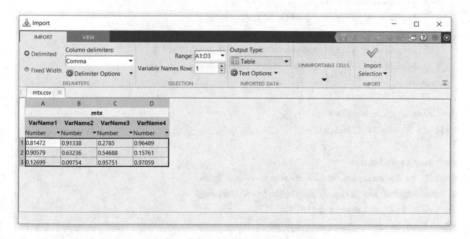

Fig. 5.9. Using `uiimport` to interactively import data from a file

Notice, however, that `uiimport` only provides support for importing data in tabular form. The output type can also be selected from the user interface (see Figure 5.9). An alternative way for loading data in tabular format is by using the function `readtable`, which is introduced at the end of exercise 5.5-4.

The final category of functions described in this section is the one corresponding to document collection management. The class `fileDatastore`, which allows for creating a datastore for a collection of files, is found within this category. This class contains a few methods for operating with and manipulating a collection of files. The available operations include reading, previewing, partitioning, and shuffling, among others. In the following example, we illustrate the use of `fileDatas-`

tore to create a datastore with some the files previously created in this chapter and copy the contents of those files into a string array.

```
>> % specifies the files to be included in the datastore          (5.56a)
>> files = ["sample_file.html","sample_file.txt","sample_file.*v"];

>> % creates the datastore object                                 (5.56b)
>> fds = fileDatastore(files,'ReadFcn',@extractFileText);

>> % loops over the files in the datastore to read their contents (5.56c)
>> strarray = [];
>> while hasdata(fds)
       textData = read(fds); % reads the content of a file
       strarray = [strarray;textData]; % copies content into string array
   end

>> % displays the contents                                        (5.56d)
>> for k=1:length(strarray)
       fname = split(fds.Files{k},'\');
       fname = fname{end};
       fcontent = strip(replace(strarray(k),newline,''));
       fprintf('\n*** %s\n%s\n',fname,fcontent);
   end


*** sample_file.html ***
This is a sample .html file

*** sample_file.txt ***
# This is an example of the feature-value file
# For details see http://svmlight.joachims.org/
+1 1:2.0 10:3.0 15:-1.2 20:8.0
-1 1:1.15 11:2.0 18:1.2 20:-1.5 23:1.0
-1 3:11.0 10:-3.0 15:1.0 21:1.1
+1 7:-1.2 15:1.0
+1 1:2.1 10:2.0 15:-1.0 20:1.0

*** sample_file.csv ***
"name","phone","birthday"
Peter,625-7725,"Feb, 27th"
Sam,717-1648,"December, 11th"
Sara,512-2345,"May, 2nd"

*** sample_file.rcv ***
1    2    250
6    3    -110
7    1    152
```

```
7    4    049
9    6    000
```

As seen form (5.56a) the file collection can be specified by a list of file names, which can also contain the wildcard `'*'`, as it was done with `"sample_file.*v"` to account for both the *.csv* and *.rcv* files. In general, the file collection definition admits either files or folder names.

In (5.56b), the datastore object is created by means of `fileDatastore`. Notice how the specific function to be used for reading the files is specified by including the key value pair `'ReadFcn'` and `@extractFileText` in the function call. This sets function `extractFileText`, see example (5.44), as the default function to be used for reading files every time the `read` method is called.

Then, in (5.56c) each file is individually read, and its content is copied into the string array variable `strarray`. Finally, in (5.56d), the file contents within `strarray` are displayed. Notice how, first, the name of each file is retrieved from the datastore object property `'Files'` and the corresponding file content (stored in `strarray`) is preprocessed to remove unnecessary newline characters, before being displayed with `fprintf`.

5.4 Further Reading

For a more detailed description on File Input/Output operations you can directly refer to the MATALB® Supported File Formats Guide (The MathWorks 2020).

A more comprehensive description on the standard specifications for text-formatting conversion characters and operators in the C-family languages can be found in any C++ reference manual. You can consider, for instance, the online version from (cplusplus.com 2021). An excellent tutorial on character encoding can be found in the free information site *IT and communication* (Korpela 2009).

Regarding file formats discussed in section 5.2, a detailed description about the CSV (comma-separated-value) format is provided by (Shafranovich 2005), and more information on the sparse-matrix format is available from (Wächter 2010). The feature-value format was introduced by (Joachims 2008). Finally, recommendations from the World Wide Web Consortium (W3C) regarding the Extensible Markup Language (XML) and from the Web Hypertext Application Technology Working Group regarding the Hypertext Markup Language (HTML) are provided in (W3C 2008) and (WHATWG 2020), respectively.

5.5 *Proposed Exercises*

1. Consider the conversion characters and operators commonly used in format-specification strings that were described in Table 5.2:

   - Create the function `printreport(rtags,ctags,values,formatstr)` for printing a table of real values.
   - The input variables `rtags` and `ctags` are string arrays, which provide tags for the rows and columns of the table, respectively.
   - The function should allow for `rtags`, `ctags` or both to be the empty string `' '`, in such a case not requiring row, column or both tag sets.
   - The input variable `values` is a matrix of numeric values, and `formatstr` specifies the format to be used for printing the matrix values.
   - Notice that if `values` is an $N \times M$ matrix, the only possible lengths for string arrays `rtags` and `ctags` are 0 or N and 0 or M, respectively.
   - Create few random matrices of different sizes, as well as some tags, and print some reports by using your created function.

2. Use the `help` command to get information about functions `fgets` and `fgetl`:

   - Create the function `docs = getdocs(fname,dtype)` for reading documents from a file in plain text format (`.txt`).
   - The input variable `fname` is a string containing the name of the file to be read, and `dtype` is a string specifying what exactly should be considered a document: `'full'` for the whole file to be considered a single document; `'line'` for each line to be considered a document; and `'paragraph'` for each paragraph to be considered a document (assume that two consecutive paragraphs are separated by an empty line).
   - The output variable `docs` should be a string array of documents.
   - Use your newly created function for reading a text data file and saving the resulting string array in a `.mat` file of the same name.

3. Use the `help` command to get information about functions `ftell`, `frewind`, `fseek`, and `ferror`.

   - Create the function `index = getindex(fname,dtype)` for creating an index of documents from a file in plain text format (`.txt`).
   - Consider the input variables `fname` and `dtype` to be exactly the same as in the previous exercise.
   - The output variable `index` should be an array of integers containing the starting positions of each document within file `fname`.
   - Add an additional entry to `index` containing the EOF (end-of-file) location. The resulting length for `index` must be $N+1$, where N is the to-

tal number of documents; so, a given document k can be located inside the file by considering the interval from `index(k)` to `index(k+1)-1`.

- Create the function `doc = getidxdoc(fname,index,n)` for retrieving the nth document from file `fname` (for which the document index has been already computed by means of function `getindex`).

4. Consider a CSV file like the one illustrated in figure 5.2, which contains both textual and numeric data:

    - Create the function `dataset = csvget(fname,fline)` able to read non-numeric CSV files, by using a similar procedure to the one illustrated in example (5.24).
    - The input variable `fname` contains the name of the file to be read, and `fline` determines whether the first line in the file should be interpreted as a normal row of data `'data'` or as a header specifying the names of the column fields `'tags'`.
    - The output variable `dataset` should be a structure array containing all the data in the input file, one row per structure in the array.
    - The fields in each structure must be named either according to the header specifications (if `fline='tags'`) or by a sequence of numbered tags: `f1`, `f2`, `f3`, etc. (if `fline='data'`).
    - Consider now the function `readtable`, which can also be used to import tabular data. However, its output is a variable of class `table` (use `help table` to get more information about `table` objects).

5. Consider the sparse-matrix format described in section 5.2.

    - Create the functions `smtxload` and `smtxsave` for reading from and writing to a specified file a sparse matrix. Consider the procedures presented in examples (5.25) and (5.28), respectively.
    - Use, for instance, the following syntaxes `smtx = smtxload(fname)` and `smtxsave(fname,smtx)`, where `fname` is the name of the file to be read/written and `smtx` is a MATLAB® sparse matrix.
    - Consider, from example (5.28c), what would happen if the last element in the sparse matrix is different from zero. Implement a solution to avoid this problem into your function `smtxsave`.

6. Consider the feature-value format described in section 5.2.

    - Create the functions `fvload` and `fvsave` for reading from and writing to a specified file a sparse matrix.
    - Consider the following syntax `[smtx,target] = fvload(fname)` for the reading function, where `fname` is the name of the file to be read, `smtx` is a sparse matrix containing the feature-value pairs (one file line

per matrix row), and `target` is a column vector containing the corresponding category values.

– Consequently, consider the syntax `fvsave(fname,smtx,target)` for the writing function, where the three input variables are the same variables involved in the `fvload` function.

7. Consider MATLAB® functions `xmlread` and `xmlwrite` for reading from and writing to an XML file a Document Object Model node.

– Use the `help` command and consult the available online documentation to learn about these functions. Implement a script for reading and writing XML documents similar to the one presented in Figure 5.6

– Use the `help` command to learn about function `xslt`. Create an XSL stylesheet[11] to create HTML code that presents in tabular format the contents of the XML file in Figure 5.6.

8. Consider the following data structure, rather than the one defined in example (5.39), for the data contained in the XML file presented in Figure 5.6:

– document.attribute.id
– doument.content.cite(n).attribute.id
– doument.content.cite(n).attribute.lang
– doument.content.cite(n).content.person
– doument.content.cite(n).content.statement
– Create scripts for writing and reading an XML file similar to the one in Figure 5.6 by using the proposed structure.
– Write another script able to parse a generic file like the one in Figure 5.6 and automatically infer the data structure proposed here.
– Alternatively, consider the function `xml2struct`.[12] What differences do you find with respect to the data structure proposed here?

9. Consider the functions for handling files and directories that were discussed in section 5.3:

– Create the function `[location,fnames] = getfilelist(path,rexp)` for getting a list of files from a specified directory `path`, such that all file names satisfy the given regular expression `rexp`.

– The input variable `path` can be a text string or an empty string. In the first case it should specify the location of the directory to be considered, in the second case the function should launch a GUI for the user to navigate to and select a directory.

---

[11] For more information about Extensible Stylesheet Language (XSL), you can refer to https:// www.w3.org/Style/XSL/ Accessed 6 July 2021

[12] Available at https://github.com/joe-of-all-trades/xml2struct Accessed 6 July 2021

- The input variable `rexp` is a string specifying a regular expression that must be matched by the file names to be listed.
- The output `fnames` must be a string array containing the list of files.
- The output `location` must be the absolute path of the specified or selected directory where the files are located.

10. Use the `help` command to get information about `webread` and `websave`.

- Select a webpage of your interest. Consider, for instance, websites that typically contain links to multiple documents or articles, such as Wikipedia, Project Gutenberg, News, Blogs, etc.[13]
- Get the content from the webpage of interest and parse it by means of `htmlTree` as previously illustrated in example (5.43b).
- Retrieve all links pointing to other content pages within the site. You can consider using functions such as `findElement` and `getAttribute` to select the links you are interested in but, first, you will need to look at the HTML code of a sample page to find out how to identify the class of links you are interested in retrieving.
- Extract the text content of all pages related to the extracted links. It is important to incorporate a delay function in your script to avoid overloading the target server with multiple consecutive requests! Consider a delay function of the form `pause(random('unif',10,20))`.
- Create a table containing relevant information about the collected pages, such as titles, URLs, etc. You can store the information in a structure first and then convert it into a table (see `help struct2table`).
- Create a new directory and save into it the table containing all relevant information about the collected pages. Save the table in a CSV file.
- Save all extracted text contents into the same directory. Create one plain text file per webpage or document collected.

5.6 References

cplusplus.com (2021) C++: Reference: C Library: cstdio (stdio.h): fprintf, http://www.cplusplus.com/reference/clibrary/cstdio/fprintf/ Accessed 6 July 2021

Joachims T (2008) SVM-light: support vector machine. Cornell University, https://www.cs.cornell.edu/people/tj/svm_light/ Accessed 6 July 2021

Korpela J (2009) A tutorial on character code issues. In IT and Communication, https://jkorpela.fi/chars.html Accessed 6 July 2021

The MathWorks (2020) MATLAB Supported File Formats, https://www.mathworks.com/help/matlab/import_export/supported-file-formats.html Accessed 6 July 2021

---

[13] You must look for specific conditions and restrictions disclosed at the websites with regards to scraping contents from them. Follow the given instructions, if any, and use the provided links or metadata files in case you intend to download a significant volume of contents from the sites.

Shafranovich Y (2005) Common format and MIME type for comma-separated-values (CSV) files. The Internet Society, http://tools.ietf.org/html/rfc4180 Accessed 6 July 2021

Wächter A (2010) Triplet format for sparse matrices. In Introduction to IPOPT, https://coin-or.github.io/Ipopt/IMPL.html Accessed 6 July 2021

W3C (2008) W3C Recommendation on Extensible Markup Language (XML) 1.0, Fifth Edition, http://www.w3.org/TR/xml/ Accessed 6 July 2021

Web Hypertext Application Technology Working Group (2020) HTML Living Standard, https://html.spec.whatwg.org/ Accessed 6 July 2021

6 The Structure of Language

This chapter introduces the structure of language, mainly focusing its attention on written language. First, in section 6.1, the different levels of linguistic phenomena are introduced and briefly described. In section 6.2, morphology and syntax are introduced and, in section 6.3, semantics and pragmatics are introduced. Some specific Text Analytics Toolbox™ functions that operate at these different levels are also described along the chapter.

6.1 Levels of the Linguistic Phenomena

Languages are systems for encoding and communicating information. The term *Natural Language* is typically used to refer to the system (or set of systems) used by human beings to communicate among themselves. As such, natural language constitutes an extremely complex system that encompasses a multiplicity of phenomena that operate at different levels of abstraction, which go from basic symbolic/encoding representations and communication means of perceptual nature to much more elaborated representations and interpretations of concepts, relationships, and beliefs of cognitive and intellectual nature. In this section, we will briefly introduce some of the different levels of the linguistic phenomena.

Given its complex nature, natural language cannot be fully described or analyzed along just a few dimensions. However, for the sake of clarity, we will present here a simplified categorization of the linguistic phenomena according to two fundamental dimensions: physical means of communication and cognitive levels of abstraction. In the first case, two main general categories of language can be distinguished according to the physical means of communication used: spoken language and written language.[1] Although certainly similar in many aspects, there are significant differences between spoken and written languages across other dimensions different from the physical means of communication described here. In the rest of this chapter (and the book), we will focus on written language.

The second dimension under consideration, the cognitive levels of abstraction, helps to understand the main components of language as we move from the physical encoding and representation of symbols (perceptual level) to the abstract representation of concepts and ideas (intellectual level). Five different structural lev-

---

[1] Notice this categorization is not complete as other different categories can be also distinguished across the physical means of communication dimension, such as sign language, for instance.

© The Author(s), under exclusive license to Springer Nature Switzerland AG 2021
R. E. Banchs, *Text Mining with MATLAB®*, https://doi.org/10.1007/978-3-030-87695-1_6

els or components are typically identified along this dimension, which are illustrated in Figure 6.1, together with the cognitive faculties involved.

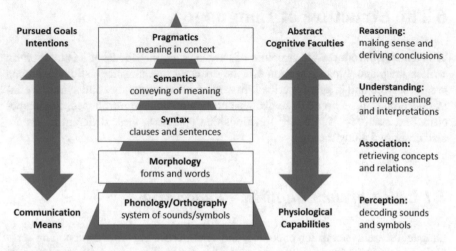

Fig. 6.1. The five structural components of language along the cognitive levels of abstraction dimension

As seen from the figure, the five structural components of language are hierarchically organized. Starting at the bottom, we encounter a pure physical layer of communication means, which requires physiological capabilities for perceiving linguistic signals and symbols. Ending at the top, we find a high-level intellectual layer of goals and intentions, which requires abstract cognitive faculties for reasoning and decision making. All the way from the bottom to the top of the hierarchy, the five presented components concur to conform the overall natural language system. These five structural components are briefly introduced next.

- Phonology/orthography. While different in nature as they are specific to spoken and written language, respectively, both phonology and orthography deal with representation systems and rules for the fundamental units of communication. In the case of phonology, it deals with the organization of sounds used in spoken language. In the case of orthography, it deals with conventions related to using graphemes in written language.

- Morphology. It deals with the structure of words and the rules that define how words are constructed in a language. This includes the creation of new words and the inflection of existent ones into their associated variants along different grammatical categories.

- Syntax. It deals with the structure of sentences and the rules and principles by which words are combined to construct valid sentences in a language. It also includes the study of the structure and sequencing of the three main components of sentences: subject, verb and object.

- Semantics. While all previous components pay attention to form, semantics pays attention to content and its relation to form. Indeed, semantics deals with the problem of understanding the meaning of language units at the different levels: words, sentences, and larger units.

- Pragmatics. It deals with putting meaning into context for the proper interpretation of the pursued intention and the information being communicated. Important elements, such as social experience and cultural background, play a fundamental role on how meaning is influenced by context.

In the following two sections, we dive into more details with regards to both the form related levels (morphology and syntax) and the content related levels (semantics and pragmatics), respectively. Along the way, we also introduce some relevant Text Analytics Toolbox™ functions used to preprocess text data across some of these dimensions.

Just before that, let us introduce two functions that are instrumental for the purposes of manipulating and visualizing text data. First, let us consider the function `tokenizedDocument`, which allows for creating an array of document objects of the same class. In this class of variables, documents are segmented and represented as a collection of tokens, in which words, punctuation and other string types are identified and extracted for analysis purposes. In the following example, we use `tokenizedDocument` for segmenting and processing a single document:

```
>> s1 = "I gave Peter my 2 theater tickets. ";                    (6.1a)
>> s2 = "He lives in San Francisco :) ";
>> strdoc = s1+s2;

>> tokdoc = tokenizedDocument(strdoc);                             (6.1b)

>> whos strdoc tokdoc                                              (6.1c)
  Name          Size            Bytes  Class               Attributes
  strdoc        1x1               278  string
  tokdoc        1x1              4199  tokenizedDocument
```

As seen from (6.1), the string scalar `strdoc` is converted into a variable of the class `tokenizedDocument`, which contains a tokenized representation of the corresponding text. A string array of all tokens used in the representation are available through the `Vocabulary` property of the resulting `tokenizedDocument` object:

```
>> tokdoc.Vocabulary                                              (6.2)
ans =
  1×14 string array
  Columns 1 through 8
    "I"     "gave"    "Peter"    "my"    "2"    "theater"    "tickets"    "."
  Columns 9 through 14
    "He"    "lives"    "in"    "San"    "Francisco"    ":)"
```

All contents and metadata within the `tokenizedDocument` object can be displayed in tabular format by using the function `tokenDetails`. This is illustrated in the following example:

```
>> tokenDetails(tokdoc)                                                            (6.3)
ans =

  14×5 table

        Token         DocumentNumber     LineNumber        Type        Language
      "I"                   1                 1          letters          en
      "gave"                1                 1          letters          en
      "Peter"               1                 1          letters          en
      "my"                  1                 1          letters          en
      "2"                   1                 1          digits           en
      "theater"             1                 1          letters          en
      "tickets"             1                 1          letters          en
      "."                   1                 1          punctuation      en
      "He"                  1                 1          letters          en
      "lives"               1                 1          letters          en
      "in"                  1                 1          letters          en
      "San"                 1                 1          letters          en
      "Francisco"           1                 1          letters          en
      ":)"                  1                 1          emoticon         en
```

Notice from (6.3) that the tokenized representation in `tokdoc` is actually an enriched representation including some metadata for each individual token. Specifically, four different metadata fields are provided: the number of the document, the number of the line within the document, the type of token (letters, digits, punctuation, etc.), and the identified language. Most of the functions presented in the remaining sections of this chapter provide a means for augmenting the representation in (6.3) with different types of linguistic information.

6.2 Morphology and Syntax

In this section, we focus our attention on the form-related components of language: morphology and syntax. As explained already, while morphology deals with the structure and rules of construction of words, syntax deals with the structure and rules of construction of sentences.

Starting with the structure of words, morphology focuses on the study of morphemes, which are the minimal meaning-bearing units in a language. Notice that morphemes and words are not the same thing. While some morphemes can stand alone and play the role of a word, others cannot and might only appear as a part of a word. In general, three main types of morphemes can be identified: free morphemes, affixes, and clitics. Free morphemes are those which can play the role of a word. A free morpheme generally constitutes the root or fundamental lexical unit of a word family. Examples of free morphemes in English include words like *explain, do, kind, nurse, septic*, etc. Notice also that free morphemes can be combined together to create new words: *pan-cake, hand-shake, break-fast*, etc.

Affixes, on the other hand, are morphemes that are combined with free morphemes to produce other words. Affixes are typically referred to as prefixes or suffixes depending on whether they attach to the beginning or the end of the root they modify, respectively. Affixes can also be classified as derivational or inflectional depending on whether they change or not the grammatical category or the meaning of the root they modify. Examples of derivational affixation include words like *explain-able, un-do, kind-ness, un-kind, nurse-ry, a-septic*, etc. Examples of inflectional affixation include words like *explain-s, explain-ed, do-es, nurse-s, pan-cake-s, break-fast-ing*, etc. Multiple affixes can also occur within a single word such as, for instance, in *un-explain-able, un-do-able*, etc.

Finally, clitics are morphemes that are similar to affixes but play a syntactic role at the phrase or sentence level. Some examples of clitics in English[2] include the possessive particle *'s*, the negative particle *n't* and auxiliary verb contractions *'ve, 's, 'm*, etc.

As in the cases of phonology and orthography, morphology varies significantly across languages. Some languages, like Chinese, have very little or practically inexistent morphology. This type of languages is referred to as isolating languages, which typically do not exhibit inflectional properties at all, i.e. each word is composed of a single morpheme. In the other extreme of the spectrum, languages exhibiting a prolific and rich morphology, like Finnish and Turkish, are referred to as agglutinative languages. In these cases, words are typically composed of a multiplicity of morphemes.

The first step of any kind of analysis of text data starts with a process called tokenization, which aims at segmenting the data by identifying word boundaries. This problem was already introduced in chapter 4; see examples (4.27), (4.32) and exercise (4.6-7). In this chapter, see example (6.1), we introduced the function `tokenizedDocument`, which automatically takes care of the tokenization problem. As seen in (6.3), each row in the table corresponds to each specific token or word resulting from the document segmentation carried out by `tokenizedDocument`.[3]

---

[2] For some of these cases, however, it does not seem to be a general consensus among linguists on whether such cases should be considered clitics rather than affixes.

[3] For English tokenization, `tokenizedDocument` uses by default the UNICODE Text Segmentation Standard https://www.unicode.org/reports/tr29/#Word_Boundaries Accessed 6 July 2021. Notice that English word segmentation is a relatively simple problem, compared to languages such as Chinese, Japanese, and Vietnamese, among others.

A common second step in text analysis, which is often referred to as normalization, consists of reducing inflectional variants of the same family of words to a single common canonical form. This process aims at reducing the sparseness of the actual language vocabulary by providing a much more compact representation of it. This is of fundamental importance for statistical methods. Two well-known processes for language normalization are *lemmatization* and *stemming*, which are introduced next and revisited later in chapter 10.

Lemmatization and stemming are methods to reduce a given word to canonical forms called lemmas and stems, respectively. While lemmas are lexicographically motivated (they are the canonical representations of words given in dictionaries), stems can be considered orthographically motivated (they are the minimal units that preserve the meaning of a word after removing inflectional affixes). Although precise definitions of these concepts might be difficult to formalize, the main difference between lemmas and stems can be easily grasped with the help of a practical example.

Consider the following example in which we use the function `normalizeWords` to reduce the inflections of the English verb *to see* to both lemmas and stems:

```
>> see_inflections = ["see","sees","saw","seen","seeing"];                    (6.4a)

>> normalizeWords(see_inflections,'Style','lemma') % Reduces to lemmas      (6.4b)
ans =
  1×5 string array
    "see"     "see"     "see"     "see"     "see"

>> normalizeWords(see_inflections,'Style','stem') % Reduces to stems        (6.4c)
ans =
  1×5 string array
    "see"     "see"     "saw"     "seen"     "see"
```

Notice, from (6.4b), that all inflected forms of the verb *to see* have the same common lemma *see*. However, as seen from (6.4c), three different stems (*see, saw* and *seen*) are obtained for the verb *to see*. While lemmatization aims at reducing all inflected forms to the basic canonical form of the word as a unit of meaning, stemming looks at the orthographical root of each inflected form, which results in different stems mainly because the verb *to see* is irregular. Indeed, for a regular verb, such as *to walk* we would expect both lemmas and stems to coincide:

```
>> walk_inflections = ["walk","walks","walked","walking"];                    (6.5a)

>> normalizeWords(walk_inflections,'Style','lemma') % Reduces to lemmas     (6.5b)
ans =
  1×4 string array
    "walk"     "walk"     "walk"     "walk"

>> normalizeWords(walk_inflections,'Style','stem') % Reduces to stems       (6.5c)
```

```
ans =
  1×4 string array
    "walk"    "walk"    "walk"    "walk"
```

Although somehow similar processes and sometimes indifferently used, lemmatization and stemming are different in nature; and, typically, lemmatization is a much more complex task than stemming. Indeed, lemmatization requires paying attention to context and determining the morpho-syntactic role of the word under consideration in order to properly reduce it to the correct lemma. Consider, for instance, example (6.4b) in which the past-tense form *saw* of the verb *to see* was reduced to its lemma *see*. It might be the case of the different word *saw*, as in "a tool for cutting wood", occurring in the text under analysis. In such a case, the correct lemma is *saw* and no *see*. Notice, however, that the stem remains the same for the two words: *saw*, as in "cutting tool", and *saw* as in past-tense of *to see*.

The kind of disambiguation needed to properly conduct lemmatization, is similar to the one needed to identify the different grammatical categories of words, which are typically referred to as parts of speech (and abbreviated as POS). Although not always necessary to conduct text mining, part of speech computation provides a means to generate linguistically enriched features that might result very useful for supporting tasks of higher levels of complexity.

The function `addPartOfSpeechDetails` allows for computing the POS category for each of the tokens in a `tokenizedDocument` object. Before seeing this function in action, let us briefly review the specific POS categories it supports. Table 6.1 lists the *17* different POS categories supported by `tokenizedDocument` along with specific examples of the corresponding category occurring in some sample sentences (in the provided examples, words belonging to the corresponding category are underlined).

Table 6.1. Part of speech categories of words and illustrative examples

POS Category	Example (words belonging to the corresponding categories are underlined)
adjective	He spent the <u>whole</u> afternoon looking at the <u>blue</u> sky .
adposition	The morning fight departing <u>from</u> Barcelona <u>to</u> London was cancelled .
adverb	All his magic tricks were <u>perfectly</u> executed .
auxiliary-verb	I <u>have</u> never seen something like this before .
coord-conjunction	They had a difficult time deciding between swimming <u>or</u> hiking .
determiner	After reading <u>the</u> statement , I still do not have <u>a</u> clue about it .
interjection	<u>Oh</u> ! I completely forgot about that !
noun	I will hand out the <u>documents</u> to the <u>layer</u> this <u>evening</u> .
numeral	I have called you more than <u>ten</u> times in the last <u>two</u> months .
particle	I have <u>n't</u> seen your parents since we graduated from highschool .
pronoun	<u>You</u> never pay attention to <u>him</u> . <u>He</u> is just trying to help <u>you</u> !
proper-noun	<u>Agnes</u> went to visit her cousins in <u>Austin</u> during the holidays .

punctuation	How are you doing? It is so nice to see you again .
subord-conjunction	I would definitely consider buying a new house if I win the lottery .
symbol	Happy faces come in different flavors and shapes: :) :-) :P ...
verb	I was looking at the pictures of our last trip. I really enjoyed that trip !
other	For more info on POS, refer to en.wikipedia.org/wiki/Part_of_speech

In general, most words tend to play the same grammatical roles most of the time. However, in some cases, the same word can be categorized into a different POS depending on the role it plays in a given context. For instance, while *water* is a noun in the sentence "I just had a glass of water", it is actually a verb in "I water the plants everyday". Similarly, while *have* plays the role of a verb in "I have a collection of books", it is an auxiliary verb in "I have read almost all books in my collection". While *blue* is typically and adjective as in "I love blue skies", it can also be a noun as in "Blue is one of my favorite colors"; and so on. In summary, POS computation is an ambiguous task, which requires contextual information (i.e. awareness of the surrounding words) in order to be properly determined.

In the following example, we use functions `addLemmaDetails` and `addPartOf-SpeechDetails` to enrich the `tokenizedDocument` object representation computed in (6.1b) with lemma and POS metadata for all the tokens in the representation.

```
>> tokdoc = addLemmaDetails(tokdoc); % Adds lemmas                          (6.6a)

>> tokdoc = addPartOfSpeechDetails(tokdoc); % Adds POS                       (6.6b)

>> doctbl = tokenDetails(tokdoc);                                           (6.6c)
>> doctbl(:,{'Token','Lemma','PartOfSpeech'})
ans =
   14×3 table
```

Token	Lemma	PartOfSpeech
"I"	"i"	pronoun
"gave"	"give"	verb
"Peter"	"peter"	proper-noun
"my"	"my"	pronoun
"2"	"2"	numeral
"theater"	"theater"	noun
"tickets"	"ticket"	noun
"."	"."	punctuation
"He"	"he"	pronoun
"lives"	"life"	verb
"in"	"in"	adposition
"San"	"san"	proper-noun
"Francisco"	"francisco"	proper-noun
":)"	":)"	symbol

It is important to notice that, in general, POS computation might affect the existent tokenization. Consider, for instance, the following example in which the POS computation changes the original tokenization:

```
>> doc = "Tokenization isn't static";                                (6.7a)

>> otk = tokenizedDocument(doc) % Original tokenization              (6.7b)
otk =
  tokenizedDocument:
   3 tokens: Tokenization isn't static

>> rtk = addPartOfSpeechDetails(otk) % Retokenized document          (6.7c)
rtk =
  tokenizedDocument:
   4 tokens: Tokenization is n't static
```

As seen from (6.7b) the tokenization performed by `tokenizedDocument` keeps the negated verb form *isn't* as a single token but, when POS is computed, it is split into two tokens: the verb part *is* and the negation particle *n't*. If needed, this behavior can be prevented by passing the argument `RetokenizeMethod` with value `none` to the function `addPartOfSpeechDetails`. This will force for the original tokenization to be preserved, although it will affect the final POS assignment:

```
>> stk = addPartOfSpeechDetails(otk,'RetokenizeMethod','none')      (6.8a)
stk =
  tokenizedDocument:
   3 tokens: Tokenization isn't static

>> stktbl = tokenDetails(stk); stktbl.PartOfSpeech'                 (6.8b)
ans =
  1×3 categorical array
     noun        other        adjective
```

Moving further up into the syntactic level, where attention is paid to the structure and rules of construction of sentences, another fundamental problem that requires disambiguation is the one of sentence segmentation. The function `addSentenceDetails` allows for splitting a `tokenizedDocument` object into sentences.[4] Let us consider adding sentence segmentation information to the `tokenizedDocument` object representation previously computed in (6.1b) and augmented in (6.6).

```
>> tokdoc = addSentenceDetails(tokdoc); % Adds sentence number      (6.9a)

>> doctbl = tokenDetails(tokdoc);                                    (6.9b)
>> doctbl(:,{'Token','SentenceNumber','Lemma','PartOfSpeech'})
```

---

[4] One of the major challenges of sentence segmentation in English and other languages is the disambiguation of punctuation characters. In this process, identifying abbreviations is critically important. Use `help addSentenceDetails` and check its documentation for more details.

```
ans =

  14×4 table

      Token        SentenceNumber       Lemma        PartOfSpeech
   _____     _____    _____     _____

   "I"                   1           "i"             pronoun

   "gave"                1           "give"          verb

   "Peter"               1           "peter"         proper-noun

   "my"                  1           "my"            pronoun

   "2"                   1           "2"             numeral

   "theater"             1           "theater"       noun

   "tickets"             1           "ticket"        noun

   "."                   1           "."             punctuation

   "He"                  2           "he"            pronoun

   "lives"               2           "life"          verb

   "in"                  2           "in"            adposition

   "San"                 2           "san"           proper-noun

   "Francisco"           2           "francisco"     proper-noun

   ":)"                  2           ":)"            symbol
```

Different from morphology, which might vary significantly from language to language, syntax tends to be more universal in the sense that most languages share similar structural building blocks with regards to sentence construction. In general, as we go up in the cognitive scale, more structural similarities can be found across different languages.

A very simplified description of the structure of sentences, in general, can be given by the distinction of the three main syntactic components of a sentence: verb (V), subject (S) and object (O). While the verb is the main action or state described in the sentence, the subject refers to the entity (person or thing) who executes or performs the action or state defined by the verb, and the object (or objects) are those entities unto which the actions or states are executed or performed. Figure 6.2 illustrates the SVO structure[5] of English by identifying verb, subject, and objects in the first sentence from (6.9).

I gave Peter my 2 theater tickets

S   V   Ind. O   Direct O

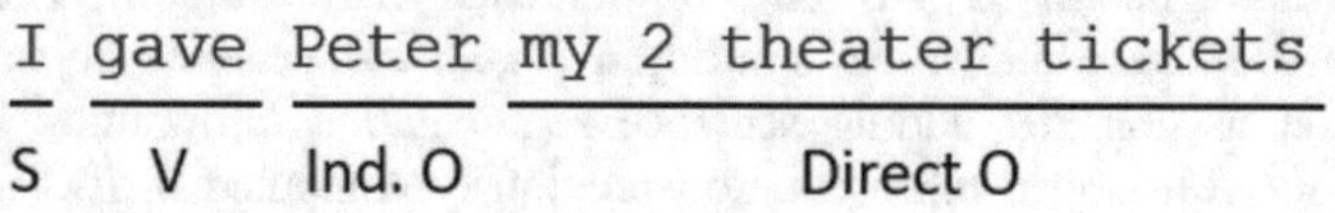

Fig. 6.2. Basic syntactic analysis of the first sentence from (6.9)

As seen from the figure, the verb (V) representing the main action performed in the statement is the word *gave*, and the subject (S) executing that action (the entity who gives) is *I*. In this particular case there are two objects (O), as the English

---

[5] Similar to English, Chinese and Spanish (among other languages) are also SVO; while other languages like Japanese and Persian, for example, exhibit an SOV structure. Interestingly, most of the languages are SVO or SOV, although a few cases exist with VSO and VOS structures.

verb *to give* actually admits two objects. The direct object *my 2 theater tickets* (the thing being given), and the indirect object *Peter* (the entity receiving the thing being given).

A more detailed and complete analysis of the structure of a sentence, generally referred to as syntactic parsing,[6] can be performed by means of two main different formalisms: constituency parsing and dependency parsing. While these two methods are different as they are based in different types of grammars, which follow different assumptions, they both aim at the same objective of extracting syntactic information about the structure of a given sentence.

Constituency parsing is based on the context-free-grammar formalism, which explains the structure of sentence as the combination of smaller chunks or phrases, referred to as constituents. These constituents can be combined according to certain rules (the grammar) to create valid sentences. These rules define, for instance, how different POS can be combined to create constituents, and those constituents can be combined to create new constituents and finally full sentences. Consider the constituency parsing of the same sentence analyzed in Figure 6.2, which is depicted in Figure 6.3.

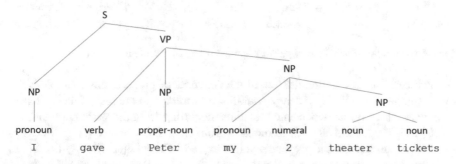

Fig. 6.3. Example of constituency parsing applied to the first sentence from (6.9)

As seen from the figure, the sentence (S) is composed of two main constituents: a noun phrase (NP), which happens to be a single pronoun (*I*), and a verb phrase (VP), which is composed of the rest of the sentence. The VP can be further decomposed into three constituents: a verb (*gave*) and two noun phrases. The first NP following the verb is composed of a single proper-noun (*Peter*), and the second NP can be further decomposed into a pronoun (*my*), followed by a numeral (*two*) and another NP, which is composed of two nouns (*theater* and *tickets*). The graphical structure depicted in Figure 6.3 is typically referred to as a constituency

---

[6] As of the time this second edition was prepared, the Text Analytics Toolbox™ did not include specific functions to perform syntactic parsing. Accordingly, the problem of syntactic parsing is beyond the scope of this book. In this section, however, we present a brief introduction to the problem along with some illustrative examples.

parse tree. This type of graph provides a visual representation of a given sentence structure in terms of its constituents.

Dependency parsing, on the other hand, is based on the dependency grammar formalism. In this formalism, the syntax of the sentence is described in terms of direct dependency relations between pairs of words. Instead of grouping words into building blocks (constituents), in the case of dependency parsing, the syntactic structure of the sentence can be seen as a "dependency signal" flowing from a root node (typically the main verb in the sentence) towards the rest of the words in the sentence. Figure 6.4 presents an example of dependency parsing for the same sentence previously analyzed in Figures 6.2 and 6.3.

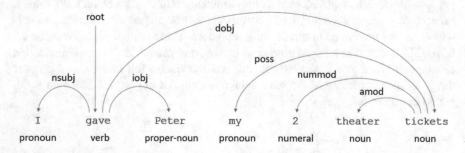

Fig. 6.4. Example of dependency parsing applied to the first sentence from (6.9)

As seen from the figure, the root of the dependency graph in this example is the verb (*gave*). From it, dependency relations are directed to the rest of the sentence. There are three direct dependency relations departing from the verb: nominal subject (nsubj), which is directed to the first pronoun in the sentence (*I*); indirect object (iobj), directed to the proper-noun (*Peter*); and direct object (dobj), directed to the last noun in the sentence (*tickets*). Notice that *I* and *Peter* are terminal nodes in the graph, while *tickets* has additional dependency relations with other words in the sentence. More specifically, the word *tickets* have three direct dependency relations: adjectival modifier (amod), directed to its preceding noun (*theater*);[7] numeric modifier (nummod), directed to the numeral (*two*); and possessive nominal modifier (nmod:poss), which is directed to the second pronoun in the sentence (*my*). The graphical structure depicted in Figure 6.4 is typically referred to as a dependency parse tree. It provides a visual representation of a given sentence structure in terms of the dependency relations among its words.

---

[7] Notice that, although its POS is of type noun, *theater* acts as a modifier with respect to *tickets*. Accordingly, we have annotated its dependency relation to be of type adjectival modifier in Figure 6.4. However, an alternative interpretation might be the one of noun compounds. In such a case, the dependency relation should be annotated as of type compound instead. For more information and examples on universal syntactic dependency relations, refer to https://universaldependencies .org/en/dep/index.html Accessed 6 July 2021.

In summary, constituency parsing and dependency parsing are analysis methods that provide complementary views of the syntactic structure of sentences in a language. While constituency parsing represents a more classical approach to syntactic analysis, dependency parsing has been gaining popularity lately due to its use in some modern language processing applications. They both are highly difficult tasks given the ambiguous nature of natural language. Indeed, many sentences typically admit more than one valid parse tree from a syntactic point of view. In such cases, valid interpretations are only possible under the light of semantics and pragmatics.

6.3 Semantics and Pragmatics

As mentioned in section 6.1, languages are systems for encoding and communicating information. In the previous section, we paid attention to the form-related components of such systems, and we saw how languages follow certain rules with respect to the construction of its fundamental structures: words and sentences. In this section, we devote our attention to the content-related components of language: semantics and pragmatics. These two components determine how the structures of language are used to convey meaning and how such meaning is affected by context, respectively.

To illustrate the importance of semantics in language, let us start by looking at a classical academic example. Consider the sentence "colorless green ideas sleep furiously".[8] Such a sentence is both morphologically and syntactically valid but completely meaningless. Indeed, although it complies to the rules defining how valid words and sentences must be constructed in English, it lacks the ability to communicate a meaningful message or, at least, anything that makes sense to a normal human being. Notice that, regardless its grammatical correctness, we cannot confidently understand or even interpret its meaning. Moreover, any attempt to understand it will make us to discover contradictions that operate at the semantic level. For instance, *colorless* refers to the lack of color and *green* is a type of color, then *colorless green* is a contradiction. Also, referring to color properties of something like an *idea*, which cannot be seen and, consequently, do not have any visual property, does not make to much sense. In the same way, we can find similar contradictions about the notions of *ideas* that *sleep*, as well as *sleep* that happens in a *furious* manner.

All the contradictions resulting from the previous example are of semantic nature, and they are responsible for the inability of a receiver of such a sentence to

---

[8] This sentence was introduced by Noam Chomsky, in his book Syntactic Structures (1957), as an example of a grammatically correct but semantically nonsensical English sentence, https://en .wikipedia.org/wiki/Colorless_green_ideas_sleep_furiously Accessed 6 July 2021.

make sense out of it. Semantics is about encoding and decoding meaning into and from language structures. This is done with the help of high-level cognitive faculties, which leverage on experience and common-sense to convey and derive meaning into a from language. Some meaningful sentences used in our daily life can be extremely ambiguous (or even contradictory, if a wrong interpretation is considered). It is actually through avoiding semantic contradictions, that understanding operates to disambiguate natural language and derive the proper meaning out of it. This process is called *grounding*[9] and it is of fundamental importance in both semantics and pragmatics.

Consider for instance the sentence "I saw the Statue of Liberty when flying to New York". Different form the previous example, this sentence makes completely sense, and it is morphologically, syntactically, and semantically valid. However, it is ambiguous as it admits two valid syntactic representations, but only one of the two is semantically valid. In one syntactic representation (the incorrect one) the *when flying to New York* part of the sentence attaches to the *Statue of Liberty* entity, literally meaning that the Statue of Liberty was flying to New York at the moment I was seeing it. In the second syntactic representation (the correct one) the *when flying to New York* part of the sentence attaches to the subject of the sentence *I*, literally meaning that when I was flying to New York, I saw the Statue of Liberty. The semantic task of selecting the proper syntactic structure to derive the correct interpretation of such an ambiguous sentence is supported by grounding.

The proper selection of the correct interpretation (understanding) in the previous example is derived from the common knowledge we share about the Statue of Liberty being an unanimated monument, which is fixed in a location (in New York) and, therefore, unable to fly. It is by avoiding the semantic contradiction of the Statue of Liberty flying that we actually validate the correct interpretation. Notice, however, that the semantic task of understanding the sentence under consideration is even more complex, as we also know that human beings do not fly. Here, common-sense also comes into play to support the selected interpretation by resolving the eventual contradiction of the subject person *I* literally *flying*. This is accomplished by using the shared assumption that people are able to fly by means of flying devices such as planes and helicopters.

The actual unambiguous restatement of the sentence in the previous example should be something like the following: "I saw the Statue of Liberty when I was in a plane and the plane was flying to New York". However, people rarely use this type of unambiguous constructions when they communicate, as the communication process also aims at being economic and efficient. In the quest for efficiency, information that is considered redundant with respect to the common knowledge shared by the people communicating is typically omitted. This makes natural lan-

---

[9] Also referred to as *common ground*, consists of the common knowledge and assumptions that must be shared by the parties communicating in order to ensure a successful communication. It operates at multiple levels and plays a fundamental role in conversational communication.

guage extremely ambiguous and non-understandable for entities, such as computers, that do not have the required common knowledge and cognitive capabilities.[10]

As deriving meaning from sentences in an automatic fashion is a complex and difficult task, for practical purposes, such a problem is typically broken down into smaller subproblems. Two of these fundamental problems, which are jointly referred to as Natural Language Understanding (NLU), are the identification of the different entities occurring in a sentence (Entity Recognition) and the identification of the general purpose of a sentence (Intent Detection).

Named Entity Recognition (NER), which is as specific type of Entity Recognition, focuses on identifying and categorizing entities that are denoted with proper names, such as persons, organizations, places, etc. This problem operates at the semantic level as it requires knowledge about the real word to assign proper categories to the different named entities occurring in a sentence. The Text Analytics Toolbox™ includes the function `addEntityDetails`, which performs Named Entity Recognition. It can be used to augment the representation of a `tokenizedDocument` object by adding entity labels to its tokens. Let us consider the following example, in which we add entity labels to the document representation from (6.9):

```
>> nerdoc = addEntityDetails(tokdoc); % Adds named entity labels       (6.10a)

>> doctbl = tokenDetails(nerdoc);                                       (6.10b)
>> doctbl(:,{'Token','SentenceNumber','PartOfSpeech','Entity'})
ans =
    13×4 table
```

Token	SentenceNumber	PartOfSpeech	Entity
"I"	1	pronoun	non-entity
"gave"	1	verb	non-entity
"Peter"	1	proper-noun	person
"my"	1	pronoun	non-entity
"2"	1	numeral	non-entity
"theater"	1	noun	non-entity
"tickets"	1	noun	non-entity
"."	1	punctuation	non-entity
"He"	2	pronoun	non-entity
"lives"	2	verb	non-entity
"in"	2	adposition	non-entity
"San Francisco"	2	proper-noun	location
":)"	2	symbol	non-entity

---

[10] The problem becomes even more complex in the case of figurative language, where the words and sentences are used to convey meanings that deviate from their conventionally accepted ones. Examples include similes and metaphors (which are commonly used in poetry and other literary works), among many others such as hyperboles, idioms, oxymorons, etc.

Notice from (6.10) that two named entities were correctly identified and categorized: *Peter* was labeled as a "person" and *San Francisco* was labelled as a "location" while the rest of the tokens were labelled as "non-entity". In general, `addEntityDetails` allows for identifying named entities in four different categories: person, location, organization and other. The latter category typically includes references to brands, products and other types of entities.

Notice also from (6.10) that the NER task has affected the original tokenization of the document. While, originally, *San Francisco* was represented by two separated tokens in (6.9), after being labeled as a location, both tokens have been combined into a single token. If needed, this behavior can be prevented by passing the argument `RetokenizeMethod` with value `none` to the function `addEntityDetails`. This will force for the original tokenization to be preserved. This is illustrated in the following example:

```
>> % Identifies named entities without changing the tokenization        (6.11a)
>> keeptok = addEntityDetails(tokdoc,'RetokenizeMethod','none');

>> doctbl = tokenDetails(keeptok);                                       (6.11b)
>> idx = doctbl.Entity ~= 'non-entity'; % gets indexes for named entities
>> doctbl.Token(idx)' % displays the tokens of the identified named entities
ans =
  1×3 string array
    "Peter"     "San"      "Francisco"
>> doctbl.Entity(idx)' % displays the labels of the identified named entities
ans =
  1×3 categorical array
    person      location        location
```

As seen from (6.11), *San Francisco* was still labeled as a location, but the original tokenization has been preserved. In this case the label "location" has been assigned to both tokens comprising the named entity.

The second fundamental problem in Natural Language Understanding is the one of Intent Detection. In this case, the objective of the task is to classify the full sentence, or utterance, according to a set of predefined categories that determine the main intention of the communicative action, such as for instance: "asking for instructions", "informing about a location", "requesting for clarification", "answering a question", etc. Typically, the Intent Detection task also involves the pragmatic level, which is described later in this section, and the specific categories used are generally application dependent. For now, let us define Intent Detection as a classification problem (text classification is discussed in more detail in chapter 11), which is of significant importance for question answering and dialogue systems (discussed in chapter 15).

Other examples of important semantic tasks include relationship extraction and event extraction. While the former focuses on the identification and categorization of semantic relationships among entities such as, for instance, cheese *is an ingredient of* pizza, vodka *is a type of* liquor, Johnathan *works at* Amazon, etc; the latter focuses on the identification and categorization of events and its attributes, aiming at identifying descriptions of something that happened, when and where it happened and who was involved.

More complex semantic analysis techniques involve the use of semantic parsing.[11] Similar to syntactic parsing, semantic parsing aims at unfolding the structure of sentences, but in this case the representation is semantically motivated. The final goal of a semantic parser is to represent sentences in logical forms that can be consumed and processed by computers. Semantic Role Labeling, sometimes also referred to as slot-filling, is considered a shallow type of semantic parsing as it focuses on identifying entities and the semantic roles they play, without addressing other more complex semantic phenomena. Figure 6.5 presents an example of semantic parsing for the same sentence from Figures 6.2, 6.3 and 6.4.

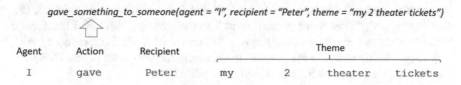

Fig. 6.5. Example of shallow semantic parsing applied to the first sentence from (6.9)

As seen from the figure, a shallow semantic parser deals with the task of identifying the values of the arguments (semantic roles) of a given predicate (the main action described). In the example, the main action is *gave something to someone* and its arguments are the agent (the entity that executed the action), the theme (the entity that was given) and the recipient (the entity receiving the theme).

Different from semantics, pragmatics deals with the problem of using context to interpret meaning in a proper way. The term pragmatics derives from the Greek word "pragma" which refers to "intentioned actions". Accordingly, pragmatics focuses on decoding the intentions being communicated beyond the pure meaning conveyed by language alone. In this process, context plays a fundamental role, as it is needed to resolve ambiguities that cannot be resolved by semantics alone.

Consider, for instance, the following example: "The police officer followed the woman with a gun in her hand". This sentence is syntactically ambiguous, as *with a gun in her hand* can be either attached to the *police officer* or *the woman*. Notice that, in this case, semantics alone cannot help selecting the correct interpretation,

---

[11] As of the time this second edition was prepared, the Text Analytics Toolbox™ did not include specific functions to perform semantic parsing. Accordingly, the problem of semantic parsing is beyond the scope of this book. In this section, however, we present a brief introduction to the problem along with an illustrative example.

as both interpretations are semantically valid. The only way of resolving this ambiguity is by having more contextual information about the situation described.

Language ambiguities can be of semantic nature too. Consider the following two sentences: "Peter and John finally met in New York. He just had a two-hour layover in New York but managed to make it possible." In this case, identifying whether the pronoun "He" in the second sentence refers[12] to "Peter" or "John" is not possible unless additional contextual information is made available such as, for instance, who lives in New York or who was travelling when they finally managed to meet in New York.

Implicatures are other example of linguistic phenomena that occur at the pragmatic level. They refer to intentions or requests that are communicated without being explicitly stated in the message. For example, in the case of the sentence "It is freezing in here!", again, it is the context what allows to determine the actual intention or request being communicated. The request in this case can be "to close an opened window", "to turn on the heater", "to get a blanket or a sweater", etc. Answering with "This is not enough food." to the question "Should I toast more bread?" is another example of implicature. While the intention is an evident "yes", it is not explicitly stated in the answer. In this case, the statement is more a sort of justification that supports the reason of the implicit intent.

Finally, Speech Act Theory is also heavily rooted in pragmatics. According to it, in language-based communications and, more specifically, in conversational interactions, individuals not only interchange information but also perform speech acts, which are specific types of actions. Similar to intents, speech acts are categories describing the type of communicative action conveyed by a sentence given its context. Some examples of speech acts include: greeting, asking, answering, complaining, requesting, explaining, confirming, etc. We discuss more about speech acts and intents in conversational systems in chapter 15.

In this chapter we have seen how natural language encompasses a multitude of linguistic phenomena, which operate at different levels of abstraction and require different types of cognitive capabilities for its proper production and understanding. In general, most of the properties and tasks described here are useful for text mining applications as they enrich the constructed text representations with linguistically informed features, some of which will be used and further discussed in the following parts of the book. However, most of the problems described in this chapter are still far from being completely solved and constitute active areas of research within the scope of Computational Linguistics and Natural Language Processing. A detailed discussion of these problems, beyond their eventual use in text

---

[12] This is a specific case of a more general problem in Computational Linguistics know as Coreference Resolution. It focuses on the identification and association of two or more references to the same entity occurring in a given passage of text. Sometimes, this problem can be handled at the syntactic level but, in other cases, it must be dealt with at the semantic level, or even the pragmatic level (as in the example presented here).

mining applications, is out of the scope of this book. You can refer to the sources cited in the following section for a more comprehensive coverage of these topics.

6.4 Further Reading

An excellent introduction and comprehensive review of Natural Language Processing and Computational Linguistics is available in (Jurafsky and Martin, 2009). For interesting discussions about the relationships between language and cognition, refer to (Hardin and Banaji, 1993) and (Harris, 2006).

Good introductions to morphology are provided in (Beard, 1995) and (Aronoff and Fudeman, 2010). For a more specific discussion on English morphology, see (Plag, 2003). The Porter stemming algorithm is presented in (Porter, 1980). An early introduction to generative grammars and syntax is given in (Chomsky, 1957). A more recent introduction to syntax can be found in (Radford, 2004). More details about constituency and context-free grammars are available in sections 12.1 and 12.2 of (Jurafsky and Martin, 2009), and dependency grammars are discussed in section 12.7 of the same book.

Introductions to semantics can be found in (Jackendoff, 1990) and (Nielson and Nielson, 1995). For more details on the problem of automatic Semantic Role Labeling, refer to (Gildea and Jurafsky, 2002). Introductions to pragmatics can be found in (Blakemore, 1990) and (Green, 1996), which also cover topics such as coreferences, implicatures, speech acts and conversational interactions. A much more recent introduction to meaning in language, covering both semantics and pragmatics, is presented in (Kroeger, 2019). For more details about grounding in communications, see (Clark and Brennan, 1991) and (Paek and Horvitz, 2000). For early works on Speech Act Theory, see (Austin, 1962) and (Searle, 1969).

For more details about the Text Analytics Toolbox™ and all its related functions, refer to the product documentation page (The MathWorks, 2020a); and for more information about the functions introduced in this chapter, refer to the section therein about preprocessing text data (The MathWorks, 2020b).

6.5 Proposed Exercises

1. Create the function `preprocess_document`, which receives a string scalar and returns a tokenized document including the following metadata for each token in the representation: document number, sentence number, line number, type of token, language, part of speech, lemma, and entity.

- Copy a paragraph of text from an online source, such a Wikipedia, save it to a plain text file, and load it into a string scalar.
- Use the function `preprocess_document` to create a first tokenized representation of the document.
- Create a second tokenized representation for the document after first lowercasing all the text in the string scalar.
- Obtain table representations for both tokenized documents by using the function `tokenDetails` and compare the counts of Part-of-Speech tags in each representation. What do you see?
- Compare the extracted entities in each case. What do you see?
- Save each table to a *.csv* file by using `writetable`. Open both files and explore the differences in more detail.
- To keep in mind: functions `addLemmaDetails`, `addSentenceDetails`, `addPartOfSpeechDetails` and `addEntityDetails` must be used before applying transformations such as `lower`, `upper`, `erasePunctuation`, `normalizeWords`, `removeWords` and `removeStopWords` as most of these transformations remove information that is needed by the former functions to do their work!

2. Compute and print frequency counts for token attributes in a given document:

- Copy a paragraph of text from an online source, such a Wikipedia, save it to a plain text file, and load it into a string scalar (you can use the same text file from the previous exercise).
- Use the function `preprocess_document` to create a tokenized representation of the document.
- Create the function `print_token_frequencies`, which should receive as input one column (attribute) of the tabular tokenized representation produced by `tokenDetails`. The function must compute and print the frequency counts for each of the attribute values (the counts should be printed in descending order, from most to least frequent).
- Print frequency counts for the following attribute types: lowercased tokens, lemmas, part-of-speech and named entities.

3. Create a function that implements a standard text preprocessing pipeline and follows the syntax: `output = text_preprocessing(input,normalization)`

- The input variables are: `input`, a string scalar or string array containing the text to be preprocessed; and `normalization`, a string specifying the type of token normalization to be used.
- The output variable `output` is a `tokenisedDocument` object after applying all preprocessing steps described next.
- Data augmentation: add the following metadata to the document representation: part-of-speech, named entity tags and sentence number.

- Remove most frequently used non-content words, commonly referred to as stopwords. You might use function `removeStopWords` (to see the list of words considered stopwords, use the command `stopWords`).
- Normalize the tokens in the document representation according to the type of normalization specified by the input variable `normalization`. The possible options are: `'lower'` for lowercasing the tokens, `'lemma'` for replacing the tokens with their lemmas, `'stem'` for replacing the tokens with their stems, or `'none'` for not doing any normalization.
- Remove punctuation. You might use function `erasePunctuation`.
- Remove short and long words. Consider short words with only one or two characters, and long those with 20 or more characters. You might use functions `removeShortWords` and `removeLongWords` for this.

4. Create the function `annotate_document`. It must take an input text file and create a new annotated version of the same file, in which each `token` in the original file is replaced with an augmented representation of the form `[token]_pos`, where `pos` is the token's corresponding part-of-speech tag.

- The function must return the `tokenisedDocument` object of the annotated document, as well as saving it into a *.txt* file in plain text format. You might want to consider using function `writeTextDocument` to write a `tokenisedDocument` object to a plain text file.
- You also want to learn about function `docfun`, which can be used to apply functions to the tokens of a `tokenisedDocument` object. It might be useful to combine tokens and POS tags into the desired format.

5. Use the `help` command to learn about the function `correctSpelling`.

- Copy a paragraph of text from an online source, save it to a plain text file (you can use the same text file from exercise 6.5-1) and manually edit the file to include few misspellings and typos.
- Load the content of the file into a string scalar, generate a tokenized document representation of it and use the function `correctSpelling` to generate a corrected version of the tokenized document.
- Display those token pairs for which the token in the original document is different from the one in the corrected document. Inspect all pairs and identify those cases, if any, in which correct words have been erroneously "corrected" into a different one.
- Based on the previous inspection, create a list of the words that have been erroneously "corrected", if any, and use the `'KnownWords'` argument to pass the list to the function `correctSpelling`.
- Use function `correctSpelling` again to generate a new corrected version of the tokenized document. Display those token pairs for which the token in the original document is different from the one in the corrected document. Inspect all pairs and verify the resulting corrections.

    – Can you find cases in which misspellings or typos have been replaced by correct words, but the correction is still wrong? Can you suggest possible ways to overcome such types of correction errors?

6. Concordances are specialized indexes that list words used in a book along with their corresponding contexts and their exact locations in the book. They constitute valuable resources for scholars as they allow for easily finding passages, as well as studying how words are used within different contexts. This exercise is about building a concordance for the King James version of the Holy Bible.[13]

    – Use the `help` command to learn about the function `context`.
    – Download the King James version of the Holy Bible from Project Gutenberg's website.[14]
    – Load the text into a string array, one verse per element in the array. Make sure you only read the text that is related to the book under consideration and not its accompanying metadata (you might want to use a modified version of function `getdocs`, created in exercise 5.5-2, to read only the paragraphs of interest).
    – Build a tokenized document representation with `tokenisedDocument`.
    – Use function `context` to get word occurrences in context for all words in the vocabulary. Store the resulting context tables for each `word` in a structure field of the form `concordance.word` and save the final `concordance` structure into a *.mat* file.
    – Compute the time it takes your program to create the full concordance (it should take about an hour or so). Similar works were conducted during the pre-computer times, each of which typically took few years to be completed!

6.6 References

Aronoff M, Fudeman K (2010) What is Morphology? Second Edition. Wiley-Blackwell
Austin JL (1962) How to Do Things with Words. Cambridge, Massachusetts. Harvard University Press
Beard R (1995) Lexeme-Morpheme Base Morphology: A General Theory of Inflection and Word Formation. Albany, NY: State University of New York Press
Blakemore D (1990) Understanding Utterances: The Pragmatics of Natural Language, Oxford: Blackwell

---

[13] One of the most famous and complete concordances of the Holy Bible is the Strong's Concordance (Strong 1890). It was manually compiled under the direction of American lexicographer James Strong, way before computers were available!

[14] The plain text version can be downloaded from https://www.gutenberg.org/files/10/10-0.txt Accessed 6 July 2021.

Clark HH, Brennan, SE (1991) Grounding in communication. In Resnick LB, Levine JM, Teasley SD (Eds.), *Perspectives on socially shared cognition*. American Psychological Association, pp. 127–149

Chomsky N (1957) Syntactic Structures. Mouton & Co.

Gildea D, Jurafsky D (2002) Automatic Labeling of Semantic Roles. *Computational Linguistics*. Vol. 28, N. 3, pp. 245–288

Green GM (1996) Pragmatics and Natural Language Understanding, Second Edition, Mahwah, New Jersey: Lawrence Erlbaum

Hardin C, Banaji MR (1993) The influence of language on thought. Social Cognition 11:3, pp. 277–308

Harris CL (2006) Language and cognition, in Encyclopedia of Cognitive Science. John Wiley & Sons

Jackendoff R (1990) Semantic Structures. Cambridge, Massachusetts: MIT Press

Jurafsky D, Martin JH (2009) Speech and Language Processing: An Introduction to Natural Language Processing, Computational Linguistics, and Speech Recognition. Second Edition. Upper Saddle River, New Jersey: Pearson Education

Kroeger P (2019) Analyzing Meaning: An Introduction to Semantics and Pragmatics. Second corrected and slightly revised edition. Berlin, Germany: Language Science Press

Nielson HR, Nielson F (1995) Semantics with Applications, A Formal Introduction. Chichester, UK: John Wiley & Sons

Paek T, Horvitz E (2000) Grounding Criterion: Toward a Formal Theory of Grounding. Microsoft Research Technical Report MSR-TR-2000-40, https://www.microsoft.com/en-us/research/wp-content/uploads/2016/02/tr-2000-40.doc Accessed 6 July 2021

Plag I (2003) Word Formation in English. Cambridge, UK: Cambridge University Press

Porter, MF (1980) An algorithm for suffix stripping, *Program*, Vol. 14, N. 3, pp 130–137, https://tartarus.org/martin/PorterStemmer/def.txt Accessed 6 July 2021

Radford A (2004) English Syntax: An Introduction. Cambridge, UK: Cambridge University Press

Searle J (1969) Speech Acts. Cambridge, UK. Cambridge University Press.

Strong J (1890) The Exhaustive Concordance of the Bible. Cincinnati: Jennings & Graham, https://archive.org/stream/exhaustiveconcor1890stro#page/n11/mode/2up Accessed 6 July 2021

The MathWorks (2020a) Text Analytics Toolbox, https://www.mathworks.com/products/textanalytics.html#import-and-visualize-text-data Accessed 6 July 2021

The MathWorks (2020b) Prepare Text Data for Analysis, https://www.mathworks.com/help/textanalytics/ug/prepare-text-data-for-analysis.html Accessed 6 July 2021

Part II: Mathematical Models

"Mathematics possesses not only truth,
but supreme beauty."

Bertrand Russell

7 Basic Corpus Statistics

This chapter opens the second part of the book, which focuses on the mathematical models used for representing text data. As already mentioned in chapter 1, the basic objective of text mining (as well as data mining, in general) can be reduced to the discovery and extraction of relevant and valuable information from large volumes of data. In this chapter we will begin our description of text models by presenting methods that are based on the observation of basic properties and regularities in large volumes of text, which are generally referred to as corpus statistics. First, in section 7.1, we describe some fundamental properties of natural language as they are reflected on the text-based representation of the language. Then, in section 7.2, we introduce the concept of word co-occurrences, as well as the commonly used measure of mutual information, for studying dependencies between words. Finally, in section 7.3, we focus on word co-occurrences at shorter distances while taking also word order into account.

7.1 Fundamental Properties

A simple exploration of a text sample in an unknown language will make evident that text sequences, although apparently random, hide a great number of regularities. Indeed, although language generation is a very prolific process and an infinite variety of valid segments of text can be produced by any person with a basic knowledge of a language, not any random combination of words or characters will produce a valid segment of text. This means that text, as a written representation of language, is subject to the same construction rules of language which operate at the morphological, syntactic, semantic, and pragmatic levels.

Perhaps, the most fundamental property of languages is the one known as Zipf's law. This law has to do with the fact that given a language's vocabulary (the set of valid words for a given language), very few words are responsible for the largest proportion of a written text, while most of the words in the vocabulary seldom occur in it. Indeed, for any language, if we plot the frequency of words versus their rank for a sufficiently large collection of text data, we will see a clear trend, which resembles a power law distribution.

Let us start our experiments on corpus statistics by illustrating Zipf's law. For this example, as well as the following ones along the chapter, we consider the data collection provided in the file `bible_en.xml`, which can be downloaded from the book's companion website. This data collection comprises the sixty-six books of

© The Author(s), under exclusive license to Springer Nature Switzerland AG 2021
R. E. Banchs, *Text Mining with MATLAB®*, https://doi.org/10.1007/978-3-030-87695-1_7

the Holy Bible, which accounts for more than *30,000* verses, distributed in more
than *1,100* chapters. Although for current standards of text collections this might
be considered a small collection, it has been proven to be a very useful resource
for text mining and natural language processing experimentation, with the main
advantage of being available in many different languages. As a small dataset, using it here will allow for conducting full experiments while maintaining computational times under control.

Although we will not take it into account for this first example, the basic structure for the XML file under consideration is as follows:

```
<collection id="bible" lang=="en">                                     (7.1)

  <book id="STR">

    <chapter id="NUM">

      <verse id="NUM"> text content of verse </verse>

      ...

    </chapter>

    ...

  </book>

  ...

</collection>
```

Let us start by copying the contents of the file into a string:

```
>> clear;                                                              (7.2a)
>> collection = string(evalc('type bible_en.xml'));
```

```
>> whos collection                                                     (7.2b)
```

Name	Size	Bytes	Class	Attributes
collection	1x1	10236198	string	

remove all XML tags:

```
>> collection = regexprep(collection,'<.*?>','');                      (7.3)
```

lowercase the text:

```
>> collection = lower(collection);                                     (7.4)
```

replace all non-alphanumeric characters with whitespaces:

```
>> collection = regexprep(collection,'\W',' ');                        (7.5)
```

and, finally, eliminate leading and trailing whitespaces:

```
>> collection = strtrim(collection);                                   (7.6)
```

Now, we are ready to segment the string **collection** into tokens or words. Notice that in this particular exercise we have already discarded all the information

related to book, chapter and verse boundaries. The whole collection has been reduced to a very long sequence of words.

```
>> words = split(collection);                                    (7.7)
```

The elements contained in the string array **words** are generally referred to as running words or tokens, which are precisely all the words or tokens (excluding punctuation marks and special characters) in the data collection. So, the total amount of running words in the collection can be computed by simply looking at the length of the string array:

```
>> running_words = length(words)                                 (7.8)
running_words =
     791420
```

As seen from (7.8), our experimental data collection comprises a total of *791,420* running words. This is actually a small collection if compared, for instance, with the European Parliament corpus, which contains more than *40* million running words.

A second important computation is the one related to the vocabulary words, i.e. the set of unique words contained in the collection. For extracting such a set, we can use the function **unique**:

```
>> vocabulary = unique(words);                                   (7.9a)

>> vocabulary_words = length(vocabulary)                         (7.9b)
vocabulary_words =
     12545
```

As seen from (7.9b), the vocabulary size of our experimental data collection is just *12,545* words, which is more than sixty times smaller than the whole dataset. This means that the complete data collection (*791,420* running words) is actually composed of repetitions of this basic set of unique words (*12,545* words). But are all the words in the vocabulary equivalently repeated in the complete data collection? The answer to this question is, obviously, no. And we are now ready to explore Zipf's law in detail.

If all words in the vocabulary were repeated approximately the same number of times in the collection, we would expect each word in the vocabulary to occur, in average, about *63* times. We can see what the actual frequency of repetition is for each vocabulary word by counting the number of times each of it occurs in the whole data collection. Instead of implementing a loop for counting the words, we can take advantage of functions **unique** and **hist** to get the frequencies as follows:

```
>> [vocabulary,void,index] = unique(words);                      (7.10a)
>> whos index
  Name            Size              Bytes   Class       Attributes
  index        791420x1           6331360   double
```

```
>> frequencies = hist(index,vocabulary_words);                              (7.10b)
>> whos frequencies
  Name              Size                  Bytes  Class      Attributes
  frequencies       1x12545              100360  double
```

In (7.10a) we have used the function **unique** to extract the vocabulary as in (7.9a), but also for getting the corresponding indexes relating both string arrays **words** and **vocabulary**. In this way, the output variable **index** contains pointers to the corresponding word in **vocabulary** for each running word in **words**. As we already computed the total amount of vocabulary words in (7.9b), we have used the function **hist** in (7.10b) to compute a histogram for **index** with as many bins as the total amount of words in the vocabulary. Then, the resulting numeric array **frequencies** contains the total number of times each word in the vocabulary occurs in the whole collection.

However, the frequency counts contained in **frequencies** are not ranked yet. Then, we should use the function **sort** for reordering the counts in **frequencies** and the words in **vocabulary** according to their rank. We do this as follows:

```
>> [ranked_frequencies,ranking_index] = sort(frequencies,'descend');    (7.11a)

>> ranked_vocabulary = vocabulary(ranking_index);                       (7.11b)
```

We are now ready for constructing the plot of word frequencies versus rank for the data collection under consideration:

```
>> set(figure,'Color',[1 1 1],'Name','Zipf''s Law');                    (7.12a)

>> loglog(ranked_frequencies,'.');                                      (7.12b)

>> xlabel('Rank'); ylabel('Word frequency');                            (7.12c)
```

The resulting plot is presented in Figure 7.1. Notice how the curve clearly approaches a linear behavior in the central portion of ranks and slightly deviates from it at both extremes. This linear behavior in the logarithmic space is characteristic to many natural phenomena and it is referred to as a power law distribution. Whether Zipf's law is exactly a power law, or not, is still debated in the scientific community, but the fact is that this sort of behavior seems to be universal for any language, and it is observable even for small datasets such as the one used here.

As seen from the figure, in the low rank extreme of the curve, there happens to be three words which are clearly separated from the rest. These are indeed the most frequently used words in our considered data collection. I suppose you can guess which these words are:

```
>> ranked_vocabulary(1:3)'                                              (7.13)
ans =
  1×3 string array
    "the"     "and"     "of"
```

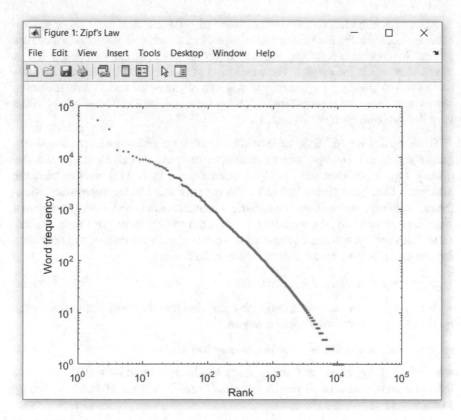

Fig. 7.1. Plot of word frequencies versus rank for the `bible_en.xml` data collection

If we just add up together the total number of occurrences for these three most
frequent words:

```
>> sum(ranked_frequencies(1:3))                                      (7.14)
ans =
      150233
```

we will realize that these three words occur *150,233* times in the data collection,
which accounts for almost *19%* of the whole dataset! Table 7.1 summarizes similar statistics for other subsets of the *n* most frequent words.

Table 7.1. Basic statistics for different subsets of the *n* most frequent words

Rank interval	1 to 7	1 to 42	1 to 249	1 to 1147
Percentage of Vocabulary	0.06%	0.33%	1.98%	9.14%
Total number of occurrences	199,795	395,047	593,882	712,352
Percentage of whole collection	25.25%	49.92%	75.04%	90.01%

Notice from the last column of the table that little less than the *10%* of the vocabulary accounts for the *90%* of the contents in the data collection. This indicates a huge disproportion regarding the way different words in the vocabulary repeat themselves within the dataset. The conclusion of this is that all words do not have the same probability of appearing in a given segment of text. There are some words with a very large probability of occurring while there are other words with a very low probability of occurrence.

At this point we are ready to define the most basic probabilistic model of language: the single word probability model, or unigram model. Based on the frequency counts obtained in (7.10b) or, alternatively, in (7.11a) we can compute maximum likelihood probability estimates for each word in the vocabulary. Maximum likelihood probabilities are empirical estimates than can be easily computed from data. In this way, the probability of occurrence of certain word w_i is estimated by counting the number of times such word occurs in the dataset and normalizing the count by the total amount of words in the dataset:

$$p(w_i) = counts(w_i) \, / \, \Sigma_j \, counts(w_j)$$
(7.15)

We do this in our current example by just dividing the word frequency vector by the total number of words in the dataset:

```
>> estimated_probabilities = ranked_frequencies/running_words;
```
(7.16)

where the validity of our probability space can be confirmed by verifying that the total probability mass of all possible events (words within the vocabulary) add up to one:

```
>> sum(estimated_probabilities)

ans =

    1.0000
```
(7.17)

From this basic probability model, we can compare now the probability of occurrence among some of the different words in the dataset. Consider the following example in which we retrieve probabilities for three different words: one we expect to appear a lot in the data, one common language word, and one less expected to occur in the kind of data under consideration:

```
>> estimated_probabilities(strcmp(ranked_vocabulary,'lord'))

ans =

    0.0101
```
(7.18a)

```
>> estimated_probabilities(strcmp(ranked_vocabulary,'water'))

ans =

    5.0037e-04
```
(7.18b)

```
>> estimated_probabilities(strcmp(ranked_vocabulary,'taxes'))

ans =

    1.2636e-06
```
(7.18c)

Although these results may confirm our intuition of what we would expect for the kind of text contents we are considering here, it is evident that these probability estimates will be very different if data collections related to other topics and domains such as politics, finance or sports are used instead. A very important warning can be issued from this observation: in general, when working with probabilistic models, special attention must be paid to the differences between the dataset from which a probability model was estimated and the dataset in which such model is intended to be used.

Additionally, a more critical problem related to this kind of models is the fact that a zero-probability value is assigned by default to any word not occurring in the dataset. Consider, for instance, the following case:

```
>> estimated_probabilities(strcmp(ranked_vocabulary,'complexity'))        (7.19)
ans =
   1×0 empty double row vector
```

In this particular case, as the word *complexity* does not occur in the corpus under consideration, a probability value has not been assigned to it in the probability vector `estimated_probabilities`. However, as seen from (7.17), all the probability mass has been already distributed among all words contained in the vocabulary. So, the only possible probability value for a word such as *complexity* (as well as for any other word not in the vocabulary) is zero. This problem is generally tackled by reducing a little bit the computed probabilities and reserving the remaining probability mass for those words that have not been seen in the dataset. We will come back to this issue in more detail in the next chapter (section 8.2).

The problems just pointed out do not mean that our basic probabilistic model derived in (7.16) is not useful in contexts different from biblical texts. In fact, although we have used a relatively small and very specific dataset for training our model, the obtained empirical results have some important implications from the linguistic point of view which are valid in any other context. Notice, for instance, which the three most common words depicted in (7.13) are. These words are not arbitrary words. It is not a coincidence that the three most common words in our experimental data collection are a determiner, a conjunction and a preposition. These are actually function words, which, different from content words, play a very specific role in language. If we take a look at the thirty most frequent words, we will find that most of them belong to the category of function words:

```
>> ranked_vocabulary(1:30)'                                               (7.20)
ans =
   1×30 string array
   Columns 1 through 8
     "the"     "and"     "of"     "to"     "that"     "in"     "he"     "shall"
   Columns 9 through 16
     "unto"     "for"     "i"     "his"     "a"     "lord"     "they"     "be"
```

```
Columns 17 through 24
   "is"     "him"    "not"    "them"    "it"    "with"    "all"    "thou"
Columns 25 through 30
   "thy"    "was"    "god"    "which"    "my"    "me"
```

As seen from (7.20), probably with the exception of 'lord' and 'god' (which are very frequent for obvious reasons), most of the frequent words shown here can be considered to be in the function word category.

On the other hand, as we move higher in the rank scale, more content words will be appearing until only content words can be seen, which are mainly composed by adjectives, verbs and nouns:

```
>> ranked_vocabulary(201:230)'                                            (7.21)
ans =
  1×30 string array
  Columns 1 through 6
    "servants"    "ever"    "might"    "gave"    "those"    "other"
  Columns 7 through 12
    "seven"    "through"    "hands"    "soul"    "another"    "would"
  Columns 13 through 18
    "life"    "cities"    "blood"    "sin"    "commanded"    "side"
  Columns 19 through 24
    "first"    "peace"    "without"    "mouth"    "sword"    "flesh"
  Columns 25 through 30
    "saul"    "work"    "gold"    "face"    "high"    "themselves"
```

Another interesting observation, also due to Zipf, which is more related to the concept of language economy, is that the most frequently used words in a language are, in average, the shortest ones. This observation can be directly verified from (7.20) and (7.21), where it is evident that the average length of the most frequent thirty words is lower than the average lengths of the thirty words within the rank range from *201* to *230*. Indeed, we can compute average lengths of words at different points in the rank scale as follows:

```
>> mean(strlength(ranked_vocabulary(1:30)))                              (7.22a)
ans =
    2.9333
```

```
>> mean(strlength(ranked_vocabulary(201:230)))                          (7.22b)
ans =
    5.2333
```

```
>> mean(strlength(ranked_vocabulary(1001:1030)))                        (7.22c)
ans =
    6.5667
```

```
>> mean(strlength(ranked_vocabulary(11001:11030)))                      (7.22d)
```

```
ans =
   6.8333
```

Examples (7.22a) to (7.22d) show the average length of the thirty words following rank positions *1, 201, 1001* and *11001*, respectively. Notice how the average length of words increases as we move higher in the ranking of words. This result is indeed a very interesting one as it provides evidence of a sort of natural "auto-compressing property" of language. Similar to code compression methods (where most frequent symbols are encoded with less bits than the average number of bits needed, and least frequent symbols are encoded with more bits), human language evolution seems to have efficiently reduced language generation costs by shortening the most commonly used words.

To see the whole picture of this natural "auto-compressing property", let us generate a cross-plot between the vocabulary word lengths and their rank for the considered data collection:

```
>> hf = figure(2);                                                    (7.23a)
>> set(hf,'Color',[1 1 1],'Name','Another Zipf''s Law');

>> semilogx(strlength(ranked_vocabulary),'.');                         (7.23b)

>> xlabel('Rank'); ylabel('Word length');                              (7.23c)
```

The resulting plot, which is presented in Figure 7.2, puts in evidence a very interesting characteristic of English, in this case. The most frequently used words in the language are also the shortest ones, while less frequent words are "allowed" to be larger but still can be short too. According to this, it seems that a sort of length-limit, which depends on the rank, exists for English words. For instance, as seen from the figure, a word in rank position *100* does not exceed *10* characters in length, and a word in rank position *1000* does not exceed *15* characters in length.

On the other hand, the opposite effect (a lower length-limit) does not happen to hold at all. As seen from the figure, for any given rank position, the words can be arbitrarily short. However, as can be observed, very short words (*1* or *2* characters in length) seldom occur.

The second fundamental property of language to be discussed here is intermittency, which is also generally referred to as *burstsiness*. This property has to do with the fact that words within a text sequence have the tendency to repeat themselves following some specific patterns that resemble *bursts* of occurrences. This means that when we first encounter a new word in a given text sequence there is a big chance to find it again relatively close in the following segments of text.

The main consequence of this intermittency property is that vocabulary words are not evenly distributed along text sequences. To better illustrate this idea, let us consider, for instance, the most frequent word in our data collection: *the*. If we assume that words are evenly distributed along a given text sequence, we would expect consecutive occurrences of the word *the* to repeat approximately every twelve words, which is basically the ratio between the total number of running words and the total number of occurrences of word *the*, just as seen below:

```
>> running_words/ranked_frequencies(strcmp(ranked_vocabulary,'the'))     (7.24)
ans =

    12.3816
```

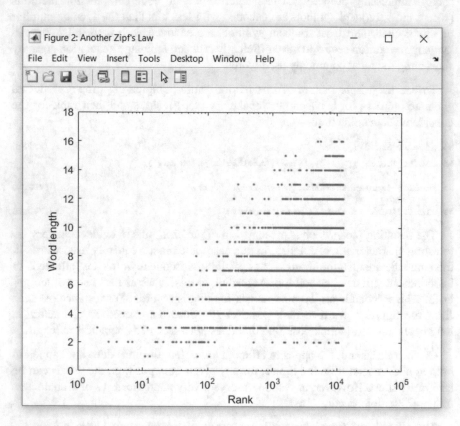

Fig. 7.2. Cross-plot between vocabulary word lengths and rank for the `bible_en.xml` data collection

Any person with a very basic knowledge of English (or any other language) would be able to tell that this assumption of evenly distributed words is wrong. Nobody would actually expect the word *the* to occur exactly every twelve or thirteen words in a given document or sequence of text. But it actually happens, in average, in our considered data collection. Then, two interesting questions we must ask here are: how does the distribution of intervals between consecutive occurrences of a given word look like? And, how do these intervals deviate from the average value?

To explore in more detail how these distributions look like, let us compute and plot the interval lengths between consecutive occurrences of the word *the* and make a histogram of such intervals. To achieve these, we will use functions `diff` and `hist` for computing the corresponding interval lengths and their histogram, respectively. Consider the following procedure, for which the resulting plots are presented in Figure 7.3:

```
>> % gets the indexes for the occurrences of the word 'the'      (7.25a)
>> locations = find(strcmp(words,'the'));

>> % computes the lengths of the repetition intervals            (7.25b)
>> intervals = diff(locations);

>> % computes a 100-bin histogram                                (7.25c)
>> histogram = hist(intervals,100);

>> hf = figure(3); % creates a new figure                        (7.25d)
>> set(hf,'Color',[1 1 1],'Name','Intermittency property');

>> % plots interval lengths between consecutive repetitions of 'the'   (7.25e)
>> subplot(2,1,1); nvals = length(intervals);
>> plot(1:nvals,intervals,'-r',[1,nvals],ones(1,2)*mean(intervals),'--k');
>> limits = axis; axis([1,nvals,limits(3:4)]); ylabel('Interval length');
>> xlabel('Consecutive occurrences of word ''the''');

>> % plots the histogram of interval lengths                     (7.25f)
>> subplot(2,2,3); bar(histogram);
>> xlabel('Interval length'); ylabel('Frequency');

>> % plots the distribution (histogram) in logarithmic space     (7.25g)
>> subplot(2,2,4); loglog(1:length(histogram),histogram,'.k');
>> xlabel('Interval length'); ylabel('Frequency');
```

As seen from the figure, occurrences of the word *the* in the considered dataset are very far from being evenly distributed. As seen from the upper panel, where the average interval length is indicated with a segmented line, many interval lengths are concentrated in the range from *1* to *12* (below the average value) while many others are distributed in a very wide range from *12* to *300* (above the average value). This can be better appreciated from the histogram of interval lengths that is depicted in the bottom-left panel of the figure. Notice from the histogram that, while most of the interval lengths tend to be concentrated on values below *20* or so, the distribution exhibits a long tail that extends beyond *100*, which is responsible for very large interval lengths.

Additionally, as seen from the interval length plot in the upper panel of the figure, such large interval lengths occur quite sporadically. Those large intervals represent sudden deviations from the mean which are followed by immediate returns to it. This phenomenon is denominated *intermittency*.

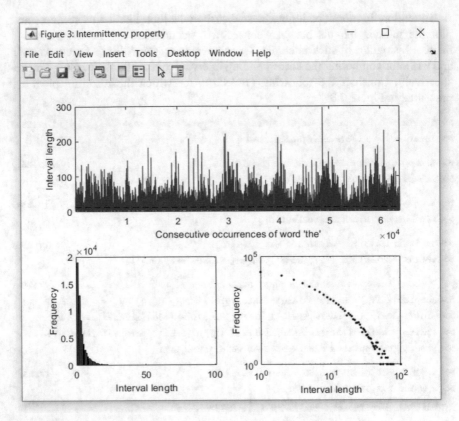

Fig. 7.3. Interval lengths between consecutive occurrences of word *the* and their corresponding histogram, in linear space (bottom-left panel) and in logarithmic space (bottom-right panel)

To exactly know where the central sample in the distribution is, we need to compute the median, as for this kind of distributions the median and the mean can actually be very far from each other:

```
>> median(intervals)                                                                (7.26)
ans =
     8
```

As derived from (7.26), the median of the distribution is actually smaller than the mean, which is a common characteristic of long-tail distributions. In this case, the half of the interval lengths are equal or less than *8*, while the other half of the interval lengths are equal or greater than *8*.

Finally, let us consider the same distribution of interval lengths but now when plotted in logarithmic space. This result is presented in the bottom-right panel of Figure 7.3. Does the shape of the curve look familiar to you? It actually resembles a lot the word frequency versus rank plot depicted in Figure 7.1. We have found again the same power-law kind of behavior when analyzing the interval lengths of word repetitions in our dataset. This *burstsiness* property, although might vary somehow from word to word, is a fundamental characteristic of human languages and other natural phenomena.

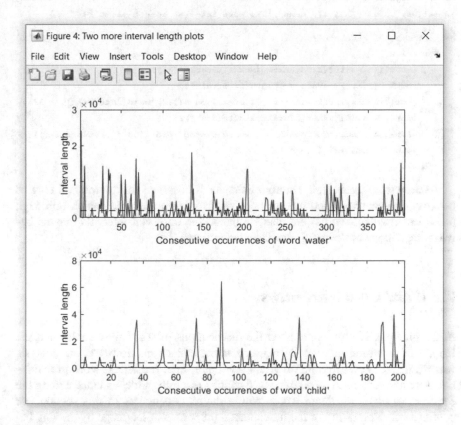

Fig. 7.4. Interval lengths between consecutive repetitions for words *water* and *child*

The following last example generates interval length plots, similar to the one in the upper panel of Figure 7.3, for the words *water* and *child*. These two words correspond to rank positions *239* and *412*, respectively. Notice how a similar pattern of behavior can be observed. The resulting plots are presented in Figure 7.4.

```
>> % writes words 'water' and 'child' into a cell array                (7.27a)
>> twowords = {'water','child'};

>> % retrieves their ranking positions                                 (7.27b)
>> find(strcmp(ranked_vocabulary,twowords{1}))
ans =
   239
```

```
>> find(strcmp(ranked_vocabulary,twowords{2}))
ans =
   412

>> hf = figure(4); % creates a new figure                            (7.27c)
>> set(hf,'Color',[1 1 1],'Name','Two more interval length plots');

>> % generates the plots                                             (7.27d)
   for k=1:2
       intervals = diff(find(strcmp(words,twowords{k})));
       subplot(2,1,k); nvals = length(intervals);
       plot(1:nvals,intervals,'-r',[1,nvals],ones(1,2)*mean(intervals),'--k');
       limits = axis; axis([1,nvals,limits(3:4)]);
       xlabel(sprintf('Consecutive occurrences of word ''%s''',twowords{k}));
       ylabel('Interval length');
   end
```

As seen from figure 7.4, a similar behavior is observed for the interval lengths between consecutive repetitions of both words *water* and *child*, which (although much less frequent words) are also similar to the observed intervals lengths between repetitions of the word *the* displayed in figure 7.3.

7.2 Word Co-occurrences

In the previous section we studied the distributions of word frequencies and the lengths of repetition intervals of words in a text collection. However, our analysis was restricted to considering only one word at a time. Individual word probabilities were derived based on simple word counts, assuming independence among the occurrences of the different words within the text sequence. In this section, we continue our exploration of the fundamental properties of language by moving beyond the single word approach and the independence assumption. More specifically, in this section we will study dependencies between pairs of vocabulary words as they occur together within different segments of text in a given data collection.

First of all, we must introduce the notion of co-occurrence. In the most general sense of the concept, we will say that two words co-occur if we observe both of them in a given unit of text. However, to precisely define co-occurrence, we must also define what we mean by "unit of text", as it can range from large-sized content units such as documents or sub-documents, to intermediate-sized grammatical units such as paragraphs or sentences, or even to small-sized and less naturally defined units such as a window of a fixed number of words. Then, in any given experimental setting, the implementation of co-occurrence analysis must always include the specification of the corresponding unit of text under consideration.

Word co-occurrence, as we will see in this and the next section, constitutes a very powerful concept for deriving useful information about words, as well as language, from large volumes of text data. For the experimental work within this section, we will be considering the verses in the Bible data collection to be the units of text for studying co-occurrences. As you might expect, different types of word and language properties will be derived from co-occurrence analyses conducted over units of text of different type and size. In our case, by considering intermediate-sized units such as the verses, which generally span from *1* to *2* or *3* sentences, we will be deriving word and language information that is mainly related to intermediate- and long-distance dependencies between words. This kind of information will be mainly of semantic nature.

Let us now start de analysis by reading again the text collection in the data file `bible_en.xml`. However, in this opportunity, we are interested in preserving boundary information at the verse level. According to this, we will create a structure array for accommodating individual verse vocabularies:

```
>> temp = string(evalc('type bible_en.xml')); % reads the file          (7.28a)

>> temp = lower(temp); % lowercases the text                            (7.28b)

>> % inserts markers at the end of each verse before removing the tags   (7.28c)
>> temp = regexprep(temp,'</verse>',' S ');

>> temp = regexprep(temp,'<.*?>',''); % removes the tags                 (7.28d)

>> temp = regexprep(temp,'\W',' '); % eliminates non-word characters     (7.28e)

>> temp = strtrim(temp); % removes leading and trailing whitespaces      (7.28f)

>> % segments the data collection into tokens                            (7.28g)
>> wordsofverses = split(temp);

>> % locates the verse boundaries                                        (7.28h)
>> limits = [0,find(strcmp(wordsofverses','S'))];

>> % creates the structure array                                        (7.28i)
>> for k=1:length(limits)-1
       verses(k).vocab = unique(wordsofverses(limits(k)+1:limits(k+1)-1));
   end
```

where each element in the structure array **verses** contains the vocabulary of each individual verse in the collection in the form of a string array of words.

Now that we have separated the verse vocabularies, we can proceed to compute word co-occurrences at the verse level. For this, we must count for every possible pair of words in the whole collection's vocabulary (**ranked_vocabulary**) the total number of verses in which they co-occur. In other to do this efficiently, we will first compute a term-document matrix, which is a binary indicator matrix with as many rows as the total number of words in the vocabulary, and as many columns as the total number of documents in the collection (verses in our case).[1]

The term-document matrix can be computed by intersecting the vocabularies of each individual verse with the vocabulary of the overall collection as follows:

```
>> % creates a sparse matrix of appropriate dimensions              (7.29a)
>> tdmtx = sparse(length(ranked_vocabulary),length(verses));

>> % updates the corresponding term-document matrix elements         (7.29b)
>> for k=1:length(verses) % this will take a while...
       [void,termindex] = intersect(ranked_vocabulary,verses(k).vocab);
       tdmtx(termindex,k) = 1;
   end
```

Finally, the term-to-term co-occurrence matrix can be computed from the term-document matrix obtained in (7.29) by transposed multiplication (you should take a moment to get convinced about this):

```
>> comtx = tdmtx*tdmtx';                                            (7.30)
```

Notice that this matrix exhibits the following properties:

- by construction, it is a square and symmetric matrix, with as many rows and columns as words in the vocabulary,
- the elements in the diagonal (i.e. $m(i,i)$ with $i=j$), represent the number of verses in which each individual vocabulary word occurs,
- the elements out of the diagonal (i.e. $m(i,j)$ with $i \neq j$), represent the number of verses in which each corresponding pair of vocabulary words co-occur, and
- as it is verified below in (7.31), it is indeed a sparse matrix.

```
>> sum(sum(comtx>0))/numel(comtx)*100                               (7.31)
ans =
    (1,1)        1.4074
```

As seen from (7.31), only *1.4%* of the elements in `comtx` are different from zero. And, because of the matrix symmetry, a large proportion of those non-zero elements are repetitions since the co-occurrence of elements i and j is exactly the same as the co-occurrence of elements j and i. What this actually indicates is that there is a huge proportion of vocabulary words that never co-occur in any verse.

Now we are ready to compute some statistics from the obtained co-occurrence matrix. But let us first introduce a very useful concept for assessing the nature of a given pair of words co-occurrence: pointwise mutual information, which is often simply referred to as mutual information. It can be defined as follows:

---

[1] Here, we will use the term-document matrix as an intermediate step for computing the term-to-term co-occurrence matrix. We will study the term-document matrix construct in more detail later in chapter 9, as it constitutes a fundamental element in the construction of geometrical models.

$$I(w_a, w_b) = p(w_a, w_b) \: / \: (\: p(w_a) \: p(w_b) \:) \tag{7.32}$$

where $p(w_a, w_b)$ denotes the joint probability for events w_a and w_b, and $p(w_a)$ and $p(w_b)$ denote the individual probabilities for the same events. In our case, events w_a and w_b refer to the occurrence of words a and b in a verse, respectively.

According to (7.32), the mutual information of two events measures how much the joint probability of such events deviates from what would be expected if they were independent. Notice that if the two events are actually independent, their joint probability would be equal to the product of the individual probabilities and the value of the mutual information in (7.32) would reduce to one. On the other hand, if the events are not independent, two different situations can happen: either a positive affinity effect, i.e. they tend to co-occur more often than expected in the independent case; or a negative affinity or rejection effect, i.e. they tend to co-occur less often than expected in the independent case. In such cases, mutual information values would be greater or less than the unit, respectively.

Let us now consider the individual and joint probabilities of words occurring and co-occurring in a verse. We can compute maximum likelihood estimates for such probabilities directly from the co-occurrence matrix `comtx` obtained in (7.30). For estimating individual word probabilities, we just need to divide the number of verses in which a given word occurs by the total number of verses in the collection. This can be done by using the diagonal elements in the co-occurrence matrix `comtx` as follows:

```
>> plword_verse = diag(comtx)/length(verses);          (7.33)
```

For estimating joint probabilities, we need to divide the number of verses in which a given pair of words co-occurs by the total number of verses in the collection. In this case, we must use the out-of-diagonal elements in the co-occurrence matrix `comtx`. However, two important considerations must be taken into account. First, as `comtx` is a symmetric matrix we should only use elements that are either above or below the diagonal. Functions `triu` and `tril` allows for extracting either group of elements, respectively.

Second, as mutual information is very sensitive to the accuracy of maximum likelihood estimates, in order to avoid artificially increased mutual information scores, we should restrict our word co-occurrence vocabulary to those cases with good empirical evidence in the dataset. What "good empirical evidence" actually refers to is that the number of samples present in the dataset should be enough to produce accurate maximum likelihood estimates. However, there is not a clear procedure to determine whether a given number of samples is enough or not, but we can just tell that the more samples we count the better estimates we will get. In any case, here we set a threshold value of 30 samples, which will not only guarantee good estimates but will also discard some uninteresting co-occurrences of singleton terms mainly due to proper names.

Then, joint probabilities for pairs of words co-occurring in more than *30* verses can be estimated as follows:

```
>> [wa,wb,rawcount] = find(triu((comtx>30).*comtx,1));                    (7.34a)

>> p2word_verse = rawcount/length(verses);                                (7.34b)
```

As seen from (7.34a), the function `find` has been used to extract from the co-occurrence matrix `comtx` all co-occurrence counts and their corresponding indexes for pairs of words with co-occurrence counts larger than *30*. Then, in (7.34b), the co-occurrence counts have been divided by the total number of verses in the collection in order to obtain joint probability estimates. Notice that joint probability estimates in (7.34b) are not provided in a matrix format but in a vector format. So, index vectors `wa` and `wb` should be used for relating joint probabilities in `p2word` with the corresponding words in `ranked_vocabulary`.

Now that we have the individual and joint probability estimates for the occurrences of words in the verses, we can compute the mutual information:

```
>> minfo_verse = p2word_verse./(p1word_verse(wa).*p1word_verse(wb));      (7.35)
```

and explore in more detail co-occurrence properties among the words within the dataset. Let us start by listing the ten pairs of words with highest mutual information scores. We do this by using the following procedure:

```
>> % sorts mutual information scores in descending order                  (7.36a)
>> [sortminfo,sortindex] = sort(minfo_verse,'descend');

>> % copies all wa into a string array according to sortindex order       (7.36b)
>> tempa = ranked_vocabulary(wa(sortindex));

>> % copies all wb into a string array according to sortindex order       (7.36c)
>> tempb = ranked_vocabulary(wb(sortindex));

>> % creates a table containing words and mutual information scores        (7.36d)
>> minfo_rank = table;
>> minfo_rank.WORDa = tempa;
>> minfo_rank.WORDb = tempb;
>> minfo_rank.PMI_Score = sortminfo;

>> % lists the first ten values for tempa, tempb and sortminfo            (7.36e)
>> disp('-> Pairs of words with largest pointwise mutual information');
>> disp(head(minfo_rank,10));
-> Pairs of words with largest pointwise mutual information
```

WORDa	WORDb	PMI Score
"blue"	"purple"	436.39
"acts"	"chronicles"	434.83
"flour"	"mingled"	346.29
"breadth"	"length"	286.19

```
"rams"        "bullocks"       284.59
"sweet"       "savour"         250.69
"linen"       "purple"         237.59
"lambs"       "rams"           224.41
"fine"        "purple"         191.99
"chariots"    "horsemen"       188.73
```

As seen from (7.36e), the pairs of words with highest pointwise mutual information have a strong semantic grounding. In most of the cases, the words within a given co-occurring pair, either belong to the same semantic category (consider, for instance, *blue* and *purple*, or *breadth* and *length*), or are related to each other by means of any other conceptual, functional, or procedural mechanism (consider, for instance, *flour* and *mingled*, *sweet* and *savour*, or *chariots* and *horsemen*).

Now, let us consider the ten pairs of words with lowest pointwise mutual information scores:

```
>> % lists the last ten values for tempa, tempb and sortminfo        (7.37)
>> disp('-> Pairs of words with lowest pointwise mutual information');
>> disp(tail(minfo_rank,10));
-> Pairs of words with lowest pointwise mutual information
```

WORDa	WORDb	PMI Score
"thy"	"you"	0.26487
"have"	"went"	0.26473
"was"	"will"	0.21907
"shall"	"were"	0.20837
"shall"	"was"	0.20645
"will"	"had"	0.2027
"shall"	"came"	0.17123
"will"	"were"	0.16465
"shall"	"went"	0.13934
"shall"	"had"	0.13893

As seen from (7.37), the pairs of words with lowest pointwise mutual information scores are composed of words that are somehow incompatible or mutually exclusive. However, these results are generally more difficult to interpret than the ones in (7.36e) as the exclusion relationships between the words can be of very different natures. For instance, most of the results in (7.37) correspond to a clear case of tense incompatibility between verbal forms. It is easy to support the idea that verses containing future tense forms such as *will* or *shall* are less likely to include also past tense forms such as *had*, *went* or *were*. However, other less evident type of exclusion relationships can be exhibited by semantically related words too. Consider, for instance, the case of words *thy* and *you*, which are indeed closely related to each other. In this case, they are mutually exclusive in the sense that in a literary style using *thy* instead of *your*, it will be much more likely to see *thou* instead of *you*.

In general, some pairs of closely related words such as synonyms and epithets are very likely to exhibit mutual information scores below the unit. This means that when one of them is used the other one is not needed as they both refer to the same thing or concept, being somehow mutually exclusive from each other.

Finally, let us list the ten pairs of words with mutual information scores closest to the unit. In this case we will expect pairs of words composed of words that are *independent* from each other. From a linguistic point of view, we can think about these words as semantically unrelated words.

```
>> % sorts mutual information scores as they depart from the unit        (7.38a)
>> [void,absindex] = sort(abs(sortminfo-1));

>> % lists values for tempa, tempb and sortminfo closest to the unit     (7.38b)
>> disp('-> Pairs of words with pointwise mutual information closest to 1.0');
>> disp(head(minfo_rank(absindex,:),10));
-> Pairs of words with pointwise mutual information closest to 1.0
```

WORDa	WORDb	PMI Score
"it"	"have"	1
"house"	"land"	0.99992
"lord"	"known"	1.0001
"and"	"reproach"	1.0001
"and"	"forasmuch"	1.0001
"of"	"you"	1.0001
"man"	"her"	0.99985
"thy"	"son"	1.0002
"therefore"	"these"	1.0002
"against"	"moses"	1.0002

As seen from (7.38b), the words in co-occurring pairs with mutual information close to the unit seem to be unrelated to each other in most of the cases. Probably, we were not expecting pairs such as *house* and *land*, or *man* and *her* in this list, as one could be conceptually expecting some positive and negative affinity between the words in those cases, respectively. Nevertheless, these are the actual statistics for such words in the specific corpus under consideration. You must be reminded at this point that, although some degree of generality can be expected when analyzing large volumes of text, all empirical information that we can extract from a dataset will be always corpus dependent.

In the same way that pointwise mutual information allows for discriminating between different types of word associations, other basic probability estimates such as the joint probability or conditional probabilities also offer important information about the relationship between two words. In Table 7.2, basic probability estimates, as well as pointwise mutual information scores, for some interesting pairs of words are presented. In the most general case, departure from independence can be evaluated by hypothesis testing (we revisit this issue in exercise 7.5-7 in section 7.5).

Table 7.2. Basic statistics and mutual information for some selected pairs of words

word$_a$	word$_b$	$p(w_a)$	$p(w_b)$	$p(w_a,w_b)$	$p(w_a|w_b)$	$p(w_b|w_a)$	$I(w_a,w_b)$
purple	blue	0.0015	0.0016	0.0011	0.6735	0.6875	436.3941
sell	buy	0.0011	0.0017	0.0003	0.1538	0.2353	140.7376
tree	fruit	0.0054	0.0059	0.0010	0.1739	0.1893	32.0072
wife	husband	0.0119	0.0035	0.0010	0.2778	0.0811	23.3506
more	less	0.0203	0.0009	0.0004	0.4074	0.0175	20.1136
your	you	0.0415	0.0645	0.0165	0.2552	0.3969	6.1539
water	river	0.0117	0.0048	0.0003	0.0676	0.0275	5.7894
fruit	seed	0.0059	0.0082	0.0003	0.0315	0.0435	5.3240
thy	thou	0.0979	0.1248	0.0454	0.3638	0.4639	3.7175
house	wife	0.0551	0.0119	0.0014	0.1189	0.0257	2.1592
say	word	0.0322	0.0217	0.0011	0.0519	0.0350	1.6111
soon	there	0.0021	0.0670	0.0002	0.0024	0.0781	1.1665
for	well	0.2282	0.0080	0.0018	0.2240	0.0079	0.9814
water	house	0.0117	0.0551	0.0005	0.0093	0.0441	0.8003
tell	word	0.0067	0.0217	0.0001	0.0044	0.0144	0.6646
thou	you	0.1248	0.0645	0.0031	0.0484	0.0250	0.3875
house	tree	0.0551	0.0054	0.0001	0.0178	0.0018	0.3223
thy	your	0.0979	0.0415	0.0011	0.0264	0.0112	0.2693
will	saw	0.0918	0.0172	0.0003	0.0150	0.0028	0.1632
had	shall	0.0570	0.1949	0.0015	0.0079	0.0271	0.1389

7.3 Accounting for Order

In the previous section, we used verses as text units for counting word co-occurrences in the `bible_en.xml` text collection. Such co-occurrences were computed by disregarding the relative position of the two words in each word pair under consideration. Although this is not necessarily always true, we assumed that semantic relationships between words are independent of their specific locations within a given segment of text. In this section, we focus our attention on units of text of a smaller size, for which the relative positions of words is otherwise very important. Accordingly, we will be taking into account word ordering when computing co-occurrences. Differently from the previous section, where we used co-occurrences to "discover" semantic relationships between words, in this section, we will focus our analysis on short-distance dependencies, from which word relationships of syntactic nature will emerge.

Let us start by considering pairs of consecutive words, i.e. we will count word co-occurrences over units of text consisting of windows of two words. Similar to (7.29) and (7.30), we will proceed to compute a co-occurrence matrix but, in this case, we will be counting co-occurrences over a sliding window of two words while taking the relative order of the two words into account. Notice that, consequently, the new resulting co-occurrence matrix is not symmetric.

Consider the following procedure for counting co-occurrences of consecutive words. It performs the computations over the string array **wordsofverses**, which was generated in (7.28g) by inserting the marker **'s'** at the end of each verse.

```
>> % creates a sparse matrix for accumulating co-occurrences            (7.39a)
>> co2mtx = sparse(length(ranked_vocabulary),length(ranked_vocabulary));

>> % creates a vector for accumulating individual word counts           (7.39b)
>> w1raw = zeros(1,length(ranked_vocabulary));

>> for k=2:length(wordsofverses) % moves along the dataset              (7.39c)
       % gets the next two consecutive words
       w1 = wordsofverses(k-1); w2 = wordsofverses(k);
       % avoids any pair of words containing the end-of-verse marker 'S'
       if w1~='S' && w2~='S'
           % gets the rank indexes of w1 and w2
           idx1 = find(ranked_vocabulary==w1);
           idx2 = find(ranked_vocabulary==w2);
           % increments the counts for word sequence w1,w2
           co2mtx(idx1,idx2) = co2mtx(idx1,idx2)+1;
           % increments the counts for each single word
           w1raw(idx2) = w1raw(idx2)+1;
       end
end % this will take a while...
```

Now that we have computed the raw counts, we can estimate maximum likelihood probabilities as we did in (7.33) and (7.34). Be reminded that, in this case, the considered text units are windows of two words and that consecutive windows overlap to each other. Notice that the total amount of windows of two words we have evaluated in our counting exercise of (7.39c) can be expressed as follows: **length(words)-length(verses)** or, alternatively, as **length(wordsofverses)-2*sum(wordsofverses=='S')** (just take a moment to convince yourself of this). Accordingly, the probability of individual words occurring in a sliding window of size two should be estimated as follows:

```
>> p1word_next = w1raw'/(length(words)-length(verses));               (7.40)
```

Similarly, we can estimate the probabilities of all consecutive pairs of words as follows:

```
>> [w1,w2,rawcount] = find((co2mtx>30).*co2mtx);
```
(7.41a)

```
>> p2word_next = rawcount/(length(words)-length(verses));
```
(7.41b)

where, again, a threshold of *30* was used for restricting probability estimates to word pairs that have been seen more than *30* times in the corpus. Notice also that subscripts *1* and *2* have been used, instead of *a* and *b*, for emphasizing the fact that word order has been taken into account. So, different from joint probability estimates in (7.34), where $p(w_a, w_b) = p(w_b, w_a)$, in this case, $p(w_1 w_2) \neq p(w_2 w_1)$.

Now, pointwise mutual information of consecutive words can be computed:

```
>> minfo_next = p2word_next./(p1word_next(w1).*p1word_next(w2));
```
(7.42)

By following similar procedures to those in (7.36), (7.37) and (7.38), we can extract the word pairs with largest, lowest and closest-to-the-unit pointwise mutual information scores. Table 7.3 summarizes these results.

Table 7.3. Pairs of consecutive words with largest, lowest and closest-to-the-unit pointwise mutual information

	Largest		*Lowest*		*Closest-to-the-unit*	
word$_1$	word$_2$	word$_1$	word$_2$	word$_1$	word$_2$
sweet	savour	they	and	put	the
fine	flour	it	the	be	thy
fig	tree	and	him	them	i
fine	linen	i	and	god	with
mercy	endureth	i	the	king	in
little	ones	they	of	and	in
unleavened	bread	thou	the	set	the
afar	off	and	them	lord	their
without	blemish	in	and	them	unto
loud	voice	the	will	and	for

As seen from the table, a clear association trend can be observed for those word pairs exhibiting the largest mutual information scores. In most of the cases, they constitute adjective-noun compounds such as *sweet savour*, *fine linen*, or *loud voice*; but they can also be other grammatical structures such as in the case of the noun-noun expression *fig tree*. In the case of word pairs with lowest mutual information, some rare word pairs such as *it the* or *in and* are observed, which can be resulting from the fact that we eliminated punctuation marks from the text, making them artificially to co-occur next to each other. However, some interesting cases can be spotted in this second group. Consider, for instance, the case of *the will*, which clearly identifies a meaning for the word *will* (noun) that is different from its most commonly used one *will* (verb). In the last case, word pairs with mutual information closest to the unit are presented. These words can be considered as "short-distance independent" words. Finally, it is worth mentioning that taking word order into account can make a big difference in the estimates for the same pair of words; while the mutual information score for *in and* is *0.12*, it is *1.01* for the reversed word sequence *and in*.

In the previous exercise we have restricted the co-occurrence analysis to the statistical dependencies between consecutive words; however, short-distance dependencies between words can actually extend to several words. Let us now consider word dependencies in a window of few words around a specific word of interest. In the following example, we compute histograms for the occurrences of a given probe word with respect to another reference word. We conduct the analysis over a window that spans from *–4* to *+4* word positions with respect to the reference word location. Histograms are presented in Figure 7.5.

```
>> % defines four reference words and their corresponding probes        (7.43a)
>> refer = ["father","shall","go","go"];
>> probe = ["house","be","to","for"];

>> hf = figure(5); % creates a figure and sets its basic properties     (7.43b)
>> set(hf,'Color',[1 1 1],'Name','Short-distance word dependencies');

>> % short-distance word dependency computation                         (7.43c)
>> for m = 1:length(refer)
       w1 = refer(m); w2 = probe(m); % gets reference and probe words
       % initializes the reference/probe word offset histogram
       histprobe = zeros(1,9);
       % loops over verses for computing co-occurrences
       for k=1:length(limits)-1
           % gets the next verse
           theverse = wordsofverses(limits(k)+1:limits(k+1)-1);
           % finds occurrences of the reference word, if there are any
           w1loc = find(theverse==w1);
           % finds occurrences of the probe word, if there are any
           w2loc = find(theverse==w2);
           % double-loop for computing all offsets between reference and probe
           for n=1:length(w1loc), for j=1:length(w2loc)
               index = w2loc(j)-w1loc(n); % computes the offset
               % updates histogram if offset is within the considered interval
               if (index>-5) && (index<5)
                   histprobe(index+5) = histprobe(index+5)+1; end;
           end; end % end of double-loop for offset computation
       end % end of loop for co-occurrence computation
       subplot(2,2,m); hb = bar(-4:4,histprobe); % plots the histogram
       temp = axis; axis([-5,5,0,temp(4)+1]); % sets axis limits
```

```
    ylabel('Frequency'); % prints the corresponding labels
    xlabel(sprintf('Location of "%s" with respect to "%s"',w2,w1));
end
```

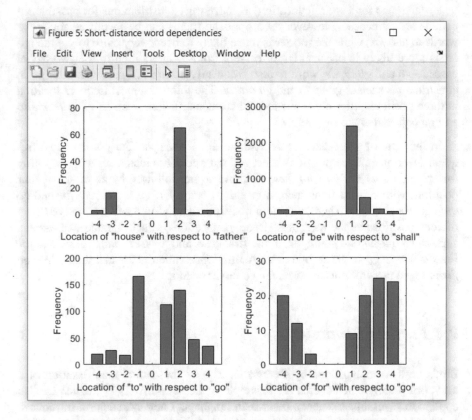

Fig. 7.5. Histograms of probe word locations with respect to given reference words

Notice from the upper half of Figure 7.5 that some clearly preferred locations are evidenced for probe words *house* and *be* with respect to reference words *father* and *shall*, respectively. In the case of word pair *father house*, the most frequent position of *house* with respect to *father* is two words after it. This corresponds, actually, to the very common construction *father's house*, which after tokenization becomes *father s house*. The second most frequent case, for which *house* is located three words before *father*, corresponds to constructions such as *house of his father*, *house of their father*, *house of my father*, and so on. In the case of word pair *shall be*, it can be seen that the most preferred location of *be* is just after *shall*. The following most preferred locations are: two words after the reference, such as in *shall it be*, *shall all be*, *shall there be*, etc; and three words after the reference, such as in *shall the end be*, *shall thy life be*, etc.

Consider now the lower half of Figure 7.5, where histograms for relative locations of probe words *to* and *for* with respect to reference word *go* are presented. Different from the previous two examples, in these two cases, it does not seem to be a preference for a specific location, as more varied distributions for possible locations are evidenced. However, two opposite trends can be identified for probe words *to* and *for*, while the former is more likely to occur very close to *go*, the latter is more likely to occur at larger offsets from it. In addition to the two trivial cases *to go* and *go to*, the word *to* also frequently occurs two positions after *go* in constructions such as *go up to* and *go out to*. The following most frequent location is three positions after the reference, where constructions such as *go with me to* and *go before it to* are encountered.

In the case of short-distance dependencies between *for* and *go*, the most frequent location is three positions after *go*, with constructions such as *go thy way for*, *go before us for*, *go with these for*, and so on; followed by an offset of four positions with constructions such as *go out to battle for*, *go to the seer for* and *go out against them for*. Then, we find a positive offset of two and a negative offset of four positions as the next most frequent locations, which include expressions such as *go out for* and *go up for*, in the first case, and *for thou shalt not go* and *for the shadow to go* in the second case. A final observation is the fact that, while *go for* is also a possible construction, *for go* never occurs.

7.4 Further Reading

Zipf's laws were explained by George K. Zipf as the natural manifestation of a universal principle he denoted the Principle of Least Effort (1949), which he also applied to several phenomenon other than language. Later, Mandelbrot introduced some refinements to the original propositions of Zipf's laws for language (1954). The specific property of *burstsiness* has been more recently studied in (Katz 1996) and (Madsen et al. 2005).

The utility of co-occurrences in the analysis of natural language is theoretically supported by the *Distributional Hypothesis*, according to which each word can be identified and defined "by the company it keeps" (Firth 1957). Particularly, co-occurrences are of great utility in specific problems such as lexical acquisition (Zernik 1991), semantic similarity estimation (Jiang and Conrath 1997) and collocation extraction (Evert 2005), as well as in more general areas of study such as statistical semantics (Furnas et al. 1983).

Pointwise mutual information (Church and Hanks 1990), as presented in (7.32), is different from the classical concept of mutual information defined in Information Theory. The latter is actually defined over probability distributions rather than over specific values of the random variables involved. For a good introduction to Information Theory, see (Cover and Thomas 1991); and for a more comprehensive overview on mutual information, see (Paninski 2003).

A commonly used procedure for validating the significance of frequent word co-occurrences is statistical hypothesis testing. We have omitted this topic in this chapter, but it is available in statistics books such as, for instance, (Ramsey and Schafer 1997) or (Lehmann and Romano 2005). For a specific example of hypothesis testing in the context of word co-occurrences see exercise 7.5-7 in the following section.

7.5 Proposed Exercises

1. Consider the basic statistics for different subsets of the n most frequent words presented in Table 7.1:

   – Compute both the percentage of vocabulary and percentage of whole collection covered for values of n varying from 1 up to total vocabulary size.
   – Generate a cross-plot between both coverage percentages. Use log scales for better visualization. What are your main observations?

2. Consider the histogram presented in the right-bottom panel of Figure 7.3:

   – As it was computed by using only 100 bins (7.25c), it is a smoothed version of the actual distribution of interval lengths.
   – Reproduce the plot by using the appropriate number of bins so the actual distribution of interval lengths observed in the data is depicted.
   – What differences do you observe between both, the smoothed and the actual, distributions?

3. Consider the procedure illustrated in (7.25a) and (7.25b) for computing the lengths of repetition intervals for a given vocabulary word.

   – Select some sample vocabulary words from different ranks and compute the length of repetition intervals for each case.
   – Use the distribution of computed repetition intervals for estimating probabilities of the form $p_r(n|word)$, where n refers to the interval length and p_r refers to the probability of repetition given the word *word*.
   – Estimate the average length of verses in the data collection and use this result along with your estimates for $p_r(n|word)$ to approximate the probability of repetition within one verse of each of the words considered.
   – Compute actual maximum likelihood estimates for the probabilities of the considered words repeating within a verse. Use the following procedure:

count the total amount of verses in which each word appears twice or more times and divide this count by the total amount of verses in the collection.

- How do your estimates from the previous two steps differ from each other? Can you explain why? Can you improve the first estimates?

4. Create a function that returns the basic co-occurrence statistics for a given list of vocabulary word pairs:

- Use the following syntax `stats = get_stats(tdmtx,vocab,wa,wb)`, where the input variables `tdmtx` and `vocab` are the term-document matrix and the vocabulary, computed in (7.29) and (7.9a), respectively; and `wa` and `wb` are the lists words under consideration.
- The output variable `stats` must a table object containing the different statistics depicted in Table 7.2 (i.e. individual word probabilities, joint probability, conditional probabilities and pointwise mutual information scores) for the different pairs of words defined by `wa` and `wb`.
- Use the created function to reproduce the results in Table 7.2 and to compute statistics for some other pairs of words selected by you.

5. Compute a co-occurrence matrix for consecutive pairs of words (use as unit of text, a window of two words) without taking word order into consideration.

- What are the differences between these co-occurrence counts and the ones computed in section 7.2 using complete verses as units of text? Select some specific word pairs for conducting your comparative analysis and try to derive general conclusions from them.
- What are the differences between these co-occurrence counts and the ones computed in section 7.3 where word order was taken into account? Again, select some specific word pairs for conducting your comparative analysis and try to derive general conclusions from them.

6. Consider the probability estimates for individual words occurring in a verse `p1word_verse` that were computed in (7.33).

- Compute the summation of probabilities `sum(p1word_verse)`. What do you observe? Can you provide an explanation for this result?
- Compute now the summation of probabilities `sum(p1word_next)` corresponding to the case of probability estimates for individual words occurring in a sliding window of size two, which were computed in (7.40).
- Should not your previous explanation apply also for this second case? Can you tell what the main differences between the two implementations are?
- Compare `estimated_probabilities` from (7.16) with `p1word_verse` and `p1word_next`. You might want to generate some cross-plots in order to conduct the comparison. What are your main observations?

7. Let us consider the statistical hypothesis testing procedure denominated *t-test*.

   - Consider the statistical independence assumption $p(w_1, w_2) = p(w_1) \, p(w_2)$ between two given consecutive words w_1 and w_2.
   - Formulate the null hypothesis that the observed sequence $w_1 w_2$ has been derived from a distribution with mean $\mu = p(w_1) \, p(w_2)$.
   - Assume a Bernulli distribution, with standard deviation $\sigma = [p(1-p)]^{1/2}$ and mean $m = p$, where p is the probability of occurrence of a given sequence of two words $w_1 w_2$.
   - Compute the *t*-value for each of the *30* word pairs $w_1 w_2$ presented in Table 7.3. Use $t = N^{1/2} \, (m - \mu)/\sigma$, where N is the total amount of running words in the data collection, and m and σ are the mean and standard deviation of the considered sequence $w_1 w_2$ derived from the data.
   - Consider the following table of confidence values versus *t*-values, which gives the critical value of t above which the *null hypothesis* (statistical independence in our case) can be rejected with the given confidence.

confidence	95%	98%	99%	99.9%
t-value	1.96	2.33	2.58	3.29

   - For all considered word pairs $w_1 w_2$, determine in which cases the *null hypothesis* can be rejected and the corresponding maximum confidence.

8. Create a script for reproducing the results presented in Figure 7.3:

   - Generate a list with the *100* word pairs with largest mutual information.
   - Manually extract from the list several subsets of word pairs according to the following criteria: word pairs with the same first word, word pairs with the same last word, word pairs of the form *adjective-noun*, word pairs of the form *noun-noun*, and any other interesting patterns you can identify.
   - What could the extracted subsets be used for?

9. Create a script for reproducing the experiment in (7.43). Use the script to explore distance dependencies between words by considering several different pairs of reference and probe words. Report your main findings.

7.6 Short Projects

1. All experiments reported in this chapter were conducted over the English version of the Bible dataset. Consider now the Spanish and Chinese versions, which are available in the companion website http://www.textmininglab.net/ Accessed 6 July 2021.

- Generate word frequency versus rank plots (as the one presented in Figure 7.1) for both the Spanish and Chinese datasets. What are the main similarities and differences with respect to the plot obtained from the English version of the dataset (presented in Figure 7.1)?
- Generate word length versus rank plots (as the one presented in Figure 7.2) for both the Spanish and Chinese datasets. What are the main similarities and differences with respect to the plot obtained from the English version of the dataset (presented in Figure 7.2)?
- Generate plots of interval length versus consecutive occurrences (as the ones presented in Figure 7.4) for some randomly selected words in both the Spanish and Chinese datasets. Take the precaution of restricting your random selection to those words that appear a minimum number of times (as, for instance, *200*) in the dataset. What are your main observations?
- Extract pairs of consecutive words exhibiting high values of mutual information, from both the Spanish and Chinese datasets. Find the corresponding translations into English of the resulting "expressions". How many of these translations are you able to find in Table 7.3? Explain your findings.
- Repeat the previous step by considering co-occurrences of words in verses.

2. Consider probability estimates for single words `p1word_next` and two consecutive words `p2word_next` that were computed in (7.40) and (7.41), respectively.

- For each probability estimate in `p2word_next` estimate conditional probabilities of the form $p(w_2|w_1)$ by using both `p2word_next` and `p1word_next`. Notice that it is possible that you will not be able to compute these conditional probabilities for all pairs of words!
- Generate random sentences by using the `p1word_next` probability estimates as follows: *step-1*, generate a random word following the probability distribution in `p1word_next`; *step-2*, repeat step-1 with probability $p=0.9$ or stop otherwise. (You might use the uniform random generator function `rand` and the cumulative distribution `cumsum(p1word_next)` for this).
- Generate random sentences by using `p1word_next` and the conditional probability estimates as follows: *step-1*, generate a random word following the probability distribution in `p1word_next`; *step-2*, generate (if possible) a random word following the conditional probability $p(w|w_s)$, where w_s is the previous generated word, if not possible (i.e. you do not have an estimate for $p(w|w_s)$ available) generate the random word following `p1word_next`; *step-3*, repeat step-2 with probability $p=0.9$ or stop otherwise. (Again, you might use the uniform random generator `rand` and cumulative distributions for $p(w|w_s)$ and `p1word_next`).
- Generate two datasets of about *1,000* sentences each by using each of the two generation methods proposed in the previous steps.

- Compute basic statistics and generate plots of frequency versus rank (as in Figure 7.1) and plots of interval lengths versus consecutive occurrences (as in Figure 7.4) for the two artificially generated datasets.
- What differences do you observe between the plots generated for each of the two datasets? What differences do you observe between these plots and the ones corresponding to real data in Figures 7.1 and 7.4?

7.7 References

Church KW, Hanks P (1990) Words association norms, mutual information, and lexicography. Computational Linguistics 16(1):22-29

Cover TM, Thomas JA (1991) Elements of Information Theory. John Wiley & Sons. New York, NY

Evert S (2005) The Statistics of Word Cooccurrences: Word Pairs and Collocations. PhD Thesis, IMS Stuttgart

Firth JR (1957) A synopsis of linguistic theory 1930-1955. In Studies in Linguistic Analysis pp. 1-32. Philological Society. Oxford, UK

Furnas GW, Landauer TK, Gomez LM, Dumais ST (1983) Statistical semantics: analysis of the potential performance of keyword information systems. Bell System Technical Journal 62(6):1753-1806

Jiang JJ, Conrath DW (1997) Semantic similarity based on corpus statistics and lexical taxonomy. In Proceedings of the International Conference on Research on Computational Linguistics

Katz SM (1996) Distribution of content words and phrases in text and language modeling. Natural Language Engineering 2:15-59

Lehmann EL, Romano JP (2005) Testing Statistical Hypotheses. Springer, New York, NY

Madsen RE, Kauchak D, Elkan C (2005) Modeling word burstiness using the dirichlet distribution. In Proceedings of the 22nd International Conference on Machine Learning. Bonn, Germany

Mandelbrot BB (1954) Structure formelle des textes et communication. Word 10:1-27

Paninski L (2003) Estimation of entropy and mutual information. Neural Computation 15:1191-1253

Ramsey FL, Schafer DW (1997) The Statistical Sleuth: A Course in Methods of Data Analysis. Duxbury Press, Belmont, CA

Zernik U (1991) Introduction. In Lexical Acquisition: Exploiting On-Line Resources to Build a Lexicon, pp. 1-26. Lawrence Erlbaum, Hillsdale, NJ

Zipf GK (1949) Human Behavior and the Principle of Least Effort. Addison-Wesley. Cambridge, MA

8 Statistical Models

In the previous chapter we studied some fundamental properties of language, as well as basic methods for analyzing dependency relations between words. Single word probabilities, as well as probabilities for pairs of words co-occurring in a given unit of text, were computed by means of maximum likelihood estimates. However, up to this point we are still not able to compute probabilities for segments of texts larger than two consecutive words. In this chapter we focus our attention on the problem of computing probabilities for larger units of text such as sentences, paragraphs and documents. More specifically, two main classes of statistical models will be considered: n-gram models and bag-of-word models. First, in section 8.1, basic n-gram models are presented. Then, in section 8.2 and 8.3, basic discounting and interpolation methods are illustrated for n-gram models. Finally, in section 8.4, statistical bag-of-word models are introduced, centering the attention on topic models.

8.1 Basic n-gram Models

Let us start our discussion by considering n-gram models. This kind of models is derived from an approximation of the probability of a sequence of words, which is based on a Markov property assumption. Let us consider, for instance, a unit of text w which consist of a sequence of words $w_1, w_2, \ldots w_m$. The probability of such a sequence can be decomposed, by means of the chain rule, in the following product of probabilities:

$$p(w) = p(w_1, w_2, \ldots w_m) \tag{8.1}$$

$$= p(w_1)\, p(w_2|w_1)\, p(w_3|w_1, w_2) \ldots p(w_m|w_1, w_2 \ldots w_{m-1})$$

A Markov process refers to a random process in which the probability of the next state only depends on the current state, and it is statistically independent on any previous states. In the specific context of word sequences, assuming the Markov property implies considering that the probability of a given word only depends on a fixed number of preceding words. According to this, n-gram models are defined by approximating the conditional probabilities in (8.1) with conditional probabilities that only depend on the previous n-1 words in the sequence (commonly referred to as the history or context). In this way, n-gram models of order one (1-gram), two (2-gram), three (3-gram), and so on, can be defined as follows:

$$1\text{-gram: } p(w) \approx p(w_1)\, p(w_2)\, p(w_3) \ldots p(w_m) \tag{8.2a}$$

© The Author(s), under exclusive license to Springer Nature Switzerland AG 2021
R. E. Banchs, *Text Mining with MATLAB®*, https://doi.org/10.1007/978-3-030-87695-1_8

$$2\text{-gram: } p(w) \approx p(w_2|w_1)\, p(w_3|w_2)\, p(w_4|w_3)\, \dots\, p(w_m|w_{m-1}) \qquad (8.2b)$$

$$3\text{-gram: } p(w) \approx p(w_3|w_2,w_1)\, p(w_4|w_3,w_2)\, \dots\, p(w_m|w_{m-1},w_{m-2}) \qquad (8.2c)$$

$$n\text{-gram: } p(w) \approx \Pi_i\, p(w_i|w_{i-1},w_{i-2}\dots w_{i-n+1}) \qquad (8.2d)$$

Maximum likelihood estimates can be easily computed for probabilities in (8.2) by using a training corpus. Also notice that, for small values of n, the probabilities in (8.2) are much easier to estimate than those in (8.1). Indeed, when long word histories are involved, the model tends to become unreliable as most of the histories are not actually seen in the training dataset and the corresponding n-gram probability estimates are not reliable. Also, notice that in the extreme case of the unigram (8.2a), the resulting model is completely independent of the order of words. Such a word-order independent model is known as bag-of-words model, which will be discussed in more detail in section 7.4. Different from the unigram case, in the bigram, the trigram and the general n-gram model (8.2b, 8.2c and 8.2d), word order is taken into account. In the bigram case, the probability of a given word depends on the word immediately before, in the trigram case, the probability of a word depends on the previous two words, and so on.

Let us now illustrate the computation of n-gram models with our experimental dataset and discuss some fundamental issues related to the practical implementation of this kind of models. First, we divide the experimental dataset into three different subsets: a train set, a development set, and a test set. This constitutes a standard practice in text mining applications, which is intended for improving the quality of the overall modeling process by reducing the risk of over-fitting, while providing an adequate framework for analysis and evaluation. The train set, which generally constitutes the largest of the three subsets, is used for training the model (in our case, estimating probabilities); the development set is used for parameter tuning and/or cross-validation purposes; and the test set is exclusively used for evaluation purposes. The test set should not be used at all in any of the phases of model computation and tuning, as it constitutes a proxy to the unknown data on which the model will be used for text mining and data analysis.

The file `datasets_ch8a.mat` contains three string arrays `trndata`, `devdata` and `tstdata` containing $27,103$, $2,000$ and $2,000$ verses, respectively, which conforms a random partition of the same dataset that was used in chapter 7. The data within each of these three string arrays is provided in the same format used for string array `wordsofverses` in (7.28); i.e. one word per string, maintaining the same word order as in the verses, and with verse boundaries delimited by the marker `"s"`.

Let us proceed to load the three datasets:

```
>> clear;                                                                     (8.3a)
>> load datasets_ch8a

>> whos                                                                       (8.3b)
```

Name	Size	Bytes	Class	Attributes
devdata	1x53451	2943458	string	
trndata	1x716711	39481306	string	
tstdata	1x52364	2884952	string	

We can compute maximum likelihood estimates for unigram probabilities as we did in section 7.1 by using (7.15). In this case, we use the train dataset for this purpose:

```
>> mask = trndata=='S'; % removes verse boundary delimiters          (8.4a)
>> [vocab,void,index] = unique(trndata(not(mask)));

>> frequencies = hist(index,length(vocab)); % gets unigram counts     (8.4b)
>> [unigram_count,ranking_index] = sort(frequencies,'descend');

>> unigram_vocab = vocab(ranking_index); % gets unigrams              (8.4c)

>> unigram_probs = unigram_count/sum(unigram_count); % computes model (8.4d)
```

With such a unigram model, the probability of a given sequence of tokens can be computed by means of (8.2a). Consider, for instance, the following example:

```
>> example1 = ["he","went","to","the","river","for","water"];        (8.5a)

>> % computes the sentence probability according to the unigram model (8.5b)
>> prob = 1;
>> for k=1:length(example1)
       prob = prob * unigram_probs(unigram_vocab==example1(k));
   end

>> disp(prob)                                                         (8.5c)
   4.0738e-17
```

Similarly, maximum likelihood estimates for bigram probabilities can be computed (just as we did at the end of section 7.3), and the resulting bigram model can be used to estimate the probability of a sequence of tokens by means of (8.2b).

Nevertheless, a very important limitation of these basic *n*-gram model implementations should be addresses: the probability of a given segment of text containing an *n*-gram that has not been seen in the training set is, in theory, equal to zero. This happens because all the probability mass has been distributed among *n*-grams whose words were seen in the training set, as in the previous unigram case:

```
>> sum(unigram_probs)                                                (8.6)
ans =
   1.0000
```

According to (8.6), the probability of any new word not contained in the vocabulary must be zero, and consequently the product in (8.2a) reduces to zero whenever a new word occurs in the text!

This problem is tackled in practice by a procedure that is referred to as smoothing, or discounting. This procedure considers reserving some amount of probability mass for unseen events, and distributing only the remaining portion among the events that are seen during training. There are many different methods for doing this, but we restrict ourselves here to the Good-Turing approach. The next section presents implementation details for this class of discounting for both unigram models and bigram models.

8.2 Discounting

There are two fundamental ideas behind Good-Turing discounting: first, it uses the notion of frequency of frequencies for smoothing counts; and second, it relies on singleton events (events that have been seen just one time) to estimate the probability mass of unseen events. (Let us recall that, in the case of the unigram model under consideration, events refer to vocabulary words. However, in the general case, event refers to whatever we are counting to estimate probabilities).

The Good-Turing approach proposes the following formula for smoothing event counts:

$$c^* = (c+1) \, E\{ N_{c+1} \} / E\{ N_c \} \qquad (8.7)$$

where c^* is the smoothed value for a given count c, $E\{ \cdot \}$ is the expectation operator, and N_{c+1} and N_c are the total number of events occurring $c+1$ and c times, respectively. For instance, the smoothed count for events that occurs twice in the training data would be equal to $(2+1)$ times the expected number of events occurring three times, divided by the expected number of events occurring twice. The expectation is introduced in the mathematical formulation as there might be some frequencies that are actually not observed in the training data. In the practice, a smoothing or interpolating function is used for estimating the values of $E\{ N_{c+1} \}$ and $E\{ N_c \}$. As you might already noticed (and if not, just take some time to convince yourself about it), the number of events occurring certain numbers of times corresponds to the frequencies of the frequencies computed in (8.4b).

The second idea behind the Good-Turing approach is the estimation of the probability mass to be reserved for unseen events. Notice that (8.7) provides us with a means for estimating a count value for unseen events, i.e. the smoothed count value for events that have been seen zero times is given by:

$$0^* = E\{ N_1 \} / E\{ N_0 \} \qquad (8.8)$$

where $E\{ N_1 \}$ is the expected number of singleton events, which can be estimated from the training data, and $E\{ N_0 \}$ is the expected number of unseen events, which we should be able to define (i.e., invent) according to some reasonable criterion.

Given the count value for unseen events 0^*, we can compute the probability of any single unseen event by dividing it by the total number of training instances:

$$p(unk) = 0^* / N = (E\{ N_1 \} / E\{ N_0 \}) / N \qquad (8.9)$$

where *unk* refers to any new unseen event and N is the total number of training instances, which, in the case of the unigram model, corresponds to the total number of running words, i.e. the normalization factor `sum(unigram_count)` used in (8.4d). According to this, the total probability mass being reserved for unseen events is $p(unk) E\{N_0\} \approx E\{N_1\} / N$, which is, actually, the same amount of probability mass devoted to singleton events in the basic unigram model computed in (8.4d). In other words, the Good-Turing approach reserves for unseen events the same amount of probability mass as the one assigned to singleton events by the maximum likelihood estimator.

Now, we are ready to apply Good-Turing discounting to the basic unigram model obtained in (8.4). First, let us compute the frequency of frequencies by constructing a histogram for the unigram counts computed in (8.4b):

```
>> freqfreq = hist(unigram_count,max(unigram_count));                    (8.10)
```

Second, we fit a smoothing function to the resulting histogram for approximating the corresponding expectations in (8.7). In order to accomplish this, four points from the distribution are selected and a spline interpolator is used:

```
>> newf(1) = 1; newff(1) = freqfreq(1); % selects singletons (c=1)       (8.11a)
```

```
>> newf(2) = 5; newff(2) = freqfreq(5); % selects the event with c=5     (8.11b)
```

```
>> % selects the event with maximum number of counts c occurring twice   (8.11c)
>> newf(3) = max(find(freqfreq==2)); newff(3) = 2;
```

```
>> % selects the event with maximum number of counts c occurring once    (8.11d)
>> newf(4) = max(find(freqfreq==1))+1; newff(4) = 1;
```

```
>> % sets slopes at the extremes for the desired smoothing function      (8.11e)
>> slope1 = (log(newff(1))-log(newff(2)))/(log(newf(1))-log(newf(2)));
>> slopeend = 0;
```

```
>> % computes the smoothing function for estimating expectations         (8.11f)
>> temp_x = log(newf); temp_y = log(newff);
>> temp_xx = log(1:length(freqfreq)+1);
>> nc = exp(spline(temp_x,[slope1,temp_y,slopeend],temp_xx));
```

Figure 8.1 illustrates the actual distribution of frequencies of frequencies, the four selected points, and the generated smoothing function. (Notice the specific commands for generating the figure are omitted here. The generation of the figure if left to you as an exercise.)

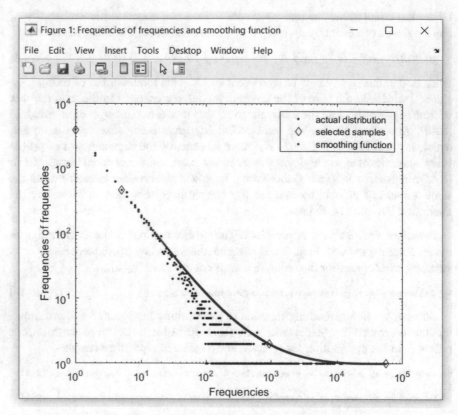

Fig. 8.1. Unigram's frequencies of frequencies and smoothing function for estimating expectations

Now, we can compute the smoothed counts by following (8.7), as well as the probability mass to be reserved for unseen events:

```
>> smoothed_counts = ((1:length(freqfreq))+1).*nc(2:end)./nc(1:end-1);    (8.12a)
```

```
>> unk_prob_mass = freqfreq(1)/sum(unigram_count);                        (8.12b)
```

With the results from (8.12), we finally have all that is needed to compute discounted unigram probabilities:

```
>> unigram_scount = smoothed_counts(unigram_count);                       (8.13a)
```

```
>> seen_prob_mass = (1-unk_prob_mass);                                    (8.13b)
```
```
>> unigram_sprobs = unigram_scount/sum(unigram_scount)*seen_prob_mass;
```

where we first get the corresponding smoothed counts for each unigram count that has been seen in the training dataset (8.13a), and then, we normalize the counts such that the probability mass defined for unseen events is actually reserved

(8.13b). We can easily verify that the total probability mass, the one reserved for unseen events plus the one assigned to seen events, adds up to one:

```
>> unk_prob_mass + sum(unigram_sprobs)                    (8.14)
ans =
    1.0000
```

One final detail is still missing! We have defined the total probability mass that all unseen events should account for, but we have not defined what the probability for a single occurrence of an unseen event is. According to (8.9), we need an estimate of the total amount of expected unseen events $E\{ N_0 \}$ in order to compute such a probability. In the particular case of the unigram model, this is equivalent to estimate how many English words are not contained in our model's vocabulary. Notice that estimating this is actually difficult, as well as arbitrary. We propose here to consider the total amount of unseen vocabulary words to be equal to the total amount of vocabulary words seen in the training dataset (if you think you can come up with a less arbitrary and better estimate than this, feel free to use your own estimate). Following this, we can estimate the smoothed count $0*$ for unseen events and the probability for the occurrence of an unseen event as:

```
>> unseenunigrams = length(unigram_vocab);               (8.15a)
```

```
>> unk_unigram_scount = freqfreq(1)/unseenunigrams       (8.15b)
unk_unigram_scount =
    0.3232
```

```
>> unk_unigram_sprobs = unk_prob_mass/unseenunigrams     (8.15c)
unk_unigram_sprobs =
    4.6873e-07
```

As seen from (8.15c), the probability assigned to an unseen event is smaller than the probability of those events that we have observed just once in the data:

```
>> unigram_sprobs(end)                                    (8.16)
ans =
    1.1500e-06
```

Now that we have completed applying Good-Turing discounting to our unigram model, we should be able to obtain better probability estimates for sentences and any other segments of texts. However, before proceeding to use our discounted model, let us analyze how these discounted probabilities differ from the original maximum likelihood estimates. To do this, we compute absolute and relative differences between the original and discounted unigram probabilities as follows:

```
>> absolute_diffs = unigram_probs-unigram_sprobs;         (8.17a)
```

```
>> relative_diffs = (unigram_probs-unigram_sprobs)./unigram_probs;   (8.17b)
```

The computed differences in (8.17) are illustrated in Figure 8.2. (Again, the commands for generating the figure have been omitted, and the generation of the plot is left to you as an exercise.)

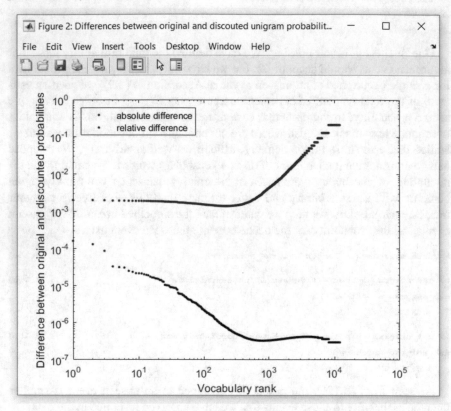

Fig. 8.2. Absolute and relative differences between original and discounted unigram probabilities

As seen from the figure, in absolute terms, more probability is discounted from the most frequent words (low vocabulary ranks) and less probability is discounted from the least frequent words (high vocabulary ranks). However, in relative terms, a much larger percentage of the probability mass is discounted from the least frequent words (around 20%) than from the most frequent ones (around 0.2%).

To see the effect of these differences when computing probabilities for a segment of text, let us reconsider the sample sentence defined in (8.5a) "*he went to the river for water*". If we estimate this sentence's probability by using the new discounted unigram model:

```
>> prob = 1; for k=1:length(example1)                          (8.18a)
       prob = prob * unigram_sprobs(unigram_vocab==example1(k));
   end
```

```
>> disp(prob)                                                           (8.18b)
   4.0233e-17
```

we will find that, compared with (8.5c), this new probability estimate is lower. This makes sense, as we have indeed discounted some probability mass from unigram probabilities and reserved it for unseen unigrams. Notice, however, that the probability reduction in (8.18b) with respect to (8.5.c) is only 1.24%, while the new discounted model has the main advantage of being able to estimate probabilities for segments of text including words which are not in the model's vocabulary.

Consider for instance, the following sentence "*bob went to the river for water*", which is very similar to the sentence considered in the previous example. As "*bob*" is not in the model's vocabulary:

```
>> find(unigram_vocab=='bob')                                           (8.19)
ans =
   1×0 empty double row vector
```

the original non-discounted model predicts a zero probability for the new sentence, while the new discounted model is able to give a non-zero probability to it:

```
>> example2 = ["bob","went","to","the","river","for","water"];         (8.20a)

>> prob = 1;                                                            (8.20b)
>> for k=1:length(example2)
        index = find(unigram_vocab==example2(k));
        if isempty(index), prob = prob * unk_unigram_sprobs;
        else, prob = prob * unigram_sprobs(index); end
   end

>> disp(prob)                                                           (8.20c)
   1.4404e-21
```

Let us now consider the bigram model defined in (8.2b), for which maximum likelihood estimates can be computed for the conditional probabilities involved in the model as follows:

$$p(w_n|w_{n-1}) = p(w_{n-1},w_n) / p(w_{n-1}) \approx c(w_{n-1},w_n) / c(w_{n-1}) \qquad (8.21)$$

where $c(w_{n-1},w_n)$ and $c(w_{n-1})$ are the numbers of times the sequence of words w_{n-1} w_n and the word w_{n-1} occur in the training dataset, respectively.

In the same way we computed smoothed counts for unigrams, we can compute smoothed counts for bigrams by using Good-Turing discounting. Notice that, in this case, we will be discounting counts of the form $c(w_{n-1},w_n)$, which corresponds to joint probabilities $p(w_{n-1},w_n)$ rather than to conditional probabilities $p(w_n|w_{n-1})$. This means that we will be discounting joint probabilities and not bigram probabilities. Nevertheless, we will be able to estimate the required conditional probabilities by using discounted counts $c^*(w_{n-1},w_n)$ and $c^*(w_{n-1})$ in (8.21).

First, we need to compute a co-occurrence matrix, similar to the one computed in chapter 7 (7.39), such that word order is taken into account. Again, we do the counting over the train dataset:

```
>> comtx = sparse(length(unigram_vocab),length(unigram_vocab));        (8.22)
>> for k=2:length(trndata)
        w1 = trndata(k-1); w2 = trndata(k);
        if (w1~='S')&(w2~='S')
            idx1 = find(unigram_vocab==w1);
            idx2 = find(unigram_vocab==w2);
            comtx(idx1,idx2) = comtx(idx1,idx2)+1;
        end
   end % coffee time!
```

Raw counts and their corresponding word indexes can be extracted from the co-occurrence matrix by using the function find as follows:

```
>> [w1,w2,rawcount] = find(comtx);                                     (8.23)
```

Then, counts must be sorted, and their corresponding bigram words obtained:

```
>> [bigram_count,bigram_index] = sort(rawcount,'descend');            (8.24a)
```

```
>> bigram_word1 = unigram_vocab(w1(bigram_index));                    (8.24b)
>> bigram_word2 = unigram_vocab(w2(bigram_index));
```

By following the same procedures already used in (8.10), (8.11), (8.12a) and (8.13a), the smoothed counts can be computed:

```
>> % computes frequencies of frequencies                              (8.25a)
>> bifreqfreq = hist(bigram_count,max(bigram_count));
```

```
>> % selects some points for guiding the interpolation                (8.25b)
>> bif(1) = 1; biff(1) = bifreqfreq(1);
>> bif(2) = 5; biff(2) = bifreqfreq(5);
>> bif(3) = max(find(bifreqfreq==2)); biff(3) = 2;
>> bif(4) = max(find(bifreqfreq==1))+1; biff(4) = 1;
```

```
>> % interpolates the smoothing function                              (8.25c)
>> slope = (log(biff(1))-log(biff(2)))/(log(bif(1))-log(bif(2)));
>> temp_x = log(bif); temp_y = log(biff);
>> temp_xx = log(1:length(bifreqfreq)+1);
>> binc = exp(spline(temp_x,[slope,temp_y,0],temp_xx));
```

```
>> % computes the smoothed count values                               (8.25d)
>> temp = binc(2:end)./binc(1:end-1);
>> smoothed_bicounts = ((1:length(bifreqfreq))+1).*temp;
```

```
>> % gets the smoothed counts for all seen bigrams                    (8.25e)
>> bigram_scount = smoothed_bicounts(bigram_count);
```

and similar to (8.15a) and (8.15b), the smoothed count *0** for unseen events can be computed. In this case, a good estimate for the total number of unseen events can be the total amount of possible word bigrams minus the total amount of bigrams seen during training:

```
>> unseenbigrams = length(unigram_vocab)^2 - length(bigram_count);     (8.26a)

>> unk_bigram_scount = bifreqfreq(1)/unseenbigrams;                     (8.26b)
```

Now we have everything we need to implement the bigram model, as described in (8.21), by using smoothed counts. Let us consider sample sentences from (8.5a) and (8.20a). Their bigram-model probabilities can be computed as follows:

```
>> for k=1:2 % computes probabilities for example1 and example2       (8.27a)
      eval(sprintf('data=example%d;',k)); % selects the sentence
      prob(k) = 1; % initializes probability value to 1
      % main loop for sentence probability computation
      for n=2:length(data)
          % step 1: computes c(w1)
          index1 = find(unigram_vocab==data(n-1)); % looks for w1
          % if w1 is unseen uses smoothed count for unseen unigrams
          if isempty(index1), c1 = unk_unigram_scount;
          % if w1 is seen uses the corresponding smoothed count for w1
          else, c1 = unigram_scount(index1); end
          % step 2: computes c(w1,w2)
          word1 = bigram_word1==data(n-1); % looks for w1
          word2 = bigram_word2==data(n); % looks for w2
          index2 = find(word1&word2);
          % if w1w2 is unseen uses smoothed count for unseen bigrams
          if isempty(index2), c2 = unk_bigram_scount;
          % if w1w2 is seen uses the corresponding smoothed count for w1w2
          else, c2 = bigram_scount(index2); end
          % step 3: updates probability with p(w2|w1) = c(w1,w2)/c(w1)
          prob(k) = prob(k) * c2/c1;
      end
  end

>> % bigram probability for example1 "he went to the river for water"   (8.27b)
>> disp(prob(1))
   1.4299e-12

>> % bigram probability for example2 "bob went to the river for water"  (8.27c)
>> disp(prob(2))
   1.4200e-13
```

As observed from (8.27b) and (8.27c), the probabilities estimated with the bigram model for the two samples sentences under consideration are higher than

those computed with the unigram model, which were reported in (8.18b) and
(8.20c). As you might be thinking already, results computed with the bigram model must be better than those computed with the unigram model. This is generally
true, as the former takes into account dependencies between consecutive words
while the latter does not. However, it is also true that the bigram model is sparser
than the unigram model, which suggests that probability estimates are more reliable in the unigram case.

From a formal point of view, the quality of an n-gram model can be quantita-
tively assessed by measuring how well the model explains or predicts unseen data.
Such a measurement is often conducted by using a function derived from information theory referred to as cross-entropy, which, under the assumptions of sta-
tionarity and ergodicity, can be computed as follows:

$$xentropy \approx -1/m \ log_2 \ p(w_1, w_2 ... w_m) \approx -1/m \ \Sigma_k \ log_2 \ p(w_k | ...) \qquad \text{(8.28)}$$

where $p(w_1, w_2, ... w_m)$ is the probability estimate generated by the model under consideration for the sequence of words $w_1, w_2, ... w_m$. Although not mandatory, generally, base-2 logarithm is used to compute cross-entropy, in which case the corresponding units are bits. As n-gram models decompose the probability of a text
sequence into the product of n-gram probabilities, the cross-entropy can be computed as the sum of n-gram log-probabilities, such as it is illustrated in the right-
hand side of (8.28).

The lower the cross-entropy obtained for a given model over a sample of unseen data, the better the model. In this way, the value of cross-entropy (when
computed for different models over the same subset of unseen data) can be used
for comparing model quality. More on this is discussed in the following section.

8.3 Model Interpolation

Before performing any cross-entropy calculation, let us introduce another important concept that is relevant to n-gram models. It is the idea of model interpolation. As just discussed before, the bigram model has the advantage of taking into
account dependencies between consecutive words, but its probability estimates
can be unreliable due to data sparseness. On the other hand, the unigram model is
unaware of word order, but its probability estimates are more reliable than bigram
model probabilities. A natural question arising here would be: is it possible to
combine both, unigram and bigram, models into one single model such that the
combined model performs better than each of the two models alone? The answer
to this question is yes!

Indeed, there are many different ways of combining statistical models like the ones we have trained in this chapter. We will restrict ourselves to illustrate a basic strategy: the simple linear combination. It can be formulated as follows:

$$p_\alpha(w_n|w_{n-1}) = (1-\alpha)\, p(w_n) + \alpha\, p(w_n|w_{n-1}) \qquad (8.29)$$

where $p(w_n)$ are the unigram model probabilities, $p(w_n|w_{n-1})$ are the bigram model probabilities, α is the combination parameter, which is also referred to as the weighting factor, and $p_\alpha(w_n|w_{n-1})$ are the probabilities of the resulting interpolated model.

Given our discussion about cross-entropy at the end of the previous section, you might be guessing now that we can select an optimal value for the combination parameter α by minimizing the measured cross-entropy over a set of unseen data (indeed, this is an example of what the development dataset can be used for). In the following exercise, we will be computing the cross-entropy over the development dataset presented in (8.3b) for the linear combination strategy proposed in (8.29) with our previously generated unigram and bigram models. To this end, we adapt the procedure described in (8.27a) for implementing the model interpolation. More specifically, the unigram model is also incorporated into the probability computation loop, and a set of probabilities is simultaneously computed for an array of different values of α aiming at finding an optimal model combination.

```
>> % step 0: initialization of variables                           (8.30)
>> alpha = 0:0.01:1; counter = 0; logprob = zeros(1,length(alpha));
>> for n = 2:length(devdata) % main loop for probability computation
        % gets the following two consecutive words
        w1 = devdata(n-1); w2 = devdata(n);
        if (w1~='S')&(w2~='S') % avoids verse boundaries
            counter = counter +1; % counts the pairs of words evaluated
            % step 1: gets unigram probabilities p(w2)
            index = find(unigram_vocab==w2); % looks for w2
            % if w2 is unseen uses smoothed probability for unk
            if isempty(index), theprob1 = unk_unigram_sprobs;
            % if w is seen uses the corresponding smoothed unigram probability
            else, theprob1 = unigram_sprobs(index); end
            % step 2: gets bigram probabilities p(w2|w1)
            % step 2.a: gets unigram smoothed counts c(w1)
            index1 = find(unigram_vocab==w1); % looks for w1
            % if w1 is unseen uses smoothed count for unseen unigrams
            if isempty(index1), countw1 = unk_unigram_scount;
            % if w1 is seen uses the corresponding smoothed count for w1
            else, countw1 = unigram_scount(index1); end
            % step 2.b: gets bigram smoothed counts c(w1,w2)
            word1 = bigram_word1==w1; % looks for w1
            word2 = bigram_word2==w2; % looks for w2
```

```
        index2 = find(word1&word2);
        % if w1w2 is unseen uses smoothed count for unseen bigrams
        if isempty(index2), countw1w2 = unk_bigram_scount;
        % if w1w2 is seen uses the corresponding smoothed count for w1w2
        else, countw1w2 = bigram_scount(index2); end
        % step 2.c: computes smoothed bigram probability
        theprob2 = countw1w2/countw1;
        % step 3: updates the log-probability sum of interpolated models
        logprob = logprob + log2((1-alpha)*theprob1 + alpha*theprob2);
    end
    % prints a dot every thousand pairs of words
    if mod(n,1000)==0, fprintf('.'); end
  end; fprintf('\n'); % this will take a while ...
```

Once log-probabilities have been computed, cross-entropies can be calculated as indicated in (8.28), and the optimal combination parameter α and its corresponding minimum cross-entropy value can be found:

```
>> xentropy = -logprob/counter; % computes cross-entropies          (8.31a)

>> [best_xentropy,best_alpha_index] = min(xentropy) % finds the minimum (8.31b)
best_xentropy =
    7.0197
best_alpha_index =
    83

>> best_alpha = alpha(best_alpha_index) % gets the optimal alpha      (8.31c)
best_alpha =
    0.8200
```

A plot of the resulting cross-entropy values versus their corresponding α values is presented in Figure 8.3 (the generation of the plot is left to you as an exercise). As seen from the figure, our original intuition about the bigram model being better than the unigram model was right. While the cross-entropy obtained for the unigram model over the development dataset was *8.7479* bits (left extreme of the curve), the cross-entropy measured for the case of the bigram model over the same dataset was *7.4593* bits (right extreme of the curve). However, the linear combination of the two models allows for further reducing the cross-entropy down to *7.0197* bits when the value of the combination parameter α is set to *0.82*.

Finally, we can evaluate and/or apply the constructed models over the test dataset, or any other experimental data. Table 8.1 reports both, log-probability and cross-entropy, values for the three constructed models (unigram, bigram and optimal linear interpolation) over the three datasets used along this chapter: test, development and train. For computing the values reported in the table, the same procedure described in (8.30) was used; with values of α set to *0.00*, *1.00* and *0.82* in order to implement each of the three models under consideration.

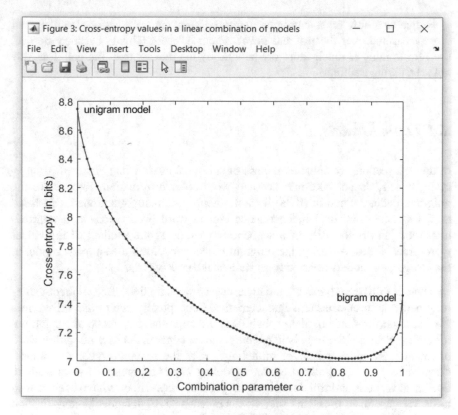

Fig. 8.3. Cross-entropy (computed over the development dataset) versus the combination parameter α in a linear combination of unigram and bigram probabilities

Notice from the table how the cross-entropy exhibits the same trend for both test and development datasets: the interpolated model exhibits the lowest cross-entropy value, followed by the discounted bigram model and, finally, the discounted unigram model. However, this is not the case for the train dataset, for which the minimum cross-entropy is achieved by the bigram model. This is basically because when all the data has been seen during the training phase, the bigram model does not benefit from its interpolation with the unigram model.

Table 8.1. Cross-entropy and log-probabilities computed over test, development and train datasets for discounted unigram, discounted bigram, and interpolated models

	Unigram Model		Bigram Model		Interpolated Model	
	log-prob	cross-entropy	log-prob	cross-entropy	log-prob	cross-entropy
tstdata	-4.25×10^5	8.7801	-3.63×10^5	7.4971	-3.41×10^5	7.0519
devdata	-4.33×10^5	8.7479	-3.69×10^5	7.4593	-3.47×10^5	7.0197
trndata	-5.79×10^6	8.7381	-4.03×10^6	6.0759	-4.13×10^6	6.2341

Notice also how cross-entropies measured over the train set are smaller than those measured over the test and development datasets. This is because the data used for training is much more predictable by the models than the data that was not seen during training.

8.4 Topic Models

In this last section, we consider a class of statistical models that, different from n-gram models, do not take into account word order information. Reconsider the unigram model defined in (8.2a) for a moment. As already mentioned, this basic model assumes statistical independence among word occurrences as it approximates the joint probability of a sequence of words as the product of individual word probabilities. Actually, the unigram model constitutes the simplest model in the category of models referred to as statistical bag-of-words.

There are different types of statistical bag-of-word models, each of them exhibiting particular properties and characteristics. Conceptually, their main differences are based on the fundamental probabilistic assumptions they make about the nature of the data and the linguistic phenomenon in general. As they all ignore word order, they are more suited for modeling text at the semantic level and, consequently, they are generally used for modeling larger segments of texts such as document sections and full documents. In practice, these models have been proven to be very useful in problems such as document search, document categorization and other applications where it is fundamentally important to assess the degree of semantic relatedness among text contents.

In this section, we will focus our attention on a general class of statistical bag-of-word models known as *topic models*. This kind of models represents document probabilities as a mixture of hidden or latent classes denominated *topics*. Inferring these latent classes from a given sample of data constitutes the fundamental problem about training this type of models. As these classes cannot be analytically derived, this is done in practice by using statistical inference methods,[1] which are beyond the scope of this book. Nevertheless, we still illustrate some fundamental concepts related to topic models by, first, implementing a very simple version of topic model and, second, looking at function `fitlda`, which implements a much more sophisticated type of topic model called latent Dirichlet allocation (LDA).

The first simple model presented here is of limited utility in practice. However, we will use it for illustrative purposes, as well as for introducing some of the fundamental concepts related to the mathematical formulation of topic models.

---

[1] More recently, deep neural network architectures have also been used to derive a wide variety of such hidden or latent spaces, which are referred to as embeddings. A more detailed discussion about embeddings is presented later in chapter 10.

Consider, for instance, the following decomposition of the probability of a given document d:

$$p(d) = \Sigma_z\, p(d,z) = \Sigma_z\, p(z)\, p(d|z) \qquad (8.32)$$

Different from (8.2a), where the probability of a document was approximated as the product of individual word probabilities (unigram probabilities), in (8.32) we represent the document probability as the result of marginalizing the joint probability distribution $p(d,z)$ over a hidden discrete variable z. This hidden or latent variable is commonly referred to as the *topic variable* within the context of topic models. According to the definition of conditional probability, we can express the joint probability between a document and a topic as the product of the topic's probability times the conditional probability of the document given the topic. This is shown in the right-hand side of (8.32).

Furthermore, we can rewrite (8.32) by approximating the document's conditional probabilities given the topics in terms of word probabilities as follows:

$$p(d) \approx \Sigma_z\, p(z)\, \Pi_n\, p(w_n|z) \qquad (8.33)$$

where the conditional probabilities of the document given a topic $p(d|z)$ have been approximated by the product of conditional probabilities of words given a topic $p(w|z)$, with the product operator running over all the words w_n contained in the document under consideration. Different from (8.2a), where statistical independence among word probabilities was assumed, topic models assume conditional independence of word probabilities given the topics. In this sense, topic models relax the independence assumption made in the unigram model case.

In order to train the model in (8.33), we need to infer both topic probabilities $p(z)$ and conditional probabilities of words given the topics $p(w|z)$ from a given set of data. We are going to do this here by means of a simplified version of an iterative procedure known as Expectation-Maximization or, in short, the EM algorithm. In the first step (E-step), we compute the conditional probabilities of the topics given de documents $p(z|d)$ from the topic probabilities $p(z)$ and the conditional probabilities of the words given the topics $p(w|z)$ as follows:

$$p(z|d) = 1/\gamma\ p(d|z)\, p(z) \approx 1/\gamma\ p(z)\, \Pi_n\, p(w_n|z) \qquad (8.34)$$

where $1/\gamma$ is just a normalization factor that ensures the resulting values of $p(z|d)$ constitute actual probabilities, i.e. $\Sigma_z\, p(z|d)=1$.

In the second step (M-step), we estimate new values for both $p(z)$ and $p(w|z)$ from the $p(z|d)$ probabilities obtained in (8.34) and word-per-document occurrences counted over the train dataset. We do this as follows:

$$p(w|z) = 1/\xi\ \Sigma_d\, c(d,w)\, p(z|d) \qquad (8.35a)$$

$$p(z) = 1/\varphi\ \Sigma_d\, \Sigma_w\, c(d,w)\, p(z|d) \qquad (8.35b)$$

where $c(d,w)$ are counts corresponding to the number of times each vocabulary word w occurs in each document d of the train dataset, and $1/\xi$ and $1/\varphi$ are normalization factors that ensure the resulting values of $p(w|z)$ and $p(z)$ are probabilities, i.e. $\Sigma_w\, p(w|z)=1$ and $\Sigma_z\, p(z)=1$, respectively.

The EM algorithm then iterates between (8.34) and (8.35) until some convergence criterion is satisfied. The main advantage of this algorithm is that its convergence is guaranteed, but its main disadvantage is that it can easily get trapped in a suboptimal solution. Notice that in the first iteration, some initial values are required for $p(z)$ and $p(w|z)$ in (8.34), so the final solution the algorithm converges to will depend on the quality of these initial guesses.

We are now ready to illustrate this procedure with a very simple dataset. Let us start by creating a toy experimental dataset of 8 documents, which spans a vocabulary of 13 words:

```
>> % defines the documents                                          (8.36a)
>> docs{1} = ["drink","water","and","wine"];
>> docs{2} = ["drink","wine","drink","water"];
>> docs{3} = ["eat","bread","eat","fruits"];
>> docs{4} = ["bread","and","fruits","are","food"];
>> docs{5} = ["use","spoon","and","fork"];
>> docs{6} = ["use","fork","use","knife"];

>> vocab = unique([docs{:}]) % extracts the vocabulary             (8.36b)
vocab =
  1×13 string array
  Columns 1 through 7
    "and"     "are"     "bread"     "drink"     "eat"     "food"     "fork"
  Columns 8 through 13
    "fruits"     "knife"     "spoon"     "use"     "water"     "wine"
```

Let us consider a model of 3 topics and randomly generate some initial values for $p(z)$ and $p(w|z)$:

```
>> % sets the random number generator to its default value2         (8.37a)
>> rng('default');

>> ntopics = 3; % sets the number of topics to 3                    (8.37b)

>> % generates an initial guess for p(z)                            (8.37c)
>> pz_ini = rand(1,ntopics);
>> pz_ini = pz_ini/sum(pz_ini);

>> % generates an initial guess for p(w|z)                          (8.37d)
```

---

[2] Setting the random number generator to its default value will allow you to reproduce the same results presented here. For the specific MATLAB™ version used in the presented examples, the default settings are the Mersenne Twister generator with seed 0.

```
>> pwz_ini = rand(length(vocab),ntopics);
>> for k=1:ntopics, pwz_ini(:,k) = pwz_ini(:,k)/sum(pwz_ini(:,k)); end
```

Let us continue by computing the counts $c(d,w)$ that are required for the computations in (8.35):

```
>> % initializes the count matrix c(d,w)                              (8.38a)
>> counts = zeros(length(docs),length(vocab));

>> % computes the counts of words within documents                    (8.38b)
>> for k = 1:length(docs)
       for n = 1:length(vocab)
           counts(k,n) = counts(k,n)+sum(docs{k}==vocab(n));
       end
   end
```

Now, we are finally ready to conduct the EM iterations. The maximum number of iterations is set to *20* but the algorithm will stop before if a convergence criterion is satisfied. In this case, convergence is achieved if the differences of $p(z)$ and $p(w|z)$ values between consecutive iterations is below a predefined threshold.

```
>> % sets initial probabilities values                                (8.39a)
>> pz = pz_ini; pwz = pwz_ini; pzd = zeros(ntopics,length(docs));

>> delta = 0.000001; % sets the convergence criterion threshold       (8.39b)

>> % implements the simplified version of the EM algorithm            (8.39c)
>> for loops=1:20
       % E-step: computes p(z|d) from p(w|z) and p(z)
       for k=1:length(docs) % loops over the documents
           for topic=1:ntopics % loops over the topics
               temp = 1; % initializes p(d|z)
               for n=1:length(docs{k}) % loops over words in current document
                   idx = find(vocab==docs{k}(n)); % gets the word index
                   temp = temp*pwz(idx,topic); % updates p(d|z)
               end
               pzd(topic,k) = temp*pz(topic); % p(z|d) <- p(d|z) p(z)
           end
           pzd(:,k) = pzd(:,k)/sum(pzd(:,k)); % normalizes p(z|d)
       end
       % M-step: estimates p(w|z) and p(z) from p(z|d) and counts c(d,w)
       newpwz = (pzd*counts)'; % p(w|z) <- sum_d {p(z|d) c(d,w)}
       for k=1:ntopics % normalizes p(w|z)
           newpwz(:,k) = newpwz(:,k)/sum(newpwz(:,k));
       end
       newpz = sum((pzd*counts)'); % p(z) <- sum_w sum_d {p(z|d) c(d,w)}
       newpz = newpz/sum(newpz); % normalizes p(z) <end of M-step>
```

```
% computes the difference between previous and new p(w|z) and p(z)
pwz_diff(loops) = sqrt(sum(sum((pwz-newpwz).^2)));
pz_diff(loops) = sqrt(sum((pz-newpz).^2));
% updates p(w|z) and p(z) with the new computed values
pwz = newpwz; pz = newpz;
% exits when convergence criterion is met
if (pwz_diff(loops)<delta) && (pz_diff(loops)<delta), break; end
end
```

Convergence of (8.39c) is actually achieved in eight iterations. This can be verified by looking at the difference between the consecutive updates computed for both $p(z)$ and $p(w|z)$ at the end of each iteration:

```
>> disp(pz_diff)                                                          (8.40a)
   Columns 1 through 8
    0.1570    0.2160    0.2232    0.0831    0.1121    0.0798    0.0009    0.0000

>> disp(pwz_diff)                                                         (8.40b)
   Columns 1 through 8
    0.4912    0.2053    0.1906    0.1612    0.1476    0.1072    0.0014    0.0000
```

To better understand what the model is doing, let us take a look at the resulting conditional probabilities of the words given the topics $p(w|z)$:

```
>> % creates a table of words and words probabilities per topic           (8.41)
>> wordprobs = table;
>> for k=1:ntopics
       [probs,idx] = sort(pwz(:,k),'descend');
       wordprobs.(sprintf('Topic%d',k)) = vocab(idx)';
       wordprobs.(sprintf('p(w|t%d)',k)) = compose('%4.2f',probs);
   end
>> disp(wordprobs)
```

Topic1	p(w|t1)	Topic2	p(w|t2)	Topic3	p(w|t3)
"use"	{'0.38'}	"bread"	{'0.22'}	"drink"	{'0.38'}
"fork"	{'0.25'}	"eat"	{'0.22'}	"water"	{'0.25'}
"and"	{'0.13'}	"fruits"	{'0.22'}	"wine"	{'0.25'}
"knife"	{'0.13'}	"and"	{'0.11'}	"and"	{'0.13'}
"spoon"	{'0.13'}	"are"	{'0.11'}	"are"	{'0.00'}
"are"	{'0.00'}	"food"	{'0.11'}	"bread"	{'0.00'}
"bread"	{'0.00'}	"drink"	{'0.00'}	"eat"	{'0.00'}
"food"	{'0.00'}	"water"	{'0.00'}	"food"	{'0.00'}
"fruits"	{'0.00'}	"wine"	{'0.00'}	"fork"	{'0.00'}
"eat"	{'0.00'}	"fork"	{'0.00'}	"fruits"	{'0.00'}
"drink"	{'0.00'}	"knife"	{'0.00'}	"knife"	{'0.00'}
"water"	{'0.00'}	"spoon"	{'0.00'}	"spoon"	{'0.00'}
"wine"	{'0.00'}	"use"	{'0.00'}	"use"	{'0.00'}

As seen from (8.41) and (8.36a), the words that mostly contribute to *Topic1* are those occurring in documents *5* and *6*, the ones in *Topic2* are those from documents *3* and *4*, and the ones in *Topic3* are from documents *1* and *2*. Indeed, the model has successfully identified the three main themes or topics within the document collection and grouped the words from the vocabulary accordingly.

Moreover, the documents in the collection can be explained in terms of the topics by means of the conditional probabilities of the topics given the documents $p(z|d)$. The distribution of topics obtained for the example under consideration is illustrated in Figure 8.4. The figure is created as follows:

```
>> hf = figure(4); % creates a new figure                                       (8.42)
>> figure_name = 'Distribution of topics per document: p(z|d)';
>> set(hf,'Color',[1 1 1],'Name',figure_name);
>> ha = axes; bar3(pzd'); % creates the axes and the bar graph
>> set(ha,'XTickLabel',compose('Topic %d',1:3)); % sets axis x labels
>> set(ha,'YTickLabel',compose('Doc %d',1:6)); % sets axis y labels
>> view([-140,35]); % rotates the view for better visualization
```

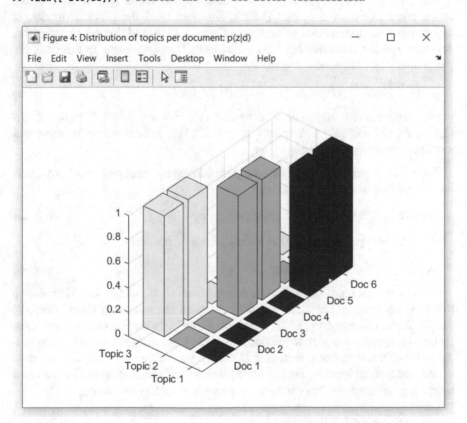

Fig. 8.4. Distribution of topics per document in the example under consideration

As seen from the figure, each document in the collection is indeed explained or represented by means of one of the three topics. For instance, documents *1* and *2* are explained by *Topic3*, documents *3* and *4* by *Topic2*, and *5* and *6* by *Topic1*. As already mentioned, it can be verified from (8.36a) and (8.41) that each document in the dataset can be derived from the words in the corresponding topic assigned to it. Accordingly, this type of models is also referred to as clustering models because the topics define a partition over the collection of documents, in which each document is associated to a single category in the partition.

The simple topic model implementation just illustrated is of very limited utility in practice. First, it always represents documents with a single topic, which significantly limits the representation power of the latent space spanned by topics. In practice, a model that is able to represent documents as mixtures of topics is more desirable and useful. Second, the implementation presented in (8.39c) becomes problematic very quickly, due to arithmetic underflow, as the number of words contained in the documents increases (see exercise 8.6-7 in section 8.6).

Next, we introduce a much more sophisticated and useful type of topic model called latent Dirichlet allocation (LDA). Different from the previous model, in which topics are represented as mixtures of words, LDA additionally assumes that documents in the collection are represented as mixtures of topics. In this case, the probability of a document d is given by:

$$p(d) = \int_\theta p(d,\theta) = \int_\theta p(\theta) \, \Sigma_z \, p(z|\theta) \, p(d|z,\varphi) \tag{8.43}$$

where θ and φ represent mixtures of topics and words that follow Dirichlet distributions $Dir(\alpha)$ and $Dir(\beta)$, respectively. α and β are referred to as the topic and word concentration parameters.

Again, by approximating document probabilities by means of word probabilities, (8.43) can be rewritten as follows:

$$p(d) \approx \int_\theta p(\theta) \, \Pi_n \, \Sigma_{zn} \, p(z_n|\theta) \, p(w_n|z_n,\varphi) \tag{8.44}$$

which, after marginalizing over the topic variable z_n, becomes:

$$p(d) \approx \int_\theta p(\theta) \, \Pi_n \, p(w_n|\theta,\varphi) \tag{8.45}$$

Two important observations can be derived from (8.44) and (8.45). First, notice that both words w_n and topics z_n vary with the word index n. So, different from the simple model described in (8.33), where all words w_n in a document are produced by the single topic assigned to the document, in (8.44) different words w_n are produced by different topics z_n according to the mixture of topics assigned to the document under consideration. Second, the probability of a document can be seen as a continuous mixture distribution with components $p(w_n|\theta,\varphi)$ and weights $p(\theta)$.

For a given document collection of D documents, training an LDA topic model implies deriving the corresponding mixture of topics θ_d for each document d and the corresponding mixture of words φ_k for each topic k. Each mixture of topics θ_d is a vector of size K (total number of topics) and each mixture of words φ_k is a vector of size M (the total number of words in the vocabulary). Accordingly, the resulting topic-mixture matrix θ_{DxK} and word-mixture matrix φ_{MxK} represent the probabilities of topics occurring within documents and the probabilities of words occurring within topics, respectively.

The Text Analytics Toolbox™ includes the function `fitlda`, which implements the latent Dirichlet allocation topic model. It generates a `ldaModel` object that includes, among others, the properties `TopicWordProbabilities` (topic-mixture matrix θ_{DxK}) and `DocumentTopicProbabilities` (word-mixture matrix φ_{MxK}). Let us now illustrate the use of `fitlda` with a practical example.

The file `datasets_ch8b.mat` includes string arrays `trndatatm` and `tstdatatm`, which contain *10* and *6* chapters, respectively, from the same dataset that was used in chapter 7. The data within each of these two string arrays is provided in a format similar to the one used for string array `wordsofverses` in (7.28). However, in this case, the marker `"s"` is used to delimit chapter boundaries. The chapters in this dataset were extracted from three specific books: *Leviticus*, *Acts* and *Psalms*. More specifically, `trndatatm` contains *5* chapters from each of the first two books, and `tstdatatm` contains *2* chapters from each of the three books.

Let us start by loading the data and computing word counts[3] for a subset of the vocabulary over the *10* chapters in `trndatatm`:

```
>> clear; load datasets_ch8b                                        (8.46a)

>> % extracts the vocabulary and selects a subset of it             (8.46b)
>> [vocab,void,index] = unique(trndatatm(trndatatm~='S'));
>> % selects vocabulary words with frequencies in the range [2-9]
>> frequencies = hist(index,length(vocab));
>> select = (frequencies>1)&(frequencies<10);
>> vocab = vocab(select);

>> % computes the matrix of frequency counts                        (8.46c)
>> limits = find(trndatatm=='S');
   counts = zeros(length(limits)-1,length(vocab));
   for k = 1:length(limits)-1
       chapter = trndatatm(limits(k)+1:limits(k+1)-1);
       for n = 1:length(vocab)
           counts(k,n) = counts(k,n)+sum(chapter==vocab(n));
       end
   end
```

---

[3] The function `fitlda` admits different types of document collection representations as input. Here we use a matrix of frequency counts, as the one computed in (8.38b). Other document collection representations such as bag-of-words and bag-of-n-grams are discussed in chapter 9.

We are now ready to train an LDA model with `fitlda`. In this example we consider *4* topics. For further exploration about number of topics and other experimental settings, refer to exercise 8.6-8 in section 8.6.

```
>> rng('default') % resets the random number generator                (8.47a)

>> ntopics = 4; % defines the number of topics                        (8.47b)
>> LDAmodel = fitlda(counts,ntopics,'Verbose',0);
```

Next, let us explore the resulting topics by listing their *20* most relevant words. For this we will rank the words according to their probabilities within the topics.

```
>> topwords = table;                                                  (8.48)
>> for k=1:ntopics
       [probs,idx] = sort(LDAmodel.TopicWordProbabilities(:,k),'descend');
       topwords.(sprintf('Topic%d',k)) = vocab(idx(1:20))';
   end
>> disp(topwords)
```

Topic1	Topic2	Topic3	Topic4
"because"	"bear"	"city"	"felix"
"brother"	"surely"	"us"	"great"
"two"	"woman"	"about"	"more"
"council"	"altar"	"forth"	"stood"
"eat"	"beast"	"our"	"caesarea"
"manner"	"bullock"	"way"	"judgment"
"spake"	"every"	"caesar"	"morrow"
"burnt"	"moses"	"hands"	"now"
"done"	"thereof"	"many"	"brethren"
"kill"	"face"	"multitude"	"hear"
"near"	"linen"	"see"	"know"
"neighbour"	"mercy"	"send"	"been"
"own"	"set"	"away"	"called"
"thyself"	"soul"	"found"	"castle"
"until"	"stranger"	"informed"	"governor"
"defile"	"wash"	"made"	"hast"
"die"	"door"	"seven"	"high"
"just"	"fire"	"such"	"laid"
"statutes"	"house"	"art"	"left"
"born"	"name"	"damascus"	"ought"

As seen from (8.48) the words in the different topics seem to be barely suggesting some different themes or subjects. At least, if we look at some of the verbs and nouns, we will be able to find some subgroups of related concepts, such as *kill*, *die* and *born* in *Topic1*, or *governor*, *castle* and *judgement* in *Topic4*, and so on. However, looking at each topic as a whole, it is not clear what each of those topics is actually referring to.

To better illustrate the utility of a topic decomposition such as the one in (8.48), let us look in more detail at the topic distributions within the documents. Figure 8.5 illustrates these distributions by means of stacked bar plots. The figure is generated as follows:

```
>> hf = figure(5); ha = axes;                                          (8.49)
>> figure_name = 'Distribution of topics (train documents)';
>> set(hf,'Color',[1 1 1],'Name',figure_name);
>> barh(LDAmodel.DocumentTopicProbabilities,'stacked')
>> xlabel("Topic probabilities")
>> set(ha,'YTickLabel',compose('Doc%d',1:size(counts,1)));
>> legend("Topic"+string(1:ntopics),...
          'Location','northoutside','Orientation','horizontal');
```

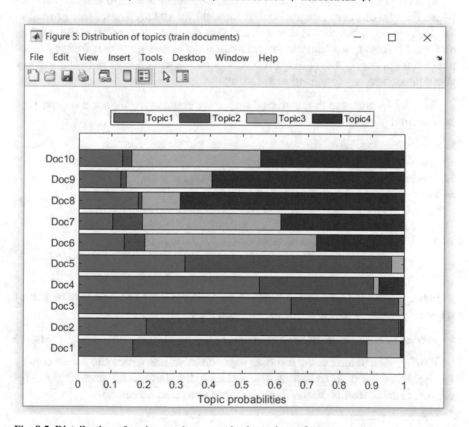

Fig. 8.5. Distribution of topics per document in the train set from `datasets_ch8b.mat`

As seen from the figure, the first 5 documents (Doc1 to Doc5) exhibit a similar topic distribution trend, which is distinctively different from the distribution trend exhibited by the last 5 documents (Doc6 to Doc10). Indeed, while the first set of documents are mainly contributed by *Topic1* and *Topic2* (*90%* or more) with a little contribution from *Topic3* and *Topic4* (*10%* or less), on the other hand, the second set of documents is mainly contributed by *Topic3* and *Topic4* (*80%* or more) with a relatively small contribution from *Topic1* and *Topic2* (*20%* or less). It is important to recall that those two groups of documents correspond to chapters of two different books! While the first *5* documents are chapters of *Leviticus*, the last *5* documents are chapters of the book of *Acts*. Notice that, even though the lists of words associated to the topics presented in (8.48) do not seem to clearly identify meaningful categories,[4] the extracted topics are actually discriminating the two different sources from which the training data was derived.

An interesting aspect of this type of topic models is that they are completely unsupervised in nature. Notice the trained LDA model is able to discriminate between the two different categories of documents we already knew were present in the training data, but the algorithm itself was completely unaware of this information. Moreover, we can use the trained model to assign topics to new documents unseen during the training process. Let us see what this model is able to say about the documents contained in the string array `tstdatatm`.

First, let us compute the matrix of frequency counts. We follow the same procedure from (8.46c):

```
>> % computes the matrix of frequency counts for testdatatm          (8.50)
>> limits = find(tstdatatm=='S');
>> counts = zeros(length(limits)-1,length(vocab));
>> for k = 1:length(limits)-1
        chapter = tstdatatm(limits(k)+1:limits(k+1)-1);
        for n = 1:length(vocab)
            counts(k,n) = counts(k,n)+sum(chapter==vocab(n));
        end
    end
```

Next, we use the `transform` function to compute topics for the new matrix of frequency counts with the previously trained model:

```
>> tsttopics = transform(LDAmodel,counts);                            (8.51)
```

Finally, we can look at the resulting topic distributions within the documents by using stacked bar plots as before. The resulting distributions are presented in Figure 8.6. (The generation of the figure is left to you as an exercise.)

---

[4] For practical purposes, *4* topics might be considered a small number indeed. This explains why the observed lists of words do not clearly represent specific semantic categories. In larger scale experiments involving larger collections of documents and vocabularies as well as a larger number of topics, most of the extracted topics tend to be more clearly related to semantic categories. Nevertheless, as seen from Figure 8.5, even apparently meaningless topics can help uncovering the underlying structure of a given document collection.

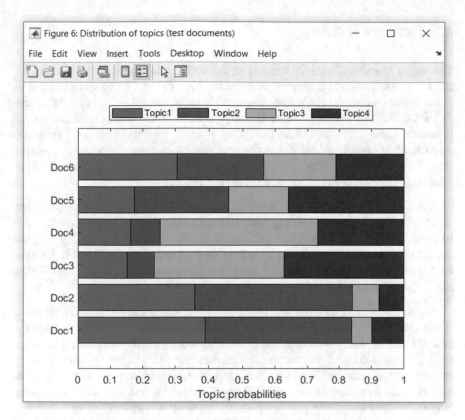

Fig. 8.6. Distribution of topics per document in the test set from `datasets_ch8b.mat`

As seen from the figure, the first two documents (Doc1 and Doc2) exhibit a similar distribution of topics to the one observed for the first set of 5 documents in Figure 8.5 (i.e. mainly represented by *Topic1* and *Topic2*). The second pair of documents (Doc3 and Doc4) exhibit a similar distribution of topics to the one observed for the second set of 5 documents in Figure 8.5 (i.e. mainly represented by *Topic3* and *Topic4*). On the other hand, the last two documents (Doc5 and Doc6) exhibit a more uniform distribution of topics, which is noticeable different from the topic distributions observed on the previous two pairs of documents.

Notice that the topic distributions in Figure 8.6 are actually discriminating the different sources that originated the documents in `tstdatatm`. Indeed, the first two documents are chapters extracted from *Leviticus*, which is the same source of the first set of documents in `trndatatm`. The second pair of documents are chapters from *Acts*, which is the same source of the second set of documents in `trndatatm`. Finally, the last two documents are chapters of the book of *Psalms*, which was not represented in `trndatatm`. However, this last pair of books exhibits a different topic distribution trend from the other two groups.

8.5 *Further Reading*

The use of Markov Chains for language modeling can be traced back to Shannon's work "A mathematical theory of communication" (1948). In this work, he proposed some empirical experiments based on *n*-gram models to estimate the entropy of the English language. Nevertheless, because of the strong criticism by rationalists during the 50s, 60s and 70s, *n*-gram models were not seriously considered in computational linguistics and natural language processing research until the late 80s, when they were finally proven to be very useful in practical applications (Jelinek 1985), (Church and Mercer 1993).

Apart from the Good-Turing discounting procedure discussed here, there are other different methods for discounting *n*-gram probabilities. These include some basic techniques such as add-one smoothing and Witten-Bell discounting. A good review on these basic techniques, including Good-Turing, can be found in Chapter 6 of (Jurafsky and Martin 2000). A more comprehensive review on Good-Turing can be found in the work of (Gale and Sampson 1995).

A non-linear interpolation technique we have not discussed here is back-off (Katz 1987), in which the probabilities of unseen *n*-grams are estimated from the probabilities of lower order *k*-grams (*k*<*n*) that were actually seen during training. Both interpolation and discounting techniques are indistinctively referred to as smoothing methods, and they are often used in combination in practice. The most popularly used method for smoothing *n*-gram probabilities is the modified Kneser-Ney (Chen and Goodman 1998), which is based on a back-off strategy that was originally proposed by (Kneser and Ney 1995).

More recently, during the deep learning era, a few deep neural network architectures have been trained with massive volumes of data giving rise to the so called pre-trained language models. This new generation of language models have been proven to provide significant gains in a large range of Natural Language Processing tasks, including language understanding and generation. For specific information about some current state-of-the-art pre-trained language models, you can look at BERT (Devlin et al. 2018), T5 (Raffel et al. 2020), GPT-3 (Brown et a. 2020), and the references within.

The basic topic model presented in section 7.4, which constitutes a mixture of unigrams, is described in detail in (Nigam et al. 2000). More advanced topic models include probabilistic latent semantic indexing (Hofmann 1999) and latent Dirichlet allocation (Blei et al. 2003). More information on the EM algorithm and statistical inference in general can be found in (Dempster et al. 1977) and (Morris 1983), respectively. For a more recent approach that incorporates the use of neural network architectures for learning topic models, refer to (Chien and Lee 2018).

Finally, although currently a little bit outdated, an excellent overview of statistical language modeling is available in (Rosenfeld 2000).

8.6 Proposed Exercises

1. Consider the smoothed bigram counts computed in (8.25) and (8.26):
   - Generate a plot showing the distribution for frequencies of frequencies and the corresponding smoothing function (just similar to the one presented in Figure 8.1 for the case of the unigram model).
   - Plot the absolute and relative differences between the original and the resulting discounted bigram probabilities (just similar to the ones presented in Figure 8.2 for the case of the unigram model).

2. Create the two functions: `get_1gram(data,file)`, for training a unigram model; and `[p,n] = use_1gram(data,file,log)`, for computing the probability of a segment of text according to a given unigram model.

   - The input variable `data` must be a string array of words containing text data in the same format used in (8.3b). It must contain either the train or the test dataset depending on whether `get_1gram` or `use_1gram` is called.
   - The input variable `file` must be a string containing the name of the file where model parameters (`unk_unigram_sprobs`, `unigram_sprobs` and `unigram_vocab`) must be either stored to or read from depending on whether `get_1gram` or `use_1gram` is called.
   - The function `get_1gram` must implement the Good-Turing discounting procedure described in section 8.2 for the case of unigram probabilities.
   - The function `use_1gram` must implement the procedure in (8.20) for estimating the probability of a segment of text by using the smoothed unigram probabilities. The input variable `log` specifies whether just regular probabilities ($log = 0$) or log-probabilities ($log = 1$) must be computed.
   - The resulting probability estimate and the total number of unigram probabilities used must be returned in the output variables `p` and `n`, respectively.

3. Create the two functions: `get_2gram(data,file)`, for training a bigram model; and `[p,n] = use_2gram(data,file,log)`, for computing the probability of a segment of text according to a given bigram model.

   - With the variables `data`, `file`, `log`, `p` and `n` as in the previous exercise.
   - The function `get_2gram` must implement the Good-Turing discounting procedure described in section 8.2 for the case of bigram probabilities.
   - The function `use_2gram` must implement the procedure in (8.27a) for estimating the probability of a segment of text by using smoothed bigram probabilities.
   - Notice that your implementation of `get_2gram` might benefit from using the function `get_1gram` already created in the previous exercise.

4. Consider the following sentence: *he went for food and water*.

   – Use the smoothed unigram and bigram models trained in this chapter to estimate its probability (you can use functions `use_1gram` and `use_2gram` if you have already completed the two previous exercises). What difference do you observe between the two estimated probabilities?
   – Consider all possible permutations of words for the given sentence and recompute the unigram and bigram probabilities again. What differences do you observe between unigram and bigram estimated probabilities?
   – Study the five sentence permutations with largest bigram probabilities and the five sentence permutations with lowest bigram probabilities. What are your observations?
   – Re-compute probabilities for all permutations by using the interpolation procedure in (8.30). Consider $\alpha = 0.82$. How different are these probabilities from your previous unigram and bigram estimates?
   – Study the five sentence permutations with largest and lowest interpolated probabilities. What differences do you observe with respect to the case in which bigram probabilities were used?

5. Different from the bigram implementation presented in this chapter, it is a common practice to include an init-of-sentence token before the first word of each sentence when both training and computing bigram probabilities.

   – Consider the procedure for computing unigram probabilities presented in (8.20b). Modify it and use it for obtaining log-probability estimates for the test and development datasets.
   – Compare these log-probabilities with the corresponding unigram-model log-probability values reported in table 8.1. Why are they different?
   – Consider the verse boundary marker `'s'` included the dataset of (8.3b) to be an init-of-sentence token. Re-compute the bigram smoothed counts and re-implement (8.30) by taking this boundary marker into account. Re-estimate test and development log-probabilities for different values of α.
   – Compare the resulting log-probabilities when $\alpha = 0$ (unigram case) with those computed in the first step of this exercise. What can you see now?
   – How different are the log-probabilities when $\alpha = 1$ (bigram case) with the corresponding ones in table 8.1? What is the new optimal value for α?

6. Let us reconsider the topic model exercise presented in section 8.4, where we only computed estimates for $p(w|z)$, $p(z)$ and $p(z|d)$.

   – Create a script for computing $p(d|z)$ from $p(w|z)$ and study the resulting distribution. What does it represent?
   – Use your computed estimates for $p(d|z)$ and the resulting estimates for $p(z)$ from (8.39) for estimating document probabilities $p(d)$ as defined in (8.32). What are your main observations?

  – Repeat the whole topic model exercise presented in section 8.4 by using
    several different initializations for $p(w|z)$ and $p(z)$. Generate bar graphs,
    similar to the one in Figure 8.4, for each case. What are your observations?

7. Extract a subset of chapters from three different books of the Bible dataset **bi-
   ble_en.xml** (you can use the ones in **datasets_ch8b.mat**).

  – Apply the following preprocessing steps to the data: remove xml tags, re-
    move punctuation marks, lowercase the text, extract vocabulary, compute
    word frequencies and retain vocabulary words within $f_{min}=2$ and $f_{max}=9$.
  – Estimate $p(w|z)$ and $p(z|d)$ by following the procedure described in section
    8.4. Consider a topic model of 3 topics. (You will need to reimplement the
    procedure by using log-probabilities to avoid arithmetic underflow!)
  – Find the 10 words with largest $p(w|z)$ probabilities for each of the 3 differ-
    ent topics. What can you conclude from your results?
  – Group the chapters based on their most probable topic according to $p(z|d)$.
    Compare the resulting grouping with the origin of the chapters (the books
    they were extracted from). What are your main observations?
  – Repeat the experiment several times with (a) different initializations for
    $p(w|z)$ and $p(z)$, (b) different values of f_{min} and f_{max}, and (c) by considering
    different number of topics. How do results vary among experiments?

8. Repeat the previous exercise by using **fitlda** to train an LDA model instead.
   What differences do you observe?

9. Look at the documentation of **fitlsa** and repeat the previous exercise by using
   an LSA model instead. What differences do you observe?

8.7 Short Projects

1. The way we had handled word counts and probabilities in this chapter is not efficient. Consider (8.20b) and (8.27a), where we first need to get the word index
   idx=find(vocab==w) before retrieving its counts **count=scount(idx)** or probabilities **prob=sprob(idx)**. An efficient way of doing this in practice is by using hash tables. A hash table is a special data structure, in which index variables (keys) are mapped into value variables (values) by means of a hash
   function. The main advantage of a hash table, with regards to the particular indexing problem under consideration, is that strings can be used as keys. So, by
   storing counts or probabilities into hash tables, they can be retrieved by using
   commands such as **scount('water')**, **scount('drink_water')**, and so on. The
   main objective of this short project is to construct efficient implementations for
   n-gram models by using this kind of data structures.

- There are a few ways for using this kind of data structures in MATLAB®. Consider, for instance, MATLAB® own class `containers.Map`. Take some time to read and learn about this class in its MATLAB® technical documentation page, https://www.mathworks.com/help/matlab/ref/containers.map.html Accessed 6 July 2021.
- Re-implement the four functions `get_1gram`, `use_1gram`, `get_2gram` and `use_2gram`, you previously implemented in exercises 8.6-2 and 8.6-3, by using the `containers.Map` class.
- Create an efficient implementation for the linear interpolation procedure presented in (8.30). Consider the following syntax when implementing the function `[p,n] = use_interp(data,file,alpha)`.
- Compute interpolated log-probabilities for `trndata` in `datasets_ch8a.mat` by considering $\alpha = 0.5$. Use both the implementation in (8.30) and the efficient implementation developed here. Compare computation times between the two implementations (you are encouraged to use functions such as `tic` and `toc`, or `cputime` in order to get accurate time estimates).

2. Create functions `get_3gram` and `use_3gram` for training a smoothed trigram model and using such a model for estimating the probabilities of a given segment of text. Use the same syntax already used for the corresponding unigram and bigram functions.

- Insert one extra verse boundary marker `'s'` at the beginning of each verse in `datasets_ch8a.mat` so that probabilities can be estimated for the first and second words of each verse, i.e. $p(w_1|\,'s',\,'s')$ and $p(w_2|w_1,\,'s')$.
- Compute counts of the form $c(w_{n-2}, w_{n-1}, w_n)$. Use the same efficient implementation you developed in the previous short project for indexing the computed trigram counts.
- Estimate discounted trigram counts by following a similar procedure to the one used in section 8.2 for discounting bigram counts.
- Estimate the trigram probabilities $p(w_n|w_{n-2}, w_{n-1})$ by considering the ratio between discounted trigram counts $c^*(w_{n-2}, w_{n-1}, w_n)$ and discounted bigram counts $c^*(w_{n-2}, w_{n-1})$.
- Compute the cross-entropy over the development and test datasets for the resulting trigram model. How does this value differ from the ones obtained for the unigram and bigram models? What can you conclude about this?
- Follow the linear interpolation procedure described in section 8.3 for combing the unigram, bigram, and trigram models into one single model. Notice that in this case you would need three combination parameters: α, β and χ, such that: α, β, $\chi \geq 0$ and $\alpha + \beta + \chi = 1$.
- Find the optimal set of combination parameters for which the interpolated model exhibits a minimum cross-entropy value over the development data set. How is this minimum value compared to the one observed for the non-interpolated trigram model?

8.8 References

Blei DM, Ng AY, Jordan MI (2003) Latent Dirichlet allocation. Journal of Machine Learning Research 3:993-1022

Brown TB, Mann B, Ryder N, Subbiah M, Kaplan J, Dhariwal P, Neelakantan A, Shyam P, Sastry G, Askell A, Agarwal S, Herbert-Voss A, Krueger G, Henighan T, Child R, Ramesh A, Ziegler DM, Wu J, Winter C, Hesse C, Chen M, Sigler E, Litwin M, Gray S, Chess B, Clark J, Berner C, McCandlish S, Radford A, Sutskever I, Amodei D (2020) Language Models are Few-Shot Learners, https://arxiv.org/abs/2005.14165v2 Accessed 6 July 2021

Chen SF, Goodman J (1998) An empirical study of smoothing techniques for language modeling. Technical Report TR-10-98, Center for Research in Computer Technology, Harvard University

Chien JT, Lee CH (2018) Deep Unfolding for Topic Models, IEEE Transactions on Pattern Analysis and Machine Intelligence 40(2):318-331

Church KW, Mercer RL (1993) Introduction to the special issue on computational linguistics using large corpora. Computational Linguistics 19:1-24

Dempster AP, Laird NM, Rubin DB (1977) Maximum likelihood from incomplete data via the EM algorithm. Journal of the Royal Society: Series B 39(1):1-38

Devlin J, Chang MW, Lee K, Toutanova K (2019) BERT: Pre-training of Deep Bidirectional Transformers for Language Understanding, https://arxiv.org/abs/1810.04805v2 Accessed 6 July 2021

Gale WA, Sampson G (1995) Good-Turing frequency estimation without tears. Journal of Quantitative Linguistics 2:217-237

Hofmann T (1999) Probabilistic latent semantic indexing. In Proceedings of the 22nd Annual International SIGIR Conference

Jelinek F (1985) Markov source modeling of text generation. In Skwirzynski, J.K. (ed.), The Impact of Processing Techniques on Communications, pp. 569-598

Jurafsky D, Martin JH (2000) Speech and Language Processing. Prentice-Hall, Inc. Upper Saddle River, NJ

Katz SM (1987) Estimation of probabilities from sparse data for the language model component of a speech recognizer. IEEE Transactions on Acoustics, Speech and Signal Processing 35:400-401

Kneser R, Ney H (1995) Improved backing-off for m-gram language modeling. In Proceedings of the IEEE International Conference on Acoustics, Speech and Signal Processing, 1:181-184

Morris C (1983) Parametric empirical Bayes inference: theory and applications. Journal of the American Statistical Association 78(381):47-65

Nigam K, McCallum A, Thrun S, Mitchell T (2000) Text classification from labeled and unlabeled documents using EM. Machine Learning 39(2/3):103-134

Raffel C, Shazeer N, Roberts A, Lee K, Narang S, Matena M, Zhou Y, Li W, Liu PJ (2020) Exploring the Limits of Transfer Learning with a Unified Text-to-Text Transformer, https://arxiv.org/abs/1910.10683v3 Accessed 6 July 2021

Rosenfeld R (2000) Two decades of statistical language modeling: where do we go from here? In Proceedings of the IEEE

Shannon CE (1948) A mathematical theory of communication. The Bell System Technical Journal, 27: 379–423

9 Geometrical Models

In the previous chapter we reviewed the basic elements of the statistical language model framework. An alternative and most commonly used modeling paradigm, which was originally developed in the field of Information Retrieval, is the geometrical framework. Within this framework, vector spaces are used for constructing mathematical representations of documents, words, and any other type of text units. Basic geometrical concepts, such as distances, angles, and projections, are then used to assess difference and similarity degrees among the units of analysis under consideration, which are modeled by means of vectors in the given vector space. In this chapter we will focus our attention on the geometrical framework of language modeling. First, in section 9.1, the term-document matrix construct is presented and described in detail. Then, in section 9.2, the vector space model approach is studied along with the popularly known TF-IDF (term frequency inverse document frequency) weighting scheme. Finally, in section 9.3, association scores and distance functions commonly used in vector model space representations are described.

9.1 The Term-Document Matrix

Consider the procedure we used for calculating the co-occurrence matrix computed in section 7.2. As an intermediate step, in (7.29b), we computed a binary matrix composed of as many rows as vocabulary words and as many columns as verses were available in the collection. Such a matrix, which is generally referred to as a term-document matrix, contains information about the occurrence of vocabulary words in the different documents (verses in our case) composing the data collection. For instance, if the matrix element m_{ij} is set to one, it means that the ith vocabulary word occurs in the jth document. On the other hand, if it is set to zero, it means that the ith vocabulary word does not occur in the jth document.

This term-document matrix construction, although very simple, is indeed a very powerful representation of the overall data collection and its corresponding distribution of words over documents. It actually constitutes a simple form of dual index, which allows for retrieving all words occurring in a given document (direct index) as well as all documents containing a given word (inverted index).

Before starting the experimentation, let us proceed to load the data file `datasets_ch9.mat`, which contains the same dataset used in section 7.2 along with its corresponding vocabulary and word counts:

© The Author(s), under exclusive license to Springer Nature Switzerland AG 2021
R. E. Banchs, *Text Mining with MATLAB®*, https://doi.org/10.1007/978-3-030-87695-1_9

```
>> clear;                                                              (9.1a)
>> load datasets_ch9

>> whos                                                                (9.1b)
  Name                    Size                    Bytes  Class    Attributes
  ranked_frequencies      1x12545                100360  double
  ranked_vocabulary       1x12545                757846  string
  verses                  1x31103              64471910  struct
```

where `ranked_vocabulary` is a string array containing all the vocabulary terms in the collection ordered by rank, `ranked_frequencies` is a numeric array containing the frequencies for each vocabulary term, and `verses` is a structure array with as many elements as verses are in the collection. This structure is composed of three fields: `verse.vocab` (a string array containing the vocabulary terms), `verse.count` (a numeric array with the frequencies of the vocabulary terms) and `verse.text` (a string containing the raw text).

Let us now reproduce the procedure for computing the term-document matrix for the considered data collection, exactly in the same way as it was done in 7.29a and 7.29b.

```
>> % creates a sparse matrix of appropriate dimensions                 (9.2a)
>> tdmtx = sparse(length(ranked_vocabulary),length(verses));

>> % updates the term-document matrix elements                         (9.2b)
>> for k = 1:length(verses) % this will take a while
       [void,termindex] = intersect(ranked_vocabulary,verses(k).vocab);
       tdmtx(termindex,k) = 1;
   end
```

As stated above, the resulting term-document matrix can be used either as a direct index or as an inverted index. In the first case, we would be interested in knowing which words occur within a given document. Consider, for instance, the words occurring in verse *400*:

```
>> word_index = find(tdmtx(:,400));                                    (9.3a)

>> ranked_vocabulary(word_index)                                       (9.3b)
ans =
  1×5 string array
    "and"     "zechariah"     "gedor"     "ahio"     "mikloth"

>> verses(400).vocab % which are indeed the words in that verse        (9.3c)
ans =
  1×5 string array
    "ahio"     "and"     "gedor"     "mikloth"     "zechariah"

>> verses(400).text                                                    (9.3d)
And Gedor, and Ahio, and Zechariah, and Mikloth.
```

On the other hand, in the second case, we would be interested in knowing which documents contain a given word. Consider, for instance, we are interested in finding the documents containing the word *spark*. The term-document matrix can be used as an inverted index to find them as follows:

```
>> doc_index = find(tdmtx(ranked_vocabulary=='spark',:))          (9.4a)
doc_index =
       14310        18334

>> verses(doc_index).text                                          (9.4b)
ans =
    "And the strong shall be as tow, and the maker of it as a spark, and they
shall both burn together, and none shall quench them."
ans =
    "Yea, the light of the wicked shall be put out, and the spark of his fire
shall not shine."
```

Similarly, the inverted index can be used to quickly answer other similar questions related to the verses containing specific terms, such as the total number of documents containing a given word:

```
>> full(sum(tdmtx(ranked_vocabulary=='spark',:)))                  (9.5a)
ans =
     2

>> full(sum(tdmtx(ranked_vocabulary=='river',:)))                  (9.5b)
ans =
   148

>> full(sum(tdmtx(ranked_vocabulary=='taxes',:)))                  (9.5c)
ans =
     1
```

or finding the documents in which two or more words either co-occur or not:

```
>> w1_idx = ranked_vocabulary=='spark';                            (9.6a)
>> w2_idx = ranked_vocabulary=='fire';

>> % 'spark' & 'fire'                                               (9.6b)
>> doc_index = find(tdmtx(w1_idx,:)&tdmtx(w2_idx,:));
>> disp(verses(doc_index).text)
Yea, the light of the wicked shall be put out, and the spark of his fire shall
not shine.

>> % 'spark' & not 'fire'                                           (9.6c)
>> doc_index = find(tdmtx(w1_idx,:)&not(tdmtx(w2_idx,:)));
>> disp(verses(doc_index).text)
And the strong shall be as tow, and the maker of it as a spark, and they shall
both burn together, and none shall quench them.
```

In general, any binary operator or combination of binary operators can be used to test the elements of a term-document matrix across its rows (vocabulary terms) and/or columns (documents), in order to search for terms and documents satisfying the specified condition or query.

An alternative, and much more efficient, way of computing the term-document matrix for a given document collection is illustrated next. The procedure makes use of specific Text Analytics Toolbox™ functions for preprocessing the documents and computing the word counts.

```
>> % copies all verses into a string array                              (9.7a)
>> for k=1:length(verses), strdocs(k)=verses(k).text; end

>> % preprocesses text by removing punctuation and lowercasing it        (9.7b)
>> strdocs = lower(erasePunctuation(strdocs));

>> tokdocs = tokenizedDocument(strdocs); % tokenizes the documents       (9.7c)

>> bowdocs = bagOfWords(tokdocs) % constructs a bag-of-words model        (9.7d)
bowdocs =
  bagOfWords with properties:

          Counts: [31103×12676 double]
      Vocabulary: [1×12676 string]
        NumWords: 12676
    NumDocuments: 31103
```

The bag-of-word model representation computed in (9.7d), which is also referred to as term-frequency counter, includes the string array **Vocabulary**, which contains the overall vocabulary of the collection, and the numeric matrix **Counts**, which contains the number of times each vocabulary word occurs in each document. There are three important observations to make with regards to the process illustrated in (9.7) as compared to the process illustrated in (9.2). First of all, due to a much more efficient implementation, the procedure in (9.7) is, by far, much faster. Second, the resulting word count matrix **bowdocs.Counts** is actually a document-term matrix (i.e. a transposed version of a term-document matrix). Each row in this matrix represents a document in the collection, and each column represents a vocabulary word. Finally, notice the vocabulary size in (9.7d) is different from the one in (9.1b). This is mainly due to differences on how the word tokenization is performed in each case. More about **bagOfWords** and other related functions is discussed in exercise 9.5-2, in section 9.5.

Before continuing our study of the geometrical nature of the language model defined by the term-document matrix representation, let us revisit the notion of co-occurrence. Consider for a moment the operation previously presented in (7.30), which implemented the right-hand side multiplication of the term-document matrix by its transpose. Here, we use the term-document matrix computed in (9.2).

```
>> comtx = tdmtx*tdmtx';                                                 (9.8)
```

This operation, as we already saw in section 7.2, accounts for word co-occurrences in the verses of the data collection. Indeed, comtx is a square matrix with as many rows and columns as vocabulary terms, with the elements in its diagonal ($m(i,i)$ with $i=j$) representing the number verses in which each individual vocabulary word occurs, and the elements out of its diagonal ($m(i,j)$ with $i \neq j$) representing the number of verses in which each corresponding pair of vocabulary words co-occur.

Similarly, an equivalent operation can be applied to account for verses or document "co-occurrences" instead of vocabulary word co-occurrences. Indeed, if we consider the left-hand side multiplication of the term-document matrix by its transpose, we will obtain a square matrix with as many rows and columns as verses in the collection, the elements of which will provide the amount of vocabulary overlap between a given pair of verses. More specifically, the elements in its diagonal ($m(i,i)$ with $i=j$) will represent the total number of vocabulary words contained in each individual verse, and the elements out of its diagonal ($m(i,j)$ with $i \neq j$) will represent the total number of vocabulary words that are common to each corresponding pair of verses.

There is only one problem with this word overlap matrix: it is not, in general, a sparse matrix. Unless you actually have a lot of memory available in your system, computing such a matrix might result in an out-of-memory error!

```
>> % be careful... an out-of-memory error might occur          (9.9)
>> ovmtx = tdmtx'*tdmtx;
>> whos comtx ovmtx
  Name        Size                     Bytes  Class    Attributes

  comtx      12545x12545            35539296  double   sparse
  ovmtx      31103x31103         14483390128  double   sparse
```

As seen from (9.9), the resulting word overlap matrix ovmtx requires around *400* times more memory (Bytes) than the word co-occurrence matrix comtx (actually, *14483390128 / 35539296 = 407.5317* times more). However, it is only about *6* times larger in size (i.e. *31103² / 12545² = 6.1470*). This suggests that ovmtx is much denser (less sparse) than comtx. Indeed, as it might be recalled from (7.31), only *1.41%* of the elements in comtx are different from zero. However, in the case of ovmtx, *93.57%* of its elements are different from zero!

```
>> sum(sum(ovmtx>0))/numel(ovmtx)*100                          (9.10a)
ans =

  (1,1)        93.5703

>> clear ovmtx                                                 (9.10b)
```

This means that, in the particular example illustrated here, the word overlap matrix ovmtx is around *66* times denser than the word co-occurrence matrix comtx.

In general, we might expect word overlap matrices to be dense. This is basically due to the fact that the most frequent words in a language are typically common to the great majority of documents in any text collection. So, the vocabulary overlap between any pair of documents has a very big chance to be greater than zero.

In practice, we do not compute a word overlap matrix like the one in (9.9) for a full term-document matrix that includes all documents and vocabulary terms. It is always necessary to consider a specific sub-collection of documents, a reduced vocabulary set, or both.[1]

Next, we illustrate the utility of the aforementioned vocabulary overlap matrix **ovmtx**. However, in order to manipulate a smaller matrix and avoid any eventual out-of-memory problems, let us recompute it for the first *10,000* verses in the data collection only:

```
>> ovmtx = tdmtx(:,1:10000)'*tdmtx(:,1:10000);                                    (9.11)
```

In the same way the word co-occurrence matrix constitutes a useful resource to identify pairs of words that are either semantically related or unrelated (see section 7.2), the vocabulary overlap matrix computed in (9.11) constitutes a very useful resource to identify pairs of documents that are either related or unrelated to each other. In this case, the degree of relatedness between the documents is defined by the total amount of vocabulary overlap between them.

Consider, for instance, the pair of verses *55* and *60*:

```
>> disp(verses(55).text)                                                          (9.12a)
These are the sons of Israel; Reuben, Simeon, Levi, and Judah, Issachar, and
Zebulun,
```

```
>> disp(verses(60).text)                                                          (9.12b)
And the sons of Zerah; Zimri, and Ethan, and Heman, and Calcol, and Dara: five
of them in all.
```

The corresponding entry in the word overlap matrix will tell us how many vocabulary words they have in common:

```
>> full(ovmtx(55,60))                                                             (9.13)
ans =
     4
```

which can be identified from (9.12) to be `'the'`, `'sons'`, `'of'` and `'and'`. Alternatively, they can also be automatically extracted by intercepting the individual vocabularies of both verses:

---

[1] The most common situation is to compute the product of a single row of the transposed term-document matrix (one single document) against the complete term-document matrix (the whole data collection). Vocabulary reduction is also a common practice for several different reasons; this issue is discussed in detail in section 10.1.

```
>> intersect(verses(55).vocab,verses(60).vocab)                    (9.14)
ans =
  1×4 string array
    "and"    "of"    "sons"    "the"
```

More interesting than the previous example, in which we just compared two predefined verses, is the possibility of finding the pair of verses with the maximum number of vocabulary overlap:

```
>> % extracts the upper triangular part of ovmtx excluding the diagonal  (9.15a)
>> upmtx = triu(ovmtx,1);

>> % gets the verses with maximum overlap                          (9.15b)
>> [v1,v2] = find(upmtx==max(max(upmtx)))
v1 =
      1777
v2 =
      3521

>> % computes the vocabulary overlap between v1 y v2               (9.15c)
>> full(ovmtx(v1,v2))
ans =
    42

>> disp(verses(v1).text)                                          (9.15d)
Therefore now, LORD God of Israel, keep with thy servant David my father that
thou promisedst him, saying, There shall not fail thee a man in my sight to
sit on the throne of Israel; so that thy children take heed to their way, that
they walk before me as thou hast walked before me.

>> disp(verses(v2).text)                                          (9.15e)
Now therefore, O LORD God of Israel, keep with thy servant David my father
that which thou hast promised him, saying, There shall not fail thee a man in
my sight to sit upon the throne of Israel; yet so that thy children take heed
to their way to walk in my law, as thou hast walked before me.
```

As seen from (9.15), we could easily find two verses which are basically a repetition of each other by searching for the maximum value within **ovmtx**.

Notice however that, in general, a high vocabulary overlap between two documents does not necessarily imply that one document is a repetition of the other. In this sense, the percentage of vocabulary overlap, or relative overlap, of one document with respect to the other provides a more reliable hint about duplicity.

```
>> % vocabulary overlap between v1 and v2 relative to v1          (9.16a)
>> full(ovmtx(v1,v2))/length(verses(v1).vocab)*100
ans =
    93.3333
```

```
>> % vocabulary overlap between v1 and v2 relative to v2          (9.16b)
>> full(ovmtx(v1,v2))/length(verses(v2).vocab)*100
ans =
    87.5000
```

Similarly, pairs of verses with a specific number of common vocabulary words can be easily identified by means of the word overlap matrix. Let us consider the cases of minimum overlap (i.e. pairs of verses with no vocabulary in common):

```
>> % find all zero elements in the vocabulary overlap matrix      (9.17a)
>> [temp1,temp2] = find(ovmtx==0);

>> % retain only those elements above the diagonal                (9.17b)
>> v1 = temp1(temp1>temp2);
>> v2 = temp2(temp1>temp2);

>> % there are quite a few verse pairs with no vocabulary overlap (9.17c)
>> whos v1 v2
  Name           Size              Bytes  Class     Attributes
  v1          2211364x1          17690912  double
  v2          2211364x1          17690912  double

>> % let us just consider one pair                                (9.17d)
>> disp(verses(v1(20)).text)
Hadoram also, and Uzal, and Diklah,

>> disp(verses(v2(20)).text)                                      (9.17e)
Adam, Sheth, Enosh,
```

After having revisited the concept of word co-occurrence, as well as introducing and illustrating the notion of vocabulary overlap, let us now go back to our term-document matrix representation in (9.2), which constitutes the very basic construction mechanism for language modeling under the geometric framework. Indeed, if we look at the rows of the term-document matrix, we will find that each unique vocabulary word in the collection is being represented by a binary vector with as many components (dimensions) as there are documents in the collection. Alternatively, if we look at the columns of the term-document matrix, we will find that each document in the collection is being represented by a binary vector with as many components (dimensions) as there are vocabulary terms.

A term-document matrix actually constitutes a dual vector space model for its corresponding document collection, which can be alternatively used as a vector space for representing words (a word-space) or a vector space for representing documents (a document-space). Although binary matrices can be constructed and used as it has been illustrated in this section, real-valued vector spaces offer a much more powerful tool for handling and operating with text data. The next section describes in detail some of the fundamental concepts related to real-valued vector space models.

9.2 The Vector Space Model

Consider a term-document matrix as the one computed in (9.2). As we just said at the end of the previous section, such a matrix provides the basic modeling mechanism for either a word-space or a document-space depending on whether we look at its rows or columns, respectively. Although both can be actually handled by means of very similar procedures, in this section we will orient our descriptions to and focus our attention on the document-space construct. According to this, we will be looking at the columns of the term-document matrix as vectors representing documents (verses in our experimental collection) that are defined over an n-dimensional vector space, where n is the total number of words in the vocabulary.

For better grasping the geometric nature of the document vector space, let us consider the very simple example depicted in Figure 9.1, which presents the three-dimensional vector space for an imaginary data collection of seven documents containing the three vocabulary words: w_1, w_2 and w_3.

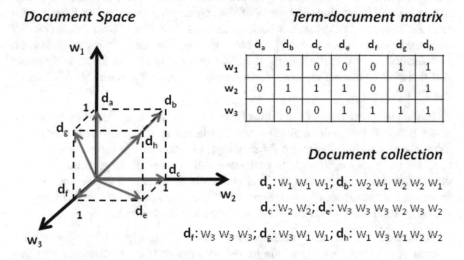

Document Space   **Term-document matrix**

	d_a	d_b	d_c	d_e	d_f	d_g	d_h
w_1	1	1	0	0	0	1	1
w_2	0	1	1	1	0	0	1
w_3	0	0	0	1	1	1	1

Document collection

d_a: $w_1\ w_1\ w_1$; d_b: $w_2\ w_1\ w_2\ w_2\ w_1$

d_c: $w_2\ w_2$; d_e: $w_3\ w_3\ w_3\ w_2\ w_3\ w_2$

d_f: $w_3\ w_3\ w_3$; d_g: $w_3\ w_1\ w_1$; d_h: $w_1\ w_3\ w_1\ w_2\ w_2$

Fig. 9.1. Illustrative example of the document space for a sample document collection

As seen from the figure, each column of the term-document matrix constitutes a binary vector representing one document in the collection. Each of these vectors has three components, which correspond to the three vocabulary terms of the collection, and it "lives" in a three-dimensional space in which the vocabulary terms constitute an orthogonal basis. Several important observations can be derived from the simple example depicted in Figure 9.1:

- Vectors that represent documents containing only one vocabulary word are collinear to the corresponding vocabulary word axis.

- Vectors that represent documents that don't contain a specific vocabulary word are orthogonal (perpendicular) to the corresponding vocabulary word axis.
- Two different documents have exactly the same vector representation in the document space if they just contain the same words in different order. (Indeed, the vector space representation constitutes a bag-of-words model, in the sense that it is completely blind to word order.)
- Two different documents have exactly the same vector representation in the document space if they are composed of the same set of vocabulary words. (Indeed, any conceivable document constructed with any combination of words from the vocabulary set $\{w_1, w_2, w_3\}$ will have one of the seven vector representations depicted in Figure 9.1.)

The first of the above observations is trivial and it might seem to be useless in practice, as it is very unlikely to find documents composed of one single vocabulary word. However, it provides a means for representing single vocabulary words into the document-space allowing for exploiting geometrical relationships between documents and individual words. The second observation happens to be very useful, and it will be exploited in some of the applications presented in future chapters. Actually, this property has been already used in (9.17) for finding pairs of documents with no vocabulary words in common. The last two observations can be thought of as severe limitations of the vector space model, as they clearly imply that this kind of models are not able to distinguish among certain sets of different documents.

In the particular case of the third observation, which is a natural consequence of the vector space model definition, word order insensitivity has been proven not to represent a serious limitation for modeling text contents in certain applications such as, for instance, information retrieval and document categorization, among others. This is mainly because the basic semantic relationships that can be inferred from word occurrences alone (independently of word order) are good enough to provide a solid framework for text data processing and analysis.

On the other hand, the fourth and last observation, which is a direct consequence of the binary nature of the term-document matrix, constitutes a very important limitation to the method. However, such limitation can be overcome by extending the proposed binary model into a real-valued vector space model. In the rest of this section, we will discuss in detail three basic weighing schemes which help to significantly improve the performance of the vector space model by assigning specific values to each non-zero element in the term-document matrix. These three schemes are: term-frequency weighting, inverse-document-frequency weighting, and vector length normalization.

The combined use of term-frequency and inverse document frequency is commonly referred to as TF-IDF weighting. Before describing how to apply such a weighting scheme to a given data collection, let us first motivate its use.

Consider for a moment a generic and basic procedure in which we are interest in assessing the semantic similarity between two documents in a collection, (indeed, such a basic procedure constitutes the core essence of many different text mining and natural language processing applications). As we already suggested it in section 9.1, with the definition of the word overlap matrix, this problem somehow reduces to accounting for common vocabulary words. Alternatively, within the vector space framework, this problem reduces to the geometrical process of computing distances among vectors in a multidimensional space, for which vocabulary words constitute an orthogonal basis.

In any case, the subset of vocabulary words that is common to a given pair of documents is of fundamental importance to determine their similarity. However, as we already saw in chapter 7, not all words are equally important within a given document or document collection, as well as their frequency distribution is not uniform at all. In this sense, the main idea behind a weighting scheme is to pay more attention to those words that are presumably more important for assessing document similarities while, simultaneously, pay less attention to those words that are not relevant for assessing document similarities.

A first intuition would tell us that the more times a certain vocabulary word occurs within a given document, the more important it should be for characterizing such a document, and this is what the term-frequency weight intends to assess. However, as we already know, the most common vocabulary words are function words, which actually convey very little or no semantic information at all. Also, as they occur in basically any document, using them alone as indicators of document similarity does not seem to be wise. In order to compensate for the negative effects induced by very frequent vocabulary words, we should look at the total number of documents in which a given vocabulary word occurs. In this case, the more documents a given vocabulary word occurs in, the less important such a word is for characterizing a document containing it, and this is what the inverse-document-frequency weight intends to assess.

The combination of both term-frequency and inverse-document-frequency weighting schemes allows for focusing the vector model's attention into those words that are common within documents but rare across documents while resting importance to those words that are either common within and across documents (very common words that are not informative) or rare within and across documents (very rare words that are informative, but do not allow to infer relationships among different documents).

To illustrate the combined effect of both the term-frequency and the inverse-document-frequency weighting schemes over the raw vocabulary counts in our sample data collection, let us consider the logarithm of the word frequencies over the complete data collection:

```
>> logfreq = log(ranked_frequencies);
```
                                                                                        (9.18)

as well as the logarithm of the inverse document frequencies:

```
>> docfreq = full(sum(tdmtx,2))';                                        (9.19a)

>> logdocs = log(size(tdmtx,2)./(docfreq+1));                            (9.19b)
```

As seen from (9.19a), the document frequency corresponding to each vocabulary word is computed by adding up the elements of the term-document matrix
along the corresponding row. Then, in (9.19b), the inverse frequencies are obtained by considering the ratio between a constant value (the total number of documents in this case) and the obtained document frequencies. Again, as in the case
of the word frequencies, the logarithm of the computed ratio is considered.

Now, let us consider the product between the term frequencies computed in
(9.18) and the inverse document frequencies computed in (9.19):

```
>> logratio = (logfreq+1).*(logdocs);                                    (9.20)
```

We added one to the word log-frequency values to avoid the overall log-ratio to
collapse to zero in those cases of words occurring only once in the data collection.
Notice that we have called the resulting quantity `logratio` as it actually represents
the ratio between the logarithm of the word frequencies and the logarithm of the
document frequencies.

For obtaining a nicer plot, we can smooth the curves by taking average values
over a sliding window of *10* vocabulary words:

```
>> wsize = 10; % defines a smoothing window of size 10                   (9.21a)

>> for k=1:length(logratio)-wsize % smoothes the curves                  (9.21b)
        sdocs(k)=mean(logdocs(k:k+wsize));
        sfreq(k)=mean(logfreq(k:k+wsize));
        sratio(k)=mean(logratio(k:k+wsize));
   end
```

Finally, we can plot the three curves corresponding to the logarithmic values
for word frequencies, inverse document frequencies and their product:

```
>> hf = figure(2); % creates a new figure                                (9.22a)
>> set(hf,'Color',[1 1 1],'Name','Word and Document Frequencies');

>> x = 1:length(logratio)-wsize; % defines a common index                (9.22b)

>> % plots the word frequency curve                                      (9.22c)
>> subplot(3,1,1); semilogx(x,sfreq,'.');
>> ax = axis; axis([0,12000,ax(3:4)]); ylabel('TF (term-freq)');

>> % plots the document frequency curve                                  (9.22d)
>> subplot(3,1,2); semilogx(x,sdocs,'.');
>> ax = axis; axis([0,12000,ax(3:4)]); ylabel('IDF (inv-doc-freq)');
```

```
>> % plots the ratio of both curves                                    (9.22e)
>> subplot(3,1,3); semilogx(x,sratio,'.');
>> ax = axis; axis([0,12000,ax(3:4)]); ylabel('Combined Effect');
>> xlabel('Vocabulary Word Rank')
```

The resulting plots are presented in Figure 9.2. As seen from the top and middle panels of the figure, the term-frequency and inverse-document-frequency curves exhibit a similar behavior but in opposite directions. In the first case, according to Zipf's law (as was seen in section 7.1), word frequencies decrease as word rank increases. In the second case, as the rank increases, inverse document frequencies increase, as the number of documents containing a given word is expected to be higher for frequent words and lower for rare words. An interesting compensation of both effects occurs when we multiply the word frequencies with their corresponding inverse document frequencies, as it is presented in the lower panel of Figure 9.2. Notice how, in this case, the resulting word weights clearly exhibits higher values for the vocabulary words in the central portion of the rank scale than in the lower and higher extremes.

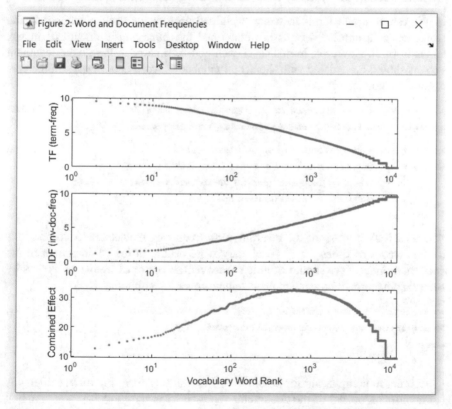

Fig. 9.2. Curves (computed over the whole document collection) corresponding to word frequencies, inverse document frequencies and the product of both

This particular way of combining term frequencies and inverse document frequencies is of fundamental importance to the vector space model implementation. As it can be seen from Figure 9.2, the term frequency to document frequency ratio does not only compensate the effect of the unbalanced frequency distribution characterized by the Zipf's law, but it also helps the model to focus its attention on those vocabulary terms that are in the middle of the rank scale, which constitute the most discriminative content words in the data collection.

Now that we have empirically motivated the inclusion of weighting mechanisms into the vector model framework, we can proceed to define the actual termfrequency weighting scheme. Differently from what we have just done in (9.18) by considering the vocabulary counts over the whole data collection, the termfrequency weighting is computed and applied locally, at the document level. Additionally, as the total number of running words occurring in a single document is significantly smaller than the number of running words in the entire collection, there is no need for considering logarithmic counts as we did in (9.18). In this sense, raw counts are typically used for computing the term-frequency weights.

We can compute term-frequency weighted vectors for each document in the collection in a similar way as we computed the binary term-document matrix **tdmtx** in (9.2). However, in this case, each non-zero entry in the given document vectors should be updated with the actual number of word occurrences in the corresponding document:

```
>> % creates a sparse matrix of appropriate dimensions                    (9.23a)
>> tfmtx = sparse(length(ranked_vocabulary),length(verses));

>> % updates term-frequencies for each document                           (9.23b)
>> for k = 1:length(verses) % this will take a while
        [void,index] = intersect(ranked_vocabulary,verses(k).vocab);
        tfmtx(index,k) = verses(k).count;
   end
```

where each element n_{ij} of the resulting term-frequency matrix **tfmtx** corresponds to the number of times the ith vocabulary word occurs in the jth document of the collection. Indeed, we should be able to recover the vector of frequency counts by summing the term-frequency matrix elements over the columns:

```
>> % compares vector sum(tfmtx,2) with ranked_frequencies                 (9.24)
>> sum(abs(ranked_frequencies-full(sum(tfmtx,2))'))

ans =

     0
```

In order to compensate for possible effects resulting from the different lengths (in number of words) of the documents represented in the model, it is a common practice to normalize the term frequencies of each document vector. One simple way of doing this is by dividing each of the term-frequencies weights by the total number of words occurring in the corresponding document. This normalized version of the term-frequency weighting can be computed as follows:

```
>> for k = 1:length(verses)                                          (9.25)
       ntfmtx(:,k) = tfmtx(:,k)/sum(tfmtx(:,k));
   end
```

Different from term frequencies, which are calculated at the document level, inverse document frequencies are computed at the collection level. For each term in the vocabulary, its inverse document frequency is basically defined as the ratio between the total number of documents in the collection and the number of documents containing the term. In practice, as document collections are commonly much larger than the sets of documents containing a typical relevant term, the logarithm is considered instead of the simple ratio.

Inverse document frequencies can be derived from the term-frequency matrix computed in (9.23) as follows:

```
>> % computes inverse document frequencies                           (9.26)
>> idfvt = log(size(tfmtx,2)./full(sum(tfmtx>0,2)));
```

As seen from (9.26), the binary term-document matrix `tdmtx` is recovered from the term-frequency matrix computed in (9.23) by using the operation `tfmtx>0`. Afterwards, it is summed up over its columns in order to obtain the total number of documents each term occurs in. Notice that the resulting document count variable is a vector, not a matrix! Then, the logarithm is computed for the ratio between the total number of columns in the term-frequency matrix (which is the number of documents in the collection) and the aforementioned vector.

Finally, the corresponding TF-IDF weight is computed for each document term by multiplying either its raw or normalized term frequency with its corresponding inverse document frequency. For instance, consider the calculation of the non-normalized TF-IDF matrix for the data collection under consideration:

```
>> % creates a sparse matrix of appropriate dimensions               (9.27a)
>> tfidfmtx = sparse(length(ranked_vocabulary),length(verses));

>> for k = 1:length(verses) % computes the TF-IDF matrix             (9.27b)
       tfidfmtx(:,k) = tfmtx (:,k).*idfvt;
   end
```

To illustrate the effects of TF-IDF normalization over different vocabulary terms, let us consider the verse *2971*:

```
>> disp(verses(2971).text)                                           (9.28)
So the priest gave him hallowed bread: for there was no bread there but the
shewbread, that was taken from before the LORD, to put hot bread in the day
when it was taken away.
```

and the corresponding TF-IDF weights assigned to the very common word *was*, the medium rank word *bread*, and the rare word *hallowed*, occurring in it:

```
>> word_rank = find(ranked_vocabulary=='was')                              (9.29a)
word_rank =
   26
>> word_tfidf = full(tfidfmtx(word_rank,2971))
word_tfidf =
   6.4450

>> word_rank = find(ranked_vocabulary=='bread')                            (9.29b)
word_rank =
   256
>> word_tfidf = full(tfidfmtx(word_rank,2971))
word_tfidf =
   13.6379

>> word_rank=find(ranked_vocabulary=='hallowed')                           (9.29c)
word_rank =
        2122
>> word_tfidf = full(tfidfmtx(word_rank,2971))
word_tfidf =
    7.3005
```

The example presented in (9.29) provides a good insight on how the TF-IDF weighting scheme works. Consider for instance the two words *was* and *bread*, each of which appears three times in verse *2971*. Although they both occur the same number of times in the document under consideration, *bread* is actually a more relevant word for characterizing this document as it is less common in the overall collection than *was*. Indeed, notice how the resulting TF-IDF weight for *bread* is more than twice the weight for *was*.

Similarly, the word *hallowed*, which is a rare word, is not given too much importance by the model as it only occurs once in the considered document. As a result, the relative importance assigned by the model to the words *was* and *hallowed* for the given document is somehow similar, but definitely much less than the importance assigned to the word *bread*.

In order to have a broader picture of the TF-IDF weights in verse *2971*, let us list, for each word contained in it, the corresponding rank, frequency (global), term frequency (TF), inverse document frequency (IDF) and TF-IDF values.

```
>> % get the indexes of the words in verse 2971                            (9.30a)
>> [void,index] = intersect(ranked_vocabulary,verses(2971).vocab);
>> [void,order] = sort(index); index = index(order);

>> % prints out rank, freq, tf, idf and tf-idf for each word               (9.30b)
>> word_weights = table;
```

```
>> word_weights.Rank = index;
>> word_weights.Freq = ranked_frequencies(index)';
>> word_weights.TF = full(tfmtx(index,2971));
>> word_weights.IDF = full(idfvt(index));
>> word_weights.('TF-IDF') = full(tfidfmtx(index,2971));
>> word_weights.Word = ranked_vocabulary(index)';
>> disp(word_weights)
```

Rank	Freq	TF	IDF	TF-IDF	Word
1	63919	4	0.25547	1.0219	"the"
4	13560	1	1.167	1.167	"to"
5	12915	1	1.1449	1.1449	"that"
6	12667	1	1.1838	1.1838	"in"
10	8971	1	1.4774	1.4774	"for"
14	7964	1	1.5281	1.5281	"lord"
18	6661	1	1.8309	1.8309	"him"
21	6129	1	1.8861	1.8861	"it"
26	4521	3	2.1483	6.445	"was"
32	3993	1	2.1155	2.1155	"but"
38	3642	1	2.3027	2.3027	"from"
41	2834	1	2.459	2.459	"when"
53	2299	2	2.7035	5.407	"there"
67	1796	1	2.939	2.939	"before"
71	1743	1	3.0029	3.0029	"day"
74	1689	1	2.9506	2.9506	"so"
88	1393	1	3.1627	3.1627	"no"
121	915	1	3.594	3.594	"away"
123	911	1	3.6128	3.6128	"put"
182	543	1	4.1754	4.1754	"priest"
204	465	1	4.2674	4.2674	"gave"
256	361	3	4.546	13.638	"bread"
269	338	2	4.5643	9.1286	"taken"
1724	31	1	6.9439	6.9439	"hot"
2122	22	1	7.3005	7.3005	"hallowed"
2464	18	1	7.4547	7.4547	"shewbread"

As seen from (9.30), IDF weights are dominating the overall tendency of the TF-IDF model. This is basically because most of the words occur only once in the verse. However, it is also evident that for those words occurring more than once, the TF weights provide an important boost to their TF-IDF values with respect to other words in the same rank range. An exception to this is the case of the word *the* that, although being the most frequent word in the verse, gets the lowest TF-IDF value due to its extremely low IDF score.

Notice also from (9.30) that the computed values of TF-IDF can exhibit a relatively large range of values, as well as their total sum can vary a lot depending on the length of the segment of text under consideration. A common practice to avoid such variability (or, at least, to reduce the possible effects resulting from it) is to normalize the TF-IDF scores for each document in the collection by using the Euclidean norm (also referred to as the *L2*-norm). This can be done as follows:

```
>> for k = 1:length(verses)                                          (9.31)
       ntfidfmtx(:,k) = tfidfmtx(:,k)/norm(tfidfmtx(:,k));
   end
```

where it can be easily verified that the resulting norm for any document vector representation is equal to one:

```
>> norm(ntfidfmtx(:,2971))                                           (9.32)
ans =

    1
```

A consequence of this type of normalization is that all document vectors are of length one in the Euclidean space in which they are contained. Moreover, as all their components are non-negative (equal or greater than zero), they all live in the same hyper-octant of the space and lay on the surface of the corresponding unitary hyper-sphere. This is particularly useful when we want to assess similarities and differences among documents by means of the angles among their vector representations. We will discuss this issue in more detail in the following section.

The Text Analytics Toolbox™ function `tfidf` provides an alternative way of computing a TF-IDF matrix for a given document collection. It is much more efficient from the computational point of view, and it also allows for using different variants of both the TF and IDF weighting schemes. More about `tfidf` and its available options is discussed in exercise 9.5-3, in section 9.5.

9.3 Association Scores and Distances

In this section we will present the association scores and distance metrics most commonly used within the vector space model framework. An association score is a measure of the degree of similarity between a given pair of vectors, while a distance is a measure of the degree of dissimilarity. In some cases, similarities can be converted into dissimilarities and vice versa by applying some simple transformation operations.

The most commonly used association scores are the Dice coefficient, the Jaccard coefficient and the cosine similarity score. In the case of the Dice coefficient, this association score computes a normalized count of all non-zero elements that are common to both vectors being compared. The used normalization factor is *2* divided by the sum of the total number of non-zero elements in each of the two vectors. This guarantees that the resulting score will be in the interval *[0,1]*. The

Dice coefficient reduces to zero when the two vectors being compared do not have any non-zero element in common, while it is equal to one when all non-zero elements are common to the two vectors. It can be defined in mathematical terms as follows:

$$dice(v_1, v_2) = 2\, n(N_1 \cap N_2) \,/\, (\, n(N_1) + n(N_2)\,) \tag{9.33}$$

where v_1 and v_2 are the two vector representations being compared (which can be either document vectors or word vectors), N_1 and N_2 represent the subsets of non-zero elements in v_1 and v_2, respectively, and $n(\,\cdot\,)$ denotes cardinality.

The Jaccard coefficient is similar to the Dice coefficient but it uses a different normalization factor. In this case, the normalization factor is 1 divided by the total number of non-zero elements in both vectors being compared. Similar to the previous case, the resulting score will be in the interval $[0,1]$, being the coefficient equal to zero when the two vectors do not have any non-zero element in common and being equal to one when all the non-zero elements are common to the two vectors. However, in this case, lower score values will result for pairs of vectors exhibiting low overlap of non-zero components. The Jaccard coefficient can be defined in mathematical terms as follows:

$$jaccard(v_1, v_2) = \ n(N_1 \cap N_2) \,/\, n(N_1 \cup N_2) \tag{9.34}$$

where, again, v_1 and v_2 are the two vectors being compared, N_1 and N_2 are the subsets of non-zero elements in v_1 and v_2, and $n(\,\cdot\,)$ denotes cardinality.

The cosine similarity score, on the other hand, is defined as the cosine of the angle between the two vectors being compared. According to this definition, the cosine scores are bounded to the interval $[-1,1]$, which extreme values correspond to angles of $180°$ and $0°$, respectively. However, as already discussed, vector components are always equal or greater than zero, which makes the maximum possible angle value between two vectors $90°$. So, in the practice, cosine similarity scores are expected to be in the interval $[0,1]$ too. The cosine similarity score is mathematically defined in terms of the inner product between normalized versions of the vectors being compared:

$$cosine(v_1, v_2) = \ <v_1, v_2> \,/\, (\|v_1\|\|v_2\|) \tag{9.35}$$

where $<\cdot,\cdot>$ denotes the scalar, or inner, product between vectors and $|\cdot|$ denotes the Euclidean norm. Notice that when using a normalized TF-IDF model, just as the one computed in (9.31), the vectors will have unitary norm. In such a case, the cosine similarity score reduces to the inner product between the vectors:

$$cosine(u_1, u_2) = \ <u_1, u_2> \tag{9.36}$$

Notice that, while the first two scores (Dice and Jaccard) will provide exactly the same values regardless we are using either the binary- or the real-valued version of a vector model, the cosine similarity score is by definition a real-valued vector score. In consequence, different score values will be obtained for the same data depending on whether we are using a binary- or a real-valued model.

To illustrate the appropriateness of association scores for assessing the degree of similarities between vectors, let us consider the following three verses from our data collection:

```
>> x1 = 392; x2 = 2419; x3 = 30063;                                          (9.37a)
```

```
>> disp(verses(x1).text)                                                     (9.37b)
Some of them also were appointed to oversee the vessels, and all the instru-
ments of the sanctuary, and the fine flour, and the wine, and the oil, and the
frankincense, and the spices.
```

```
>> disp(verses(x2).text)                                                     (9.37c)
And Eli said unto her, How long wilt thou be drunken? put away thy wine from
thee.
```

```
>> disp(verses(x3).text)                                                     (9.37d)
And cinnamon, and odours, and ointments, and frankincense, and wine, and oil,
and fine flour, and wheat, and beasts, and sheep, and horses, and chariots,
and slaves, and souls of men.
```

As can be seen from the examples presented in (9.37), although all three verses are different, there is clearly a higher degree of similarity between the first example (9.37b) and the third example (9.37d), than between any of them and the second one (9.37c). Next, let us compute the three association scores described above for each of the three possible pair-wise combinations of verses in (9.37):

```
>> x = [x1,x2,x3,x1];                                                        (9.38a)
>> for k=1:3
       % gets the pair of vectors to be compared
       va = full(tfidfmtx(:,x(k)));
       vb = full(tfidfmtx(:,x(k+1)));
       % computes the association scores for the given vector pair
       dice(k) = 2*sum((va>0)&(vb>0))/(sum(va>0)+sum(vb>0));
       jaccard(k) = sum((va>0)&(vb>0))/sum((va>0)|(vb>0));
       cosine(k) = (va'*vb)/norm(va)/norm(vb);
   end
```

```
>> [dice;jaccard;cosine]                                                     (9.38b)
ans =
     0.1081      0.1143      0.3684
     0.0571      0.0606      0.2258
     0.0701      0.0597      0.3030
```

Notice that each row of values in (9.38b) represents each of the considered association scores: Dice, Jaccard, and cosine; and each column represents one of the three pair-wise comparisons: (x1,x2), (x2,x3), and (x3,x1). As observed from the results, all three metrics agree on assigning a higher degree of similarity to the latter verse pair (x3,x1), i.e. the last column in (9.38b). On the other hand, much lower degrees of similarity are assigned to the other two verse pairs, (x1,x2) and (x2,x3), i.e. the first and second columns in (9.38b).

Let us now consider the distance metrics. Differently from association scores, which accounts for similarity, distances account for the opposite concept: dissimilarity. Generally, similarity and dissimilarity can be derived from each other by means of simple mapping operations such as $d = 1 - s$ or $d = 1/s$, where d stands for dissimilarity and s for similarity. According to this, for instance, we can define the "cosine distance" in terms of the previously defined cosine similarity score as follows: $cosine_{dist}(v_1, v_2) = 1 - cosine(v_1, v_2)$.

The most commonly used distance functions (apart from the cosine distance just defined above) are the Hamming distance, the Euclidean distance, and the city block distance. This latter one is also known as the Manhattan distance.

The first of these three distances is defined for binary-valued vectors only, and it is computed as the total number of different elements between both vectors being compared. We can actually think about the Hamming distance as a binary-vector implementation of the Levenshtein distance previously discussed in section 4.1. In this case, as both vectors being compared must have the same number of elements, neither insertion nor delete operations are needed; so, only replacement operations are required. Additionally, as binary-valued vectors are considered, the replacement operation reduces to a "toggle" operation between the two values 1 and 0. However, notice that the Hamming distance can be also computed for real-valued vectors just by considering any non-zero element to be "1". Then, the Hamming distance can be defined as follows:

$$hamming(v_1, v_2) = n(N_1 \cap Z_2) + n(N_2 \cap Z_1) \qquad (9.39)$$

where v_1 and v_2 are two the vectors being compared, N_1 and N_2 represent the subsets of non-zero elements in v_1 and v_2, Z_1 and Z_2 represent the subsets of zero elements in v_1 and v_2, and $n(\cdot)$ denotes cardinality.

The Euclidean distance between two points is simple defined as the length of the line segment connecting the two points. This is probably the most intuitive and general notion of distance used in everyday's life.[2] If we state this definition in terms of vectors, the Euclidean distance between two vectors corresponds to the Euclidean norm of the difference vector:

$$euclidean(v_1, v_2) = \| v_1 - v_2 \| = \left[\sum_i (v_1(i) - v_2(i))^2 \right]^{1/2} \qquad (9.40)$$

---

[2] It was proposed by the Greek mathematician Euclid of Alexandria in his fundamental treaty on geometry "The Elements", about 23 centuries ago! The definition of the Euclidean distance is based on the Pythagoras' theorem.

where v_1 and v_2 are the two vectors being compared, and $\| \cdot \|$ denotes the Euclidean norm, which is defined as the square root of the sum of the squared differences between the individual vector elements.

Differently from the Euclidean distance, the city block distance or Manhattan distance is computed by summing the absolute differences between the corresponding vector elements. In other words, the city block distance between two vectors corresponds to the $L1$-norm of the difference vector:

$$cityblock(v_1,v_2) = \| v_1 - v_2 \|_1 = \sum_i | v_1(i) - v_2(i) | \qquad (9.41)$$

where v_1 and v_2 are the two vectors being compared, $\| \cdot \|_1$ denotes the $L1$-norm, and $| \cdot |$ is the absolute value operator.

Notice from (9.39), (9.40) and (9.41) that distances do not include normalization factors in their definitions. Indeed, in the three specific definitions provided here, the distance between two vectors will always depend on the lengths of the two vectors involved in its computation. Nevertheless, if we are interested in reducing the effect of vector lengths in distance computations, we may consider any of the following two alternatives: the use of normalized vector models, as the one computed in (9.31), or the definition of an appropriate normalization factor for the specific distance function under consideration.

Let us now illustrate the computation of distance values with the same three verse examples from (9.37). In this case we will consider the normalized TF-IDF model computed in (9.31):

```
>> for k=1:3                                                        (9.42a)
       va = full(ntfidfmtx(:,x(k)));
       vb = full(ntfidfmtx(:,x(k+1)));
       hamming(k)  = sum(not((va>0)==(vb>0)));
       euclidean(k) = norm(va-vb);
       cityblock(k) = sum(abs(va-vb));
   end

>> [hamming;euclidean;cityblock]                                    (9.42b)
ans =
    33.0000   31.0000   24.0000
     1.3638    1.3714    1.1806
     7.1208    7.2139    5.5644
```

Notice now that the most similar verse pair `(x3,x1)`, represented by the third column in (9.42b), is the one exhibiting the shortest distance values for all three considered distance functions. Also notice the big differences in relative magnitudes among the three considered distance functions, i.e. the rows in (9.42b).

In general, the use of association scores and distance functions in the context of vector space model representations is especially useful for quantifying basic notions such of similarity and ordering. These scores are commonly used in practice for quantifying the degree of similarities and dissimilarities among contents in a collection of documents, making it possible to generate rankings and ordered lists of the contents. This is actually the basic operational strategy behind commonly used search engines, as well as text classification systems.

Let us consider, for instance, the problem of computing vector similarities between a given verse and all other verses in our data collection. If we consider the normalized TF-IDF model computed in (9.31), cosine similarity scores between a document, let us say verse *392*, and the rest of the collection can be easily obtained as follows:

```
>> cosine_scores = ntfidfmtx(:,392)'*ntfidfmtx;                    (9.43)
```

We can now plot a histogram of the resulting scores to see the distribution of similarities, with respect to verse *392*, across the whole collection:

```
>> hf = figure(3); % creates a new figure                         (9.44a)
>> set(hf,'Color',[1 1 1]);
>> set(hf,'Name','Distribution of similarities with respect to verse 392');

>> % computes a histogram of similarities (20 beams)              (9.44b)
>> [freq,vals] = hist(cosine_scores,20);

>> % generates a bar plot considered log frequencies              (9.44c)
>> hb = bar(vals,log(freq)+1);

>> % adds the figure labels                                       (9.44d)
>> xlabel('Cosine Similarity Scores');
>> ylabel('Log-frequencies + 1');
```

The resulting histogram is presented in Figure 9.3. Notice from the figure that the overall range of similarities *[0,1]* has been divided in *20* beams. As the lowest similarity beam contains a huge number of verses:

```
>> freq(1)                                                        (9.45)
ans =
      29931
```

we have plotted the logarithm of the frequencies instead of the frequencies themselves, see (9.44c). We have also added *1* to the log-frequency value in order to be able to identify in the plot the only sample with similarity *1* (maximum similarity), which corresponds to the same verse *392* when it is compared to itself. As observed from the figure, all verses in the data collection exhibit cosine similarity scores below *0.35*. Moreover, as seen in (9.45), *29,931* verses have similarities below *0.05*. If we consider that our data collection has a total of *31,103* verses, this means that *96%* of the verses in the collection have cosine similarity scores below *0.05* with respect to our probe verse *392*.

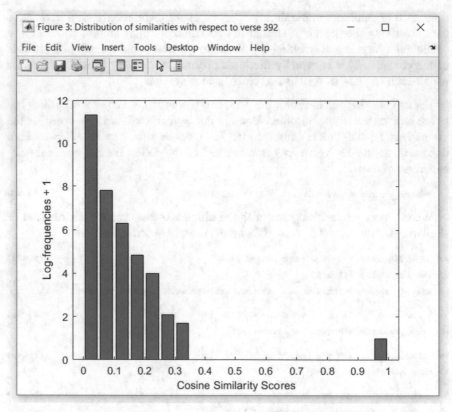

Fig. 9.3. Distribution of cosine-based similarities with respect to verse 392 for all verses in the data collection (bar amplitudes correspond to log-frequency values)

If we recompute (9.43) and (9.44) for every single verse in the data collection, we will find that the resulting distributions of similarities will follow, in general, the same trend observed in Figure 9.3. If we think about this highly skewed distribution of similarities in terms of distances, the conclusion is that in our vector space model representation every single verse is very far away from the rest of the verses in the collection. This phenomenon is typical of high-dimensional spaces (be reminded that the dimensionality of the vector space under consideration is equal to the vocabulary size of the data collection, i.e. *12,545* dimensions). This phenomenon is generally referred to as the "curse of dimensionality" and it is a geometrical manifestation of the sparse nature of the dataset.

The problem of data sparseness is tackled in practice by means of dimensionality reduction techniques. These methods attempt to reduce data sparseness by projecting the model space into a space of a lower dimension with the hope that distance or similarity distributions among the data samples (such as the one presented in Figure 9.3) become more uniform, i.e. data samples get closer to each other. The next chapter is fully devoted to dimensionality reduction techniques.

9.4 Further Reading

Geometry is one of the oldest fields of mathematics and, interestingly, it still continues to be one of the most intuitive and generic frameworks for approaching and modeling a great variety of problems in different areas of knowledge in modern science. Indeed, the vector space framework presented in this chapter for modeling text is totally a geometrical one. For a wonderful introduction to geometry as a natural framework for modeling language and meaning at their different abstraction levels (Widdows 2004) is strongly recommended.

Although the use of binary-valued vector models, which are generally referred to as feature spaces, has a long tradition in Artificial Intelligence and Cognitive Sciences, the introduction of weighted vector space models in Natural Language Processing firstly occurred within the context of Information Retrieval applications. The term-frequency and the inverse-document-frequency weighting factors, as we know them today, are the result of many years of empirical work and contributions from several authors during the 1960s and 1970s, but mainly from G. Salton and K. Spärck Jones (Salton 1971), (Spärck Jones 1972), and (Salton et al. 1975). More recently, it has been shown that the empirically observed "optimality" of the TF-IDF weighting scheme can be theoretically supported on statistical basis (Roelleke and Wang 2008).

The term-document matrix construct described in this chapter is not the only useful artifact for building vector space model representations of text contents. Here, we have focused our attention into looking at the relationships among different documents in a collection (the columns of the term-document matrix). Alternative constructs exist, such as the word-context matrix and the pair-pattern matrix, that are useful for studying other different aspects of text contents, such as the similarities among words or the similarities among relationships between pairs of words (Turner and Pantel 2010).

For a more complete overview on association scores and their use in automatic document and keyword clustering and classification, you can refer to chapter 3 of (Rijsbergen 1979). For the specific use of association scores in problems related to lexical acquisition, you can refer to chapter 9 of (Manning and Schütze 1999).

9.5 Proposed Exercises

1. Consider the binary-valued term-document matrix `tdmtx` constructed in (9.2), and an arbitrary given list of vocabulary terms `term_list`.

- Create the function `indexes = binary_search(tdmtx,term_list)` which returns an array containing the indexes of all documents containing one or more of the vocabulary terms in the list.
- Modify the previous function such that it also returns the number of different vocabulary terms in the list that are contained in each of the documents identified: `[indexes,nterms] = binary_search(tdmtx,term_list)`

2. Consider the function `BagOfWords` provided in the Text Analytics Toolbox™:

- Reproduce the results presented in (9.7) and explore the differences in vocabulary with respect to (9.2). Can you isolate the specific causes that explain the observed larger vocabulary in (9.7)?
- Study some of the methods available for `BagOfWords`, such as `topkwords`, `removeWords`, `removeInfrequentWords`, and `removeDocument`. Apply some of these operations to your previously computed representation.
- Study the function `BagOfNgrams`, which similar to `BagOfWords` computes frequency counts but for n-grams of words instead of individual words. Generate bag-of-n-gram representations with $n=2$ and $n=3$ for the same data collection used in (9.7).

3. Consider example (9.30), where the corresponding TF, IDF and TF-IDF values were listed for each of the vocabulary terms in a given verse. In that example, the TF-IDF weights were derived from `tfmtx`, the non-normalized version of TF computed in (9.23).

- Compute a new set of values NTF-IDF by using `ntfmtx`, the normalized version of TF computed in (9.25).
- Apply the *L2*-norm based normalization described in (9.31) to both TF-IDF and NTF-IDF weights.
- Similar to (9.30), list the corresponding values of TF, NTF, IDF, TF-IDF, NTF-IDF, L2-TF-IDF and L2-NTF-IDF for each vocabulary term in a given verse. Explore the differences among the different sets of weights.
- Consider the function `tfidf` provided in the Text Analytics Toolbox™. Study the different name-value pair arguments that allow for using different TF, IDF and normalization methods.
- Compute weights for vocabulary terms in your previously considered verse by using `tfidf` with different combinations of arguments. Compare these new results with your previous results.

4. Let us revisit example (9.38), where pair-wise comparisons were conducted among three different verses by using the three discussed association scores: Dice, Jaccard and Cosine. In (9.38), TF-IDF weights were used.

- Repeat example (9.38) but use the binary-valued term-document matrix `tdmtx` constructed in (9.2) instead.

– What are the main differences between the new results and those presented in example (9.38)? Could you explain these results?

5. Consider the rows of the real-valued term-document matrix **tfidfmtx** constructed in (9.27). This set of vectors constitutes a vector space model for the words in the vocabulary.

– Use the *L2*-norm based normalization, as described in (9.31), for normalizing these row vectors.
– Use the cosine similarity implementation described in (9.36) for identifying the five most similar vocabulary terms to each of the following ones: **'bread'**, **'water'**, **'desert'**, **'cattle'** and **'wine'**.
– Repeat the same experiment by using the binary-valued matrix **tdmtx** constructed in (9.2) instead.
– What differences and/or similarities do you observe?

6. Study the MATLAB® function **pdist**, which allows for computing different types of distances metrics among vectors.

– Consider specifically the following distance types: **'jaccard'**, **'cosine'**, **'hamming'**, **'euclidean'** and **'cityblock'**.
– For each case, find the corresponding mapping function (if any) between the values returned by **pdist** and the ones resulting from our definitions in (9.34), (9.35), (9.39), (9.40) and (9.41).

7. Search for the definition of the Chebyshev distance, which is also referred to as the *L∞*-norm metric.

– Write a script implementing this distance function.
– Repeat example (9.42) and include the Chebyshev distance computation.
– What differences and/or similarities do you observe with respect to the previously studied distances?

8. Consider that all vectors in a given vector space model representation have been normalized with respect to their *L2*-norm, by following the procedure described in (9.31).

– Propose a normalization factor for each of the studied distance functions, such that any distance measurement between two vectors in the model are guaranteed to be in the interval *[0,1]*.
– Repeat exercise (9.42), apply your proposed normalization factors, and obtain association scores by applying the mapping: *score = 1 − distance*.
– Compare the obtained results with those reported in example (9.38). What are your observations?

9. Consider Figure 9.3, which illustrates the results from example (9.44).

- Repeat these computations and generate the corresponding figures, for several randomly selected verses of the data collection.
- What are the main differences and/or similarities among the different generated figures?

9.6 Short Projects

1. Create the function `getvectormodel` for constructing a vector space model representation for a given collection of documents. The function should comply with the following specifications:

   - `[mtx,params] = getvectormodel(dataset,vocab,type,idfvt)`
   - `dataset`: input variable containing the data collection for which the vector model is to be constructed. It must be a structure array similar to `verses`, the structure presented in (9.1), containing at least the following three fields for each document in the collection: `vocab` (a string array containing the document's vocabulary terms), `count` (a numeric array containing the term frequencies within the document) and `text` (a string containing the raw text of the document).
   - `vocab`: input variable containing the specifications of the overall collection vocabulary to be used for constructing the model. It can be either a string array of words or an empty string. In the first case, it will contain the specific vocabulary words to be used for constructing the model. In the case of an empty string, the vocabulary should be extracted directly from `dataset`.
   - `type`: input variable containing the specifications of the weighting schemes to be applied. It should provide options for using binary-valued entries, TF, NTF (normalized TF), IDF, and *L2*-norm based normalization, as well as any valid combination of them. Select the variable type of your preference for your implementation.
   - `idfvt`: input variable containing the IDF vector to be used for IDF weighting. It only makes sense in the case the vocabulary is provided. If an empty vector is given, the IDF vector should be computed from `dataset`.
   - `mtx`: output variable returning the corresponding weighted term-document matrix for the given data collection (as many rows as vocabulary terms and as many columns as documents). It should be a sparse matrix.
   - `params`: output variable returning the main parameters used for generating the vector model representation. It must be a structure containing at least the following fields: `vocab` (a cell array containing the vocabulary terms used in the construction of the vector space model, in the same order as vocabulary dimensions or rows appear in the model), `idfvt` (the vector of IDF weights for the given data collection) and `type` (the specifications of

the applied weighting schemes). Notice that this output variable is of fundamental importance for being able to compute compatible vector representations for new documents that were not present in the data collection from which the original parameters and the original vector space representation were computed.

2. Consider the book "Oliver Twist" by Charles Dickens, which is available in digital form from the Project Gutenberg website https://www.gutenberg.org/files/730/730-0.txt Accessed 6 July 2021.

- Download the plain text version of the book from the website.
- Eliminate the front matter and back matter from the text file. You should conduct this step manually.
- Segment the complete book into sentences as best as you can. (You might consider using function splitSentences for this).
- Create a string array for the dataset containing one sentence per element.
- Tokenize and lowercase the complete sentence collection. Compute a TF-IDF matrix representation for the collection.
- Consider the rows of the TF-IDF matrix and use some of the studied association scores to find word-pairs with similar vector representations.
- Consider the columns of the TF-IDF matrix and use some of the studied association scores to find sentence-pairs in the data collection with similar vector representations.

9.7 References

Manning CD, Schütze H (1999) Foundations of Statistical Natural Language Processing. The MIT Press, Cambridge, MA

Roelleke T, Wang J (2008) TF-IDF Uncovered: A Study of Theories and Probabilities. In Proceedings of the 31st Annual International ACM SIGIR Conference, pp.435-442

Salton G (ed) (1971) The SMART Retrieval System – Experiments in Automatic Document Retrieval. Prentice Hall Inc., Englewood Cliffs, NJ

Salton G, Wong A, Yang CS (1975) A vector space model for information retrieval. Communications of the ACM, 18(11):613-620

Spärck Jones K (1972) A statistical interpretation of term specificity and its application in retrieval. Journal of Documentation, 28(1):11-21

Turney PD, Pantel P (2010) From Frequency to Meaning: Vector Space Models of Semantics. Journal of Artificial Intelligence Research, 37(1):141-188

van Rijsbergen CJ (1979) Information Retrieval. Butterworths, London

Widdows D (2004) Geometry and Meaning. CSLI Publications, Center for the Study of Language and Information

10 Dimensionality Reduction

We closed the previous chapter by introducing the "curse of dimensionality", which refers to the sparseness problem that typically affects models involving a very large number of variables, i.e. high-dimensional spaces. This problem is alleviated in practice by using dimensionality reduction techniques, which aim at reducing the sparseness of the data representation by projecting the original model into a new space of lower dimensionality. There exist several different approaches to dimensionality reduction, which constitutes a very common practice in data mining applications. Indeed, almost every standard data mining method or procedure involves some sort of dimensionality reduction. In this chapter we focus our attention on methods for dimensionality reduction within the context of text mining applications. First, in section 10.1, vocabulary pruning and merging methods are presented. Then, in section 10.2, the linear transformation approach to dimensionality reduction is introduced and, in section 10.3, non-linear projection methods for dimensionality reduction are described. In section 10.4, a more detailed presentation of commonly used reduced space representations for language, which are commonly referred to as embeddings, is provided. Finally, some relevant references to the presented methods, as well as some other methods, are provided in the *Further Reading* section at the end of the chapter.

10.1 Vocabulary Pruning and Merging

A common practice in text mining applications is to remove some of the vocabulary terms from a given data collection before constructing a model representation for it. This procedure is referred to as vocabulary pruning, and it is generally applied to those vocabulary terms that are considered to be the least discriminative or relevant for the specific analysis under consideration. According to this, two different types of terms are typical candidates to be considered for pruning: the most frequent and the least frequent vocabulary terms.

In the case of the most frequent vocabulary terms, two different pruning criterions can be used: either a relative frequency threshold F (with $0<F<1$) is defined or, alternatively, an absolute number N of terms is defined. In the first case, all vocabulary words in the dataset with relative frequencies above the predefined threshold value F are removed from the corpus. In the second case, the most frequent N vocabulary terms are removed from the corpus. On the other hand, in the case of the least frequent vocabulary terms, a common practice is to remove all

© The Author(s), under exclusive license to Springer Nature Switzerland AG 2021
R. E. Banchs, *Text Mining with MATLAB®*, https://doi.org/10.1007/978-3-030-87695-1_10

unique terms in the data collection, i.e. those terms that only occur once in the entire data collection. Such terms are generally referred to as singletons.

Before continuing with the description of vocabulary pruning methods, let us load the data file **datasets_ch10a.mat**, which contains some relevant datasets and models from previous chapters:

```
>> clear;                                                                    (10.1)
>> load datasets_ch10a
```

Consider, for instance, the terms which have relative frequencies above *0.1* in the Bible data collection:

```
>> % gets the indexes of terms with relative frequencies above 0.1          (10.2a)
>> hf_terms = (ranked_frequencies/max(ranked_frequencies))>0.1;

>> ranked_vocabulary(hf_terms) % lists the terms                            (10.2b)
ans =
  1×20 string array
  Columns 1 through 8
    "the"    "and"    "of"    "to"    "that"    "in"    "he"    "shall"
  Columns 9 through 16
    "unto"    "for"    "i"    "his"    "a"    "lord"    "they"    "be"
  Columns 17 through 20
    "is"    "him"    "not"    "them"

>> % computes the percentage of running words they represent                (10.2c)
sum(ranked_frequencies(hf_terms))/sum(ranked_frequencies)*100
ans =
    38.1762
```

Notice from (10.2) that although they account for almost the *40%* of the running words in the entire data collection, they are only *20* vocabulary terms! So, for the effects of dimensionality reduction, eliminating these vocabulary terms will only discard *20/12545*100 = 0.16%* of the full space dimensions.

On the other hand, if we consider dimensionality reduction by means of eliminating the singletons in the collection, we will find the following situation:

```
>> % gets the indexes of all singletons                                     (10.3a)
>> lf_terms = ranked_frequencies==1;

>> sum(lf_terms) % compute the total amount of these terms                  (10.3b)
ans =
      3937

>> % computes the percentage of running words they represent               (10.3c)
>> sum(ranked_frequencies(lf_terms))/sum(ranked_frequencies)*100
ans =
```

        0.4975

Notice that, in this case, *3,937* singletons accounts for *0.50%* of the running words, which is actually a very small percentage of the text contained in the collection. However, it corresponds to *3937/12545*100 = 31.38%* of the vocabulary set! So, eliminating these vocabulary terms will definitively have a great impact on the model's dimensionality.

In practice, both high and low frequency vocabulary terms are commonly removed from the model, but implementations will vary depending on the specific applications under consideration. In the specific case of our previously computed TF-IDF model, we can remove these terms from the model by using the indexes computed in (10.2) and (10.3) as follows:

```
>> pruned_tfidfmtx = tfidfmtx(not(hf_terms|lf_terms),:);                    (10.4)
```

The two criterions of vocabulary pruning discussed so far can be regarded as statistically motivated methods. In both cases, word frequencies were used to determine whether a vocabulary term was to be removed or not. An alternative or complementary common practice for vocabulary pruning is the one based in the use of stopwords, in which a predefined set of vocabulary terms is removed for modeling purposes. This pruning criterion is more linguistically motivated as the corresponding lists of terms are manually handcrafted and adapted to the specific analysis or procedures under consideration.

Stopword lists are language dependent, but they generally include the same classes of function words such as:[1] determiners (*that, the, these*), pronouns (*I, you, she, her*), prepositions (*across, after, about*), auxiliary verbs (*am, are, should, would*), etc. Indeed, most of the words commonly included in stopword lists are precisely those vocabulary terms exhibiting the highest frequencies in data collections. If we consider, for instance, the *20* most frequent words in (10.2b) we will find that almost all of them are part of English standard stopword lists. Only one word (*lord*) for sure and probably another one (*not*), depending on the specific application under consideration, will not be found in an English stopword list.

In practice, more than an alternative procedure, the use of stopword lists constitutes a pruning method that complements frequency-based methods. However, as we have already mentioned, the use of specific pruning criterions will always depend on the specific application the computed model is expected to serve. For instance, in an automatic categorization scenario, removing singletons constitutes a very good strategy as it results in an important reduction of the model dimensionality, and words appearing once in the corpus (and, therefore, in a single docu-

---

[1] The Text Analytics Toolbox™ function **stopWords** returns a string array containing the *225* most commonly used English stopwords. Several stopword lists for many different languages can be also found online. For instance, for a general-purpose English stopword list, you can see http://www.textfixer.com/resources/common-english-words.txt Accessed 6 July 2021.

ment) will not be useful for establishing relationships among documents. On the other hand, in a document search scenario, removing singletons is not wise at all because these terms are actually the best pointers to the documents they appear in, which can be contained in user-defined queries.

Different form vocabulary pruning, where vocabulary terms are removed from the dataset for modeling purposes, vocabulary merging implies the combination of groups of two or more vocabulary terms into categories or classes of terms. In this case, instead of eliminating individual rows in the TF-IDF matrix, we should combine different groups of rows into single rows, one row per group. Typically used vocabulary merging techniques include lemmatization and stemming, both of which were already introduced in chapter 6.

In the case of lemmatization, the objective is to reduce all different forms of morphologically related words to their common root or lemma. Consider for instance words such as *water*, *waters*, *watery*, *waterless* and *wateriness*. Lemmatization should reduce them all to their basic lemma form: *water*. In the case of stemming, the main objective is to reduce different forms to a common root or stem, but differently from lemmatization, the stem of a word does not necessarily match its morphological root or lemma. In this case, the criterion is more orthographical than morphological, making this process much simpler than lemmatization from the computational point of view. For illustrating the main difference between lemmatization and stemming, in addition to the list of words depicted above, a stemmer will also map words such as *watermelon*, *waterfall* and *watergate* to the same root: *water*, while a lemmatizer should not do so.

Typical implementations for stemmers include *ad hoc* rules that are, in most of the cases, language dependent. However, language independent implementations are also possible by using simple procedures such as the one illustrated next. First, consider stems obtained by retaining the first n characters of each word and, second, regroup all original vocabulary terms into the new generated vocabulary of stems. Let us illustrate this algorithm for the case of $n = 6$:

```
>> % converts the vocabulary string array into a character array      (10.5a)
>> char_vocabulary = char(ranked_vocabulary{:});

>> % retains the first six characters of each term and reconverts     (10.5b)
>> % the truncated character array back into a string array
>> stems = string(cellstr(char_vocabulary(:,1:6))');

>> % groups vocabulary terms into the new vocabulary of stems         (10.5c)
>> [stems_vocabulary,void,pointer] = unique(stems);

>> length(stems_vocabulary) % gets the stem vocabulary size           (10.5d)
ans =
     9383
```

As seen from (10.5), the simple stemming strategy based on retaining the first *6* characters of each word has reduced the original vocabulary size from *12,545* terms to *9,383* stems. The index `pointer` computed in (10.5c) allows for grouping different rows of the term-document matrix into their new corresponding stem rows as follows:

```
>> % creates a sparse matrix of appropriate dimensions                (10.6a)
>> stems_tfmtx = sparse(length(stems_vocabulary),length(verses));

>> % updates the term frequency for each new stem category            (10.6b)
>> for k=1:length(ranked_vocabulary) % this will take a while
      stems_tfmtx(pointer(k),:) = stems_tfmtx(pointer(k),:)+tfmtx(k,:);
   end
```

Two important remarks must be made regarding (10.6). First, notice that the updating procedure was conducted over the TF matrix and not over the TF-IDF matrix. Different from pruning, where rows were just dropped from the matrix, in merging we must add up contributions from all terms being merged into the same stem category. The implications of this on the computation of TF and IDF are actually different. While TF weights can be updated by adding together individual TF of merged words, IDF cannot be updated this way. Consider for instance two words that are merged into the same stem category, one occurring in N documents and the other occurring in M documents. As they can co-occur in, let us say, K documents, the actual document count for their combination is $N+M-K$. According to this, the IDF weight vector `stems_idfvt` must be directly derived from the stem TF matrix `stems_tfmtx`. Afterwards, the corresponding TF-IDF matrix can be computed.

The second important remark about (10.6) has to do with the fact that in the resulting stem TF matrix, stems are not ordered by frequency ranks. This is basically because, the stem vocabulary obtained in (10.5c) is not ordered by ranks either, as the function `unique` returns the elements in alphabetical order. In order to get stems ordered by rank you must derive the global frequency counts for the stems and rank them by using the function `sort`. Then, you should apply the resulting reordering pattern to both `stems_vocabulary` and the rows of `stems_tfmtx`.[2]

To better illustrate how stemming groups vocabulary words into stem categories, let us consider a simple example. Let us first identify which are the stems that group more than *8* vocabulary terms. For this we can proceed as follows:

```
>> % creates a histogram of pointers from terms to stems              (10.7a)
>> nterms = hist(pointer,length(stems_vocabulary));

>> % identifies those stems grouping more than 8 terms               (10.7b)
>> index = find(nterms>8)
```

---

[2] This is actually the same procedure we followed when we computed `ranked_frequencies` and `ranked_vocabulary` for the first time in chapter 7.

```
  index =
 Columns 1 through 6
          87          1130         1722         1724         1792         2106
 Columns 7 through 12
        2897          4203         4697         6397         6492         8379
 Columns 13 through 14
        8530          8568
```

Notice that there are *14* stems, each of which groups together more than *8* different terms. Let us consider the first of them:

```
>> stems_vocabulary(index(1))                                              (10.8a)
ans =

    "accept"

>> % gets all vocabulary terms that have been reduced to 'accept'          (10.8b)
>> ranked_vocabulary(pointer==index(1))
ans =
  1×9 string array
  Columns 1 through 5
    "accepted"      "accept"      "acceptable"      "accepteth"      "acceptation"
  Columns 6 through 9
    "acceptably"      "acceptance"      "acceptest"      "accepting"
```

Notice that, in this case, our simplistic stemming algorithm is doing a great job; but this might not be always the case, as unrelated words sharing the same first *6* characters will be inevitably reduced to the same stem. Larger dimensionality reductions can be achieved by considering, instead of *6*, the first *5*, *4*, or *3* characters shared by the vocabulary words. However, this would be at the expense of making the model more inaccurate as the resulting stems will contain more unrelated words as the required number of character matches is reduced.

In this section we have focused on the study of dimensionality reduction techniques by means of vocabulary pruning and merging. This conveys the implicit assumption that the words are the variables of the model, and the documents are the data samples under study. Alternatively, we can be interested in representing words (the samples) as vectors of documents (the variables). However, common practices for "document pruning" or "document merging" are not standard and commonly available. Projection methods, as the ones described in the following sections, are standard procedures which can be used in both cases.

10.2 The Linear Transformation Approach

The linear transformation approach to dimensionality reduction is based on a matrix factorization procedure known as Singular Value Decomposition, also referred to as SVD. This matrix factorization allows for decomposing a given $n \times m$ matrix H into an associated set of singular values and two orthogonal bases of singular vectors, as follows:

$$H = U S V^T \tag{10.9}$$

were U and V are unitary matrices (i.e. $U^T = U^{-1}$ and $V^T = V^{-1}$) of dimensions $n \times n$ and $m \times m$, S is an $n \times m$ diagonal matrix (i.e. $s_{ij} = 0$ if $i \neq j$), and T denotes transposition (or conjugate transposition in the case of complex-valued matrices). The diagonal elements of S are known as the singular values of H, and the columns of U and V are known as the "left" and "right" singular vectors of H, respectively.

Notice that SVD does not constitutes a space reduction technique *per se*. It is actually a factorization. What is interesting about this factorization is that it generates optimal orthogonal bases for the given dataset. Indeed, the n columns of U constitute an orthogonal basis for the vector space spanned by the rows of H. Be reminded that the rows of a TF-IDF matrix correspond to vocabulary terms, so if H is a TF-IDF matrix, the columns of U would constitute an orthogonal basis for the associated document space. Similarly, the m columns of V constitute an orthogonal basis for the vector space spanned by the columns of H. In this case, as the columns of a TF-IDF matrix correspond to documents, V would provide an orthogonal basis for the associated word space.

These orthogonal bases are optimal in the sense that they concentrate the variability of the data in as few dimensions as possible. In other words, the first singular vector is aligned with the direction in which the data exhibits its maximal variability, the second singular vector is aligned with the orthogonal direction (with respect to the first singular vector) in which the data exhibits maximal variability, the third singular vector is aligned with the orthogonal direction (with respect to both the first and second singular vectors) in which the data exhibits maximal variability, and so on.

In the SVD decomposition defined in (10.9), singular values are always nonnegative, and their relative magnitude is an indicator of the amount of variability exhibited by the data along the direction of its corresponding singular vectors. It is a convention to provide the singular values in S ordered from largest to smallest, i.e. $s_{11} \geq s_{22} \geq s_{33} \geq s_{44} \ldots$

In order to project the columns of the data matrix H (documents in the case of a TF-IDF matrix) into the orthogonal basis generated by the SVD procedure, we should left-multiply (10.9) with U^T, as follows:

$$D = U^T H = S V^T \tag{10.10}$$

The columns of the resulting matrix D are the corresponding projections of the columns of H into the orthogonal basis given in U. Notice that D is an $n{\times}m$ matrix just like H. So, at this point, no dimensionality reduction has actually occurred. What we have achieved so far is to project the columns of H into a new set of axes which guaranties that the maximum variability along the data samples is concentrated in the first dimensions (or components) of this new representation. So, in order to perform the dimensionality reduction, we must drop certain amount k of the low variability components of D, i.e. its last k rows. Alternatively, we can discard the columns of U that are associated to the k smallest singular values in S.

This approach to dimensionality reduction constitutes a very common practice in data mining and exploratory data analysis and is generally known as Principal Component Analysis (PCA). In the specific context of text mining and natural language processing, this procedure (when applied to the columns of a TF-IDF matrix as in (10.10), which means that document vectors are projected into a new space of reduced vocabulary dimensions) is called Latent Semantic Indexing.

Now, let us illustrate the use of this methodology for reducing the dimensionality of the document space model for our experimental dataset. First of all, let us apply vocabulary pruning for removing the most and least frequent words in the vocabulary. We do this to maintain computational time manageable, as the computational cost of SVD can be significantly increased when considering large document collections with large vocabularies. More specifically, let us remove those vocabulary terms with relative frequencies above *0.2*, as well as those occurring less than five times in the entire collection:

```
>> hf_terms = (ranked_frequencies/max(ranked_frequencies))>0.02;        (10.11a)
>> lf_terms = ranked_frequencies<=5;

>> pruned_tfidfmtx = tfidfmtx(not(hf_terms|lf_terms),:);                 (10.11b)
>> whos pruned_tfidfmtx
```

Name	Size	Bytes	Class	Attributes
pruned_tfidfmtx	4689x31103	4518240	double	sparse

Then, we normalize the resulting document vectors:

```
>> pruned_ntfidfmtx = sparse(size(pruned_tfidfmtx,1),length(verses));    (10.12)
>> for k=1:length(verses) % applies L2-norm normalization
       temp = norm(pruned_tfidfmtx(:,k));
       if temp>0 % avoids divisions by zero
           pruned_ntfidfmtx(:,k) = pruned_tfidfmtx(:,k)/temp;
       end
   end
```

Before conducting the SVD and the subsequent dimensionality reduction, let us revisit examples (9.43) and (9.44). In those examples we computed cosine association scores between one verse and the rest of the collection and generated a histogram of the resulting scores. Indeed, we are going to repeat the same exercises here, and for the same verse *392*, with two objectives in mind: first we will be able to compare this new histogram with the one depicted in Figure 9.3 and observe the effects of vocabulary pruning on the distribution of similarity scores; and second, we will compute a similar histogram after applying SVD-based dimensionality reduction and study the effects of this kind of dimensionality reduction on the distribution of similarity scores. Figure 10.1 presents the resulting histogram of cosine similarity scores. The generation of the plot, by applying the same procedures in (9.43) and (9.44) to `pruned_ntfidfmtx`, is left to you as an exercise.

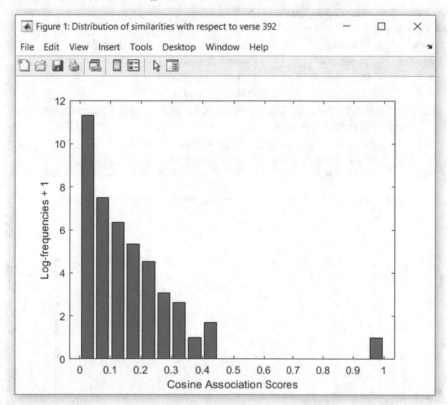

Fig. 10.1. Distribution of cosine similarities with respect to verse *392* for all verses in the collection after vocabulary pruning (bar amplitudes correspond to log-frequency values)

As seen from the figure, the resulting distribution of cosine similarity scores is similar to the one depicted in Figure 9.3. However, while all verses in the data collection exhibited cosine similarity scores within the interval *[0,0.35]* when all the *12,545* vocabulary terms were taken into account (see Figure 9.3), in this case, the observed interval for cosine similarities has been extended to *[0,0.45]*. This means that some document vectors have become a little bit closer after pruning the vo-

cabulary. Notice, however, that this small increment of the maximum score resulted from throwing away more than half of the vocabulary terms! As seen in
(10.11b), the vocabulary size has been reduced to only *4,689* terms after pruning.

Next, let us apply the SVD to our data model `pruned_ntfidfmtx`. For this, we
will use MATLAB® function `svds`, which allows for computing a specified number of singular values and vectors. In this way, `[u,s,v] = svds(H,k)` will return
the k largest singular values of H along with their corresponding right and left singular vectors (notice also that `svds` admits matrices in sparse format). Let us only
consider the *100* largest singular values:

```
>> [u,s,v] = svds(pruned_ntfidfmtx,100);                              (10.13)
>> whos u s v
   Name           Size              Bytes  Class       Attributes

   s              100x100           80000  double
   u              4689x100        3751200  double
   v              31103x100      24882400  double
```

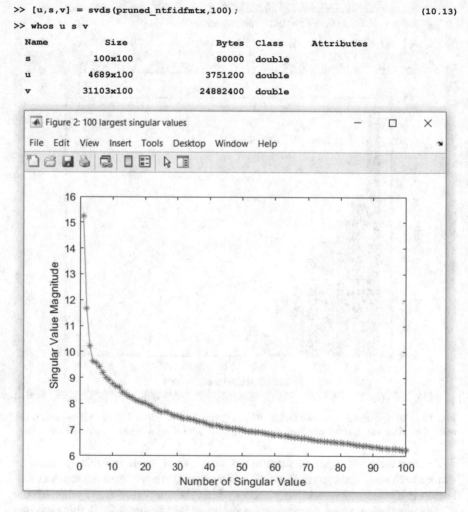

Fig. 10.2. Largest singular values for the pruned-vocabulary TD-IDF model in (10.12)

Figure 10.2 depicts the *100* largest singular values obtained in (10.13). Again, the generation of the plot is left to you as an exercise (you might want to use function `diag(s)` for getting the singular values from the diagonal matrix S).

As can be verified from the figure, the first singular values are actually the largest ones, and their magnitudes drop very quickly.

In order to see how things look like in very low dimensionality spaces, let us now generate a new histogram of cosine similarity scores, just like the one previously shown in Figure 10.1. However, in this case, we will be retaining the first *20* components (dimensions) only.[3]

```
>> % computes the projected document vectors in 20 dimensions        (10.14a)
>> projected_docs = u(:,1:20)'*pruned_ntfidfmtx;

>> norm_projected_docs = sparse(20,length(verses));                  (10.14b)
>> for k=1:length(verses) % applies L2-norm normalization
       temp = norm(projected_docs(:,k));
       if temp>0 % avoids divisions by zero
           norm_projected_docs(:,k) = projected_docs(:,k)/temp;
       end
   end

>> % computes cosine similarity scores with respect to verse 392     (10.14c)
>> cosine_scores = norm_projected_docs(:,392)'*norm_projected_docs;
```

Figure 10.3 presents the resulting histogram of cosine association scores with respect to verse *392* in *20* dimensions (again, the generation of the plot is left to you as an exercise). As seen from the figure, the distribution of similarities looks totally different from those previously presented in Figures 10.1 and 9.3. Different from the previous cases, the new obtained distribution is much more uniform, exhibiting scores in the whole interval *[0,1]*. This puts in evidence how in very low dimensional spaces, on average, data samples come much closer to each other. Such an "improved" distribution of similarities and, consequently, of distance distributions among data samples is of great utility in many different text mining and natural language processing applications.

Another important observation from 10.3 is the fact that some negative similarity scores have also appeared. This means that there are some document vectors forming angles greater than *90* degrees with respect to the vector of verse *392*. In other words, the projected vectors are not restricted to the positive hyper-octant as the original set of vectors in the TF-IDF matrix was. This is because, resulting from the SVD, the new document vectors can contain some negative-valued entries. Although this has been regarded as an important drawback of the vector

---

[3] In practical applications, with probably a few exceptions, it is not common to use such a low number of dimensions. Generally, reduced space dimensionalities will be around *100* to *500* dimensions. Here, we are using an exaggeratedly low dimensionality for illustrative purposes.

space model framework from a theoretical point of view, in most practical applications this issue does not seem to represent any problem at all.

If for any reason the resulting low-dimensional similarity scores need to be restricted to the interval *[0,1]*, these similarities can be either truncated or normalized. In the first case, all those similarity scores with negative values are forced to be zero, i.e. $score_{xy} = 0$ *if* $score_{xy} < 0$ (where x and y refer to the two documents being compared). In the second case, a standard normalization formula, such as the following one, can be used:

$$norm\_score_{xy} = (score_{xy} - min\{score\}) / (1 - min\{score\})$$ (10.15)

where *min{score}* refers to the minimum similarity score observed among document pairs in the data collection.

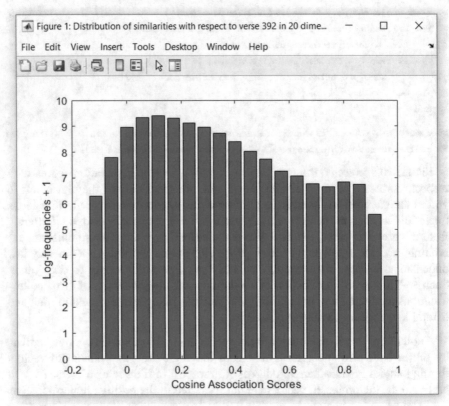

Fig. 10.3. Distribution of cosine similarities with respect to verse *392* for all verses in the collection after applying SVD-based dimensionality reduction; the dimensionality of the reduced space is *20* (bar amplitudes correspond to log-frequency values)

Similar to (10.10), where we projected the columns of the data matrix *H* (documents in the case of a TF-IDF matrix) into the orthogonal basis generated by the

SVD, we can think about projecting the rows of the data matrix H (vocabulary terms in the case of a TF-IDF matrix) into their corresponding orthogonal basis. For achieving this, we should right-multiply (10.9) with V, as follows:

$$W = HV = US \qquad (10.16)$$

Now, the rows of the resulting matrix W are the corresponding projections of the rows of H into the orthogonal basis given in V. Notice that W is also an $n \times m$ matrix, just like H. In order to perform the dimensionality reduction, we must drop the less informative components of W, i.e. its last k columns. Alternatively, we can discard the columns of V that are associated to the k smallest singular values in S before right multiplying H.

In the specific context of text mining and natural language processing, performing dimensionality reduction to the rows of a TF-IDF matrix by means of SVD, as in (10.16), is referred to as Latent Semantic Analysis. Different from the projections discussed before, in this case, word vectors are projected into a new word space of reduced "document-dimensions".

Let us now conduct some experiments to illustrate the main implications of dimensionality reduction over the vector space model of words. First of all, we need to apply $L2$-norm normalization to the rows of the **pruned_tfidfmtx** matrix computed in (10.11.b). We do this to be able to compute cosine similarities among rows, instead of columns:

```
>> % applies L2-norm normalization to rows of pruned_tfidfmtx          (10.17)
>> norm_terms = sparse(size(pruned_tfidfmtx,1),size(pruned_tfidfmtx,2));
>> for k=1:size(pruned_tfidfmtx,1)
       temp = norm(pruned_tfidfmtx(k,:));
       norm_terms(k,:) = pruned_tfidfmtx(k,:)/temp;
   end
```

Notice that, as we have pruned some vocabulary terms in (10.12) and we are interested in looking at the terms now, we need to obtain the corresponding subset of vocabulary terms that is currently represented in the model:

```
>> pruned_vocabulary = ranked_vocabulary(not(hf_terms|lf_terms));     (10.18)
```

Next, we can proceed to perform the dimensionality reduction by following the same procedure we used in (10.13). Again, we set the dimensionality of the reduced space to *100* and we normalize the projected vectors:

```
>> [nu,ns,nv] = svds(norm_terms,100); % performs SVD                  (10.19a)

>> % computes the projections (be reminded that here we are          (10.19b)
>> % projecting the rows of the matrix instead of its columns)
>> projected_terms = norm_terms*nv;

>> % applies L2-norm normalization to the projected matrix rows       (10.19c)
```

```
>> for k=1:length(pruned_vocabulary)
       temp = norm(projected_terms(k,:));
       norm_projected_terms(k,:) = projected_terms(k,:)/temp;
   end
```

Now, we are ready to explore some interesting differences between the original word-space model and the dimensionality-reduced one. Consider, for instance, the two words: *star* and *shining*. It happens to be the case that these two words never co-occur in any verse of the data collection. This can be easily verified as follows:

```
>> % gets the indexes of both words                                     (10.20a)
>> w1 = find(pruned_vocabulary=='star');
>> w2 = find(pruned_vocabulary=='shining');

>> % computes the number of verses in which these two words co-occur     (10.20b)
>> full(sum((norm_terms(w1,:)>0)&(norm_terms(w2,:)>0)))
ans =
     0
```

Consequently, the cosine similarity score for these two words is zero:

```
>> full(norm_terms(w1,:)*norm_terms(w2,:)')                              (10.21)
ans =
     0
```

Nevertheless, this is not what we would expect for this specific pair of words, because these two words are supposed to be semantically related somehow. Indeed, we could argue that *shining* is one of the properties of a *star*, but as they do not co-occur in any verse of the corpus, our model is not able to infer any relationship between these two words. This problem can be solved by taking into account the relationships among the other words that are commonly seen to co-occur with both, *star* and *shining*. In other words, we can infer that these two words are related if they tend to co-occur with the same set of words, i.e. if they tend to occur within similar contexts. The good news is that this is precisely the effect of dimensionality reduction! In the reduced space, the dimensions are indeed a sort of "weighted combinations" of the different verses in the collection. So, when we compare two word vectors in the reduced space, we are not simply looking at word co-occurrences within verses. We are actually looking at context similarities among the different contexts in which those two words tend to occur.

If we compute the cosine similarity score between *star* and *shining* in the reduced space, we will notice that it is not zero in this case:

```
>> full(norm_projected_terms(w1,:)*norm_projected_terms(w2,:)')          (10.22)
ans =
     0.8166
```

Even more, notice from (10.22) that the cosine similarity between the words under consideration is relatively high, which means that these two words are effectively related to each other.

By playing around with the data, we can find more examples like this one. Consider, for instance, the following word pairs: *love-obedience*, *thirst-thirsty*, *hungry-poverty*, *feet-knees*, *fish-creature*; all of which have a cosine similarity score equal to zero in the original high-dimensional space, but a relatively high positive score in the low-dimensional space. Let us print a report displaying both scores for each of these word pairs:

```
>> % enters the word pairs into two string arrays                        (10.23a)
>> list1 = ["love","thirst","hungry","feet","fish"];
>> list2 = ["obedience","thirsty","poverty","knees","creature"];

>> % computes cosine similarity scores                                   (10.23b)
>> for k=1:5
      w1 = find(pruned_vocabulary==list1(k));
      w2 = find(pruned_vocabulary==list2(k));
      hd_score(k) = norm_terms(w1,:)*norm_terms(w2,:)';
      ld_score(k) = norm_projected_terms(w1,:)*norm_projected_terms(w2,:)';
   end

>> % generates the report                                                (10.23c)
>> headers = {'Word 1','Word 2','HD Score','LD Score'};
>> disp(table(list1',list2',hd_score',ld_score','VariableNames',headers))
```

Word 1	Word 2	HD Score	LD Score
"love"	"obedience"	0	0.68122
"thirst"	"thirsty"	0	0.65798
"hungry"	"poverty"	0	0.6109
"feet"	"knees"	0	0.60132
"fish"	"creature"	0	0.65451

In the same way as dimensionality reduction allows for extracting hidden or latent relationships among terms that do not co-occur in any document of the data collection, it also helps (in general terms) to improve the quality of the observed word associations among terms that co-occur in documents. Let us illustrate this by considering a different exercise. In the following example we will obtain the *10* terms that are closest to the vocabulary term *drink*, in both the original high-dimensional space and the reduced low-dimensional one. Along with the cosine similarity scores, we also report the number of word co-occurrences (within verses) for each case:

```
>> w = find(pruned_vocabulary=='drink'); % gets the word index           (10.24a)

>> % computes and sorts similarity scores in high-dimensional space      (10.24b)
>> scores = norm_terms*norm_terms(w,:)';
```

```
>> [sortedscores,order] = sort(scores,'descend');
>> hd_sc = full(sortedscores(1:10));
>> hd_wd = pruned_vocabulary(order(1:10))';

>> for k=1:10 % gets co-occurrences and scores for 10 closest words    (10.24c)
       temp = sum((norm_terms(w,:)>0)&(norm_terms(order(k),:)>0));
       hd_cc(k) = full(temp);
   end

>> % computes and sorts similarity scores in low-dimensional space     (10.24d)
>> scores = norm_projected_terms*norm_projected_terms(w,:)';
>> [sortedscores,order] = sort(scores,'descend');
>> ld_sc = full(sortedscores(1:10));
>> ld_wd = pruned_vocabulary(order(1:10))';

>> for k=1:10 % gets co-occurrences and scores for 10 closest words    (10.24e)
       temp = sum((norm_terms(w,:)>0)&(norm_terms(order(k),:)>0));
       ld_cc(k) = full(temp);
   end

>> % generates the report                                              (10.24f)
>> hds = {'HD Co-oc','HD Score','HD Words','LD Co-oc','LD Score','LD Words'};
>> disp(table(hd_cc',hd_sc,hd_wd,ld_cc',ld_sc,ld_wd,'VariableNames',hds))
```

HD Co-oc	HD Score	HD Words	LD Co-oc	LD Score	LD Words
327	1	"drink"	327	1	"drink"
67	0.26691	"wine"	7	0.74875	"drunk"
105	0.19174	"eat"	67	0.74574	"wine"
65	0.16685	"water"	3	0.7433	"drank"
24	0.15785	"cup"	8	0.70889	"drunken"
50	0.15041	"offering"	3	0.68198	"bottle"
6	0.14694	"pitcher"	105	0.65958	"eat"
53	0.14654	"meat"	5	0.64362	"vinegar"
5	0.1459	"vinegar"	24	0.62898	"cup"
20	0.12207	"strong"	7	0.6275	"thirsty"

Notice from the results reported in (10.24) that, although both lists contain several terms in common, the two lists are quite different. Apart from this, two important observations can be derived from (10.24). First, observe how the ranking of the terms seems to be more dependent on co-occurrence counts in the case of the high-dimensional space than in the case of the low-dimensional one.[4] Second, notice how cosine similarity scores are higher in the reduced dimensionality space than in the original one. This verifies the fact that, in low-dimensional spaces, an-

---

[4] Although this can vary significantly from word to word, the general trend is to observe a more direct dependency on pair-wise co-occurrences in the high-dimensional space. On the other hand, in low-dimensional spaces, similarity scores among terms depend more on context similarities than on simple word co-occurrences.

gles among term vectors are smaller on average and, consequently, the terms are closer to each other than in the original high-dimensional space.

Regarding the quality of both lists of terms in (10.24), it is actually very difficult to be assessed as the obtained sets of words do not clearly belong to any specific category of semantic relationships. We could probably argue that the set of terms obtained in the original space is worse because it contains some terms such as *offering*, *meat* and *strong* that are clearly not related to *drink*. However, we could also challenge the merits of the set of terms obtained in the reduced space as it does not include the most obvious term: *water*. A further evaluation of both term lists will bring us back to the idea of the full space model being prone to rely on co-occurrences, giving more preference to those words that tend to occur close to *drink*, such as *eat*, *wine* and *water*. On the other hand, the low-dimensional space model seems to be paying more attention to the similarity among the contexts in which the words occur, giving more preference to those words that tend to be used "instead of" *drink*, such as *drunk*, *drank* and *drunken*.

Finally, let us consider some additional examples on word associations, similar to the ones presented in (10.24). In Table 10.1, we present the 5 closest terms, in both the original high-dimensional space and the reduced one, for some different terms. Notice from the table that, regardless the high degree of variability observed in these results, the lists of closest terms obtained from low-dimensionality scores tend to be more appropriate. However, this is very subjective.

Table 10.1. Lists of the 5 closest words to a given vocabulary term; proximity is computed by means of cosine similarities in both the original high-dimensional space (*31,103* dimensions) and the reduced low-dimensional space (*100* dimensions)

Rank	High-dimensional	Low-dimensional	High-dimensional	Low-dimensional
-	moon	moon	fury	fury
1	sun	stars	anger	fierceness
2	stars	darkened	poured	furious
3	darkened	sun	oppressor	indignation
4	light	shineth	accomplish	anger
5	shining	shining	pour	jealous
-	rain	rain	cup	cup
1	showers	clouds	drink	drunken
2	latter	latter	outside	drink
3	rained	showers	testament	drunk
4	clouds	hail	brim	wine
5	former	heaven	drunken	fill

10.3 Non-linear Projection Methods

The dimensionality reduction methods described in the previous section are based on the use of linear transformations of the data space. Indeed, the idea was to use a linear transformation to project the data into a new basis, such that the variability of the data is concentrated in as few dimensions as possible. These transformations are linear in the sense that they can be represented in terms of a matrix operator (i.e. internal products among vectors). In this section we will discuss the use of non-linear projections for the same purpose we used linear transformations in the previous section: dimensionality reduction. Although there are several different approaches to dimensionality reduction by means of non-linear projections, here we will focus our attention into a particular family of methods regarded to as multidimensional scaling. You can get more information about other alternative methods within the references provided in the further reading section.

Different from linear projections, non-linear projections attempt to construct low-dimensional embeddings that preserve, as much as possible, the structure of the given data collection. While we can think about a linear projection as the "shadow" produced by the data into a low-dimensional space, a non-linear projection would be more like a "distorted shadow" which aims at preserving as much as possible, in the reduced space, the original "shape" of the dataset in the high-dimensional space. Figure 10.4 illustrates this idea.

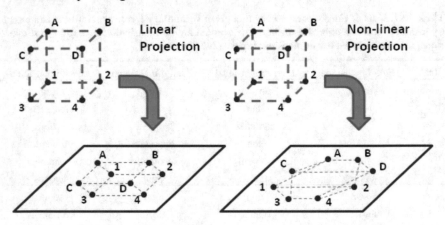

Fig. 10.4. Linear and non-linear projections of a cube into a 2-dimensional space

A first glance at Figure 10.4 might suggest that the linear projection is more reliable than the non-linear one because we still are able to distinguish the faces of the cube in it; while, in the case of the non-linear projection, the cube faces have been completely distorted. This appreciation, however, is wrong! It is based on our natural intuition about projections as shadows, and it is actually missing a very important and desirable property of reduced space representations. If you examine the right-hand side of the figure more carefully, you will realize that the non-linear projection has managed to maintain far from each other those pairs of points which constitute opposite vertices in the original cube; while the linear projection has not been able to do it. In fact, in the linear projection, the closest points to *D* are *1* and *4*. While it is fine for *4* to be close as it is a consecutive vertex to *D*, *1* should not be close. Actually, point *1* should be the farthest point from *D* because it is the opposite vertex to it. On the other hand, in the case of the non-linear projection, *1* is indeed the farthest point from *D*; and this is also true for all opposite pairs of vertices. This inherent capacity of non-linear projection methods for better preserving data structure at reduced dimensionality spaces make them very attractive, as well as useful, in several data mining and text mining applications.

In this section, we will devote our attention to a specific family of non-linear projection methods, which is called multidimensional scaling (MDS). This specific type of non-linear projection was originally conceived as a visualization technique, which main objective was to construct two- or three-dimensional representations for datasets defined in a higher dimensionality space (a large number of variables). As humans cannot visualize more than three dimensions, visualizing a dataset with more than three variables is not viable. In this sense, MDS was conceived with the objective of constructing low-dimensional embeddings for high-dimensional datasets such that they can be visualized, while preserving as much as possible the original structure of the dataset.

There are two different classes of MDS algorithms: metric and non-metric. In the case of metric MDS, distances among data points in the low-dimensional embedding are required to match as much as possible the corresponding distances, or dissimilarities, among data points in the original high-dimensional space. In the case of non-metric MDS, which constitutes a relaxed version of the metric case, distances among data points in the embedding are only required to match a monotonic transformation of the corresponding distances in the original space.

MDS is implemented in practice as an optimization problem, in which an objective function, generally referred to as the *stress function*, is to be minimized. Different variants of MDS rely in different implementations of this objective function, but all of them are designed to measure in one way, or another, the amount of distortion observed in the distances among the data samples in the reduced space with respect to the distances or dissimilarities in the original space. In other words, MDS projections are computed by minimizing the differences between the distances among data points in both the reduced and the original space. The general form of the *stress function* is as follows:

$$stress\_function = [\alpha \, \Sigma_x \, \Sigma_y \, (f(dh_{xy}) - d_{xy})^2]^{1/2} \qquad (10.25)$$

where dh_{xy} refers to the distance, or dissimilarity value, between data points *x* and *y* (that are either document vectors or word vectors in our TF-IDF matrix scenario) as it is measured in the original high-dimensional space; d_{xy} refers to the distance between the same pair of data points in the low-dimensional embedding; $f(\cdot)$ is the monotonic transformation used to relax the matching criterion in the case of non-metric MDS; and α is a normalization factor.

A classical example that is frequently used to illustrate the power of MDS is the one about constructing a map given the matrix of distances between the elements in the map. Consider, for instance, some major U.S. cities and the approximate distances (in kilometers) among them:[5]

```
>> cities = {'Atlanta','Chicago','Denver','Houston','Los Angeles',...    (10.26a)
        'Miami','New York','San Francisco','Seattle','Washington'};

>> % matrix of approximate inter-city distances in kilometers            (10.26b)
>> d = [ 0   1503   3103   1795   4956   1546   1915   5476   5586   1390
       1503      0   2355   2406   4467   3041   1825   4756   4447   1528
       3103   2355      0   2250   2127   4419   4175   2429   2614   3825
       1795   2406   2250      0   3517   2478   3635   4211   4841   3123
       4956   4467   2127   3517      0   5988   6275    888   2455   5888
       1546   3041   4419   2478   5988      0   2796   6641   6999   2363
       1915   1825   4175   3635   6275   2796      0   6582   6164    525
       5476   4756   2429   4211    888   6641   6582      0   1736   6252
       5586   4447   2614   4841   2455   6999   6164   1736      0   5962
       1390   1528   3825   3123   5888   2363    525   6252   5962      0];
```

The matrix of inter-city distances is actually a matrix of distances or dissimilarities among the data points in a dataset, which in this particular example happens to be originally a two-dimensional dataset. So, in this experiment we will not be conducting any dimensionality reduction at all. However, as you probably noticed already, a matrix as the one in (10.26b) can be computed for any dataset regardless the dimensionality of the space containing it. Indeed, we can obtain a dissimilarity matrix, just like this one, for either documents or words by computing the dissimilarities among the columns or rows, respectively, of a TF-IDF matrix.

Continuing with our example on city distances, we now apply MDS to the matrix of inter-city distances in (10.26). As these distances are already measured in a two-dimensional space and we will be using MDS for creating a two-dimensional embedding for the dataset, no distortion is expected to occur, and the actual map locations of the cities should be generated. Of course, to be able to put the cities into an actual U.S. map, we will need to adjust some rotation and scaling factors in the resulting embedding. MDS will satisfy the inter-city distance constrains when placing the cities in the embedding, but it does not have any way to know what the scale and orientation of it must be.

_____

[5] A similar example can be found in http://www.analytictech.com/borgatti/mds.htm Accessed 6 July 2021.

```
>> % creates a new figure and displays a U.S. map                    (10.27a)
>> hf = figure(5);
>> set(hf,'Color',[1,1,1],'Name','Placing cities into a map with MDS');
>> [imagen,cmap] = imread('map_usa.gif','gif');
>> image(imagen); colormap(cmap);

>> % applies metric MDS to the inter-city distance matrix in (10.26b)  (10.27b)
>> [coord,stress] = mdscale(d,2,'criterion','metricstress');

>> % rotates and rescales to fit the resulting embeding into the map   (10.27c)
>> rotang = pi/17; scale1 = 325; scale2 = 200; offset1 = 20; offset2 = 30;
>> rotcoord = -coord*[cos(rotang),-sin(rotang);sin(rotang),cos(rotang)];
>> minval = min(rotcoord); maxval = max(rotcoord);
>> normcoord(:,1) = (rotcoord(:,1)-minval(:,1))/(maxval(:,1)-minval(:,1));
>> normcoord(:,2) = (rotcoord(:,2)-minval(:,2))/(maxval(:,2)-minval(:,2));
>> mapcoord(:,1) = normcoord(:,1)*scale1 + offset1;
>> mapcoord(:,2) = (1-normcoord(:,2))*scale2 + offset2;

>> % plot city locations and names into the map                       (10.27d)
>> hold on; plot(mapcoord(:,1),mapcoord(:,2),'*b'); hold off;
>> text(mapcoord(:,1),mapcoord(:,2)+7,cities,'Color',[0,0,1],'FontSize',10,...
        'HorizontalAlignment','center');
```

The U.S. map with the resulting city locations on it is shown in Figure 10.5. As can be verified from the figure, MDS has been able to infer the actual city coordinates (with some rotation, offsets and scaling factors required)[6] from the inter-city distance matrix. Although this result might seem impressive, actually it is not. The MDS algorithm has just generated coordinates for placing objects into a two-dimensional space from a set of distances that was also computed in two dimensions. As the original dataset "lives" in two dimensions, what we are seeing in Figure 10.5 is the actual structure of the dataset, which MDS has been able to recover from its corresponding dissimilarity matrix.

Nevertheless, MDS becomes really impressive when it is used for placing into two dimensions a dataset that "lives" in a very high-dimensional space. Reconsider, for instance, the word vector space model we constructed in (10.17), which accounts for the vocabulary terms in (10.18). In this model, *4,689* vocabulary terms are represented in a vector space of *31,103* dimensions. In the next example, we are going to use MDS for projecting a subset of these vocabulary term representations into a two-dimensional map and see how the resulting map looks like. Let us consider the following list of terms: *bird, cloud, field, fish, flock, goat, lighting, mountain, rain, river, sea, sheep, sky, storm, thunder* and *wind*.

---

[6] Different versions and/or initializations of the `mdscale` algorithm can produce different rotation, offsets and scaling factors; if you are not able to reproduce the results shown in Figure 10.5, you will have to play for a while with the values of variables `rotang`, `scale1`, `scale2`, `offset1` and `offset2` until you can properly fit the cities into the map.

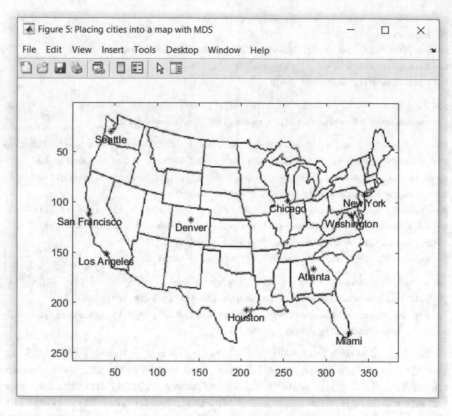

Fig. 10.5. Generating city coordinates from inter-city distances by means of MDS

We now apply MDS to the proposed list of words:

```
>> % Creates a string array with the selected vocabulary terms        (10.28a)
>> list = ["bird","cloud","field","fish","flock","goat","lightning",...
     "mountain","rain","river","sea","sheep","sky","storm","thunder","wind"];

>> % Gets the corresponding index for each word in the list            (10.28b)
>> for k=1:length(list)
       idx(k) = find(pruned_vocabulary==list(k));
   end

>> % Computes the cosine distance matrix for the subset of words       (10.28c)
>> thedist = pdist(full(norm_terms(idx,:)),'cosine');

>> % Applies MDS to the computed cosine distance matrix               (10.28d)
>> limit = statset('MaxIter',500); % Sets the maximum number of iterations
>> [map,stress] = mdscale(thedist,2,'criterion','stress','options',limit);
```

Next, we plot the resulting map of words. The results are shown in Figure 10.6.

```
>> hf = figure(6); % Creates a new figure                              (10.29a)
>> set(hf,'Color',[1,1,1],'Name','Constructing a map of words with MDS');

>> plot(map(:,1),-map(:,2),'*'); % Plots the resulting word locations  (10.29b)
>> axis([-0.6,0.6,-0.3,0.3]);

>> % Prints each word below its corresponding location                 (10.29c)
>> text(map(:,1),-map(:,2)-0.02,list,'HorizontalAlignment','center');
```

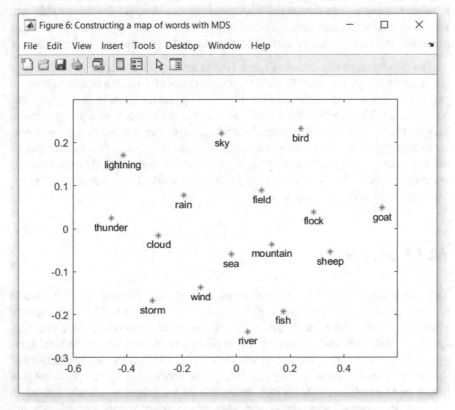

Fig. 10.6. Two-dimensional map for a selected subset of words generated by applying MDS to their original vector space representation in *31,103* dimensions

A careful exploration of Figure 10.6 will reveal a very interesting arrangement of the terms in the generated map. First of all, let us consider the horizontal axis. Notice how living things appear towards the right-hand side of the map, while non-living things are located towards the central portion and left-hand side of the map. Moreover, an interesting difference can be observed between the terms referring to non-living things in these two zones. The terms appearing in the central portion are habitats or physical locations (*river*, *sea*, *mountain*, *field*, and *sky*), while the terms in the left-hand side of the map refer to meteorological phenomena (*storm*, *wind*, *rain*, etc.).

If we now consider the vertical axis, we can also discover and interesting arrangement of the terms in this direction. Notice, for instance, how the habitats are arranged from bottom to top as water, land, and sky. You can now understand why we intentionally reverted the direction of the second coordinate in (10.29b), just to put *sky* on top. Additionally, you can see from the figure how the living things are also arranged in the vertical direction according to their corresponding habitats. Regarding meteorological phenomena, although they affect all habitats similarly, it is interesting to see that *storm* and *wind* appear towards the lower side and somehow closer to *sea*, while *cloud* and *rain*, for instance, appear higher towards *sky*. Also interesting is the proximity of *thunder* and *lightning* to *cloud* and *rain*.

The arrangement of words in Figure 10.6 is certainly impressive if we consider the nature of both the model and the space reduction procedure used, where no explicit linguistic information has been used at all. However, it is particularly important to clarify here that this impressive result is the consequence of considering only a reduced and carefully selected set of terms. In fact, you can verify how easily this map gets distorted and starts losing its semantic appeal just by incorporating few more words into the analysis. And, of course, this is totally reasonable as we cannot expect the complexity of a *31,103*-dimensional space model to be perfectly mapped in two dimensions!

10.4 Embeddings

The semantic properties of the low-dimensional representations for documents and words presented in the previous sections happen to be very useful in practical applications. These constructs, commonly referred to as embeddings, are probably the most important and interesting accomplishment in the field of Natural Language Processing during the last decade. Basically, embeddings are projections of the sparse discrete forms of language into dense continuous spaces. In these projected representations, fundamental properties of language, which are hidden in the original representation space, are surfaced. In other words, embeddings are tools for alleviating the "curse of dimensionality" problem in Text Mining and Natural Language Processing applications.

In this section, some specific implementations of word embeddings[7] that are available within the Text Analytics Toolbox™ are presented, and some syntactic and semantic properties of these constructs are explored. We also devote some attention to the problem of constructing a document embedding for the data collection we have been experimenting with along the chapter.

---

[7] Most recent approaches to embedding computation are based on Deep Learning architectures that are trained over massive volumes of text data. Details on these recent implementations are beyond the scope of this book. Relevant references are provided in the Further Reading section.

Let us start by exploring some of the morpho-syntactic properties of word embeddings. For this, we use `fastTextWordEmbedding`, which is a pretrained word embedding provided with the Text Analytics Toolbox™. This function returns a word embedding of *300* dimensions with a vocabulary coverage close to a million English words.[8] The pretrained word embedding can be loaded as follows:

```
>> word_embedding = fastTextWordEmbedding                    (10.30)
word_embedding =
  wordEmbedding with properties:
    Dimension: 300
    Vocabulary: [1×999994 string]
```

In the loaded model, words in the English vocabulary are represented as vectors of 300 dimensions. Most of the syntactic and semantic properties of the embedding are manifested by means of vector offsets, as illustrated in Figure 10.7.

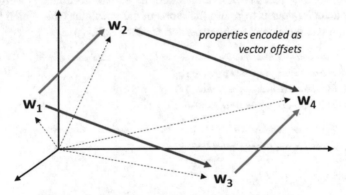

Fig. 10.7. Linguistic properties as vector offsets in embedded representations of language

In a word embedding, linguistic properties typically tend to manifest as vector offsets, as illustrated in the figure. According to this, the linguistic relationship between a given pair of words w_1 and w_2 is expected to be the same as the one between the pair of words w_3 and w_4 as far as the linguistic relationship between the pair w_1 and w_3 is consistent with the one between w_2 and w_4 and vice versa. This would be the geometrical analogous to word relationship assessment questions of the form: "*man* is to *men* as *woman* is to *women*". Notice that, in this example, there are two linguistic properties that govern the relationships among the four words under consideration: gender and number. While the gender property drives the relationships of the pairs *man* vs. *woman* and *men* vs. *women*, the number property drives the relationships of the pairs *man* vs. *men* and *woman* vs. *women*.

---

[8] The `fastTextWordEmbedding` function requires a support package. If the package is not installed in your system, a link will be provided the first time you execute the function. Click the link and follow the instructions to perform the installation process.

It happens that in a word embedding representation, these classical word relationship assessment questions can be reduced to basic vector arithmetic. For instance, the vector difference *men − man* would be equal to the vector difference *women − woman*. So, the plural form *women* can be obtained by a vector operation of the form: *women = men − man + woman*. Notice that this vector operation admits two different but completely equivalent interpretations. First, the number offset inferred from the vector subtraction *men − man* is added to *woman* in order to derive the plural form. Second, the gender offset inferred from *woman − man* is added to *men* in order to obtain the feminine form.

Let us conduct the actual computations of this example with the pretrained word embedding loaded in (10.30). Here, we will use functions `word2vec` and `vec2word` to map word forms into and back from embedding vectors, respectively. Notice that, in the general case, the result of vector operations in the embedding result into vectors that are not associated to actual vocabulary words. What `vec2word` actually does is to return the word in the vocabulary for which its vector representation is the closest to the provided vector result.

```
>> % Gets embedded vector representations for man, men and woman      (10.31a)
>> w1 = word2vec(word_embedding,"man");
>> w2 = word2vec(word_embedding,"men");
>> w3 = word2vec(word_embedding,"woman");

>> % Gets the closest word in the embedding to vector men-man+woman   (10.31b)
>> w4 = vec2word(word_embedding,w2-w1+w3)
w4 =
     "women"
```

A similar example exploring relationships among verb forms is presented next:

```
>> w1 = word2vec(word_embedding,"sell");                              (10.32a)
>> w2 = word2vec(word_embedding,"sold");
>> w3 = word2vec(word_embedding,"buy");

>> w4 = vec2word(word_embedding,w2-w1+w3)                             (10.32b)
w4 =
     "bought"
```

Even more interesting are the variety of semantic relationships also encoded as vector offsets in word embeddings. Consider for instance the following examples:

```
>> wds(1,:) = ["Portugal","Lisbon","Spain"];                         (10.33a)
>> wds(2,:) = ["dinner","p.m.","breakfast"];
>> wds(3,:) = ["nurses","hospital","teachers"];

>> for n=1:size(wds,1)                                                (10.33b)
       for k=1:3, wv(k,:) = word2vec(word_embedding,wds(n,k)); end
       w4 = vec2word(word_embedding,wv(2,:)-wv(1,:)+wv(3,:));
```

```
        disp(compose("'%s' - '%s' + '%s' = '%s'",[wds(n,[2,1,3]),w4]))
    end
'Lisbon' - 'Portugal' + 'Spain' = 'Madrid'
'p.m.' - 'dinner' + 'breakfast' = 'a.m.'
'hospital' - 'nurses' + 'teachers' = 'school'
```

As seen from the examples in (10.33) some more complex relationships of semantic nature are also encoded as vector offsets in the word embedding under
consideration. Specifically, we have illustrated semantic relationships of the type
country to *capital*, *meal type* to *time of day*, and *profession* to *work place*. In general, this sort of property encoding is what makes embeddings powerful tools for
language processing and analysis.

As it might be expected, the examples shown in (10.31), (10.32) and (10.33)
have been cherry picked as embeddings do not always perform as nicely as in
these examples. However, in general, you might expect meaningful results, most
of the times, when exploring the neighborhoods of the resulting vector operations.
Indeed, with **vec2word** you can specify how many words to retrieve from the
neighborhood of the provided embedding location. Consider for instance the following example:

```
>> w1 = word2vec(word_embedding,"nurses");                          (10.34a)
>> w2 = word2vec(word_embedding,"hospital");
>> w3 = word2vec(word_embedding,"judges");

>> w4 = vec2word(word_embedding,w2-w1+w3,5)'                         (10.34b)
w4 =
  1×5 string array
    "judges"    "judge"    "court"    "judicial"    "courts"
```

In this particular case, the correct response corresponds to the third closest
word vector to the provided vector operation result. For a more comprehensive
exploration of embedding neighborhoods, see exercise 10.6-7 in section 10.6.

In addition to the possibility of loading a pretrained embedding, the Text Analytics Toolbox™ also offers the functionality to train a word embedding from a
given dataset. This can be done by means of function **trainWordEmbedding**, which
implements two specific methods for word embedding calculation: the skip-gram
model (the default option) and the continuous bag-of-words model.[9] Once embeddings are trained, they can be saved to and read from files by means of functions
writeWordEmbedding and **readWordEmbedding**, respectively.

Next, we will train and visualize a word embedding from the Bible dataset used
in the previous sections of this chapter. This is done for illustrative purposes only
as the volume of text in this dataset is significantly smaller than the volumes of

---

[9] We are omitting here details on how these models are trained. For more detailed information
about them, you can refer to the original publications (Mikolov *et al.* 2013a and 2013b).

text typically used to train word embeddings. In practice, training a word embedding requires much larger volumes of data such as, for instance, the entire Wikipedia. Just for reference, the pretrained embedding loaded in (10.30) was trained over a *16*-billion token dataset, while our dataset under consideration contains only a total of *791,420* tokens.

First, let us start by reloading and preprocessing the dataset:

```
>> % loads the dataset and creates a string array of verses          (10.35a)
>> clear; load datasets_ch10a
>> for k=1:length(verses), docs(k) = verses(k).text; end

>> % preprocess the data                                             (10.35b)
>> tokdocs = tokenizedDocument(docs);
>> tokdocs = erasePunctuation(tokdocs);
>> tokdocs = lower(tokdocs);
>> tokdocs = removeStopWords(tokdocs);
```

Now, we are ready to train the embedding. In this particular example, we set the dimensionality of the resulting vector space to *30*, we include only those words that occur *10* or more times in the dataset, and we increase the number of training epochs (from its default value of *5*) to *20*.

```
>> word_embedding = trainWordEmbedding(tokdocs,...                   (10.36)
                     'Dimension',30,'MinCount',10,'NumEpochs',20)
Training: 100% Loss: 2.12945  Remaining time: 0 hours 0 minutes.
word_embedding =
  wordEmbedding with properties:
     Dimension: 30
     Vocabulary: [1×3451 string]
```

As seen from (10.36) the resulting word embedding is a *30*-dimensional vector space with a coverage of *3,451* vocabulary words. We can now visualize the embedding. First, we use function **tsne**[10] to construct a 2D map for the embedding, followed by **textscatter** to create an interactive plot of the 2D map:

```
>> words = word_embedding.Vocabulary;                               (10.37a)
>> vectors = word2vec(word_embedding,words);
>> word_map = tsne(vectors);

>> hf = figure(8);                                                  (10.37b)
>> set(hf,'Color',[1,1,1],'Name','Word Embedding Visualization');
>> textscatter(word_map,words)
```

---

[10] T-distributed Stochastic Neighbor Embedding (TSNE) is a non-linear projection technique, similar to MDS, that aims at building low-dimensional representations that preserves the original high-dimensional structure of the data. Different from MDS, which focuses on minimizing the difference between pairwise data sample distances, TSNE focuses on minimizing the difference between the overall data distributions (i.e. Kullback-Leibler divergence).

The resulting plot is depicted in Figure 10.8, which provides a high-level overview of the structure of the word embedding trained in (10.36). As seen from the figure, some different neighborhoods or regions of semantic nature seems to appear in the map. Consider for instance the proximity of words like *fire* and *light*, *idols* and *sin*, *wife* and *sons*, *house* and *work*, etc.

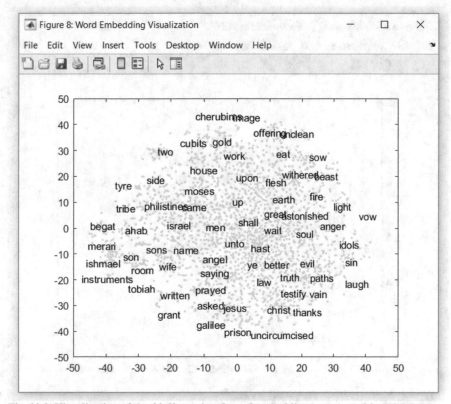

Fig. 10.8. Visualization of the *30*-dimensional word embedding constructed in (10.36)

The map in the figure can be interactively explored by zooming in and out, as well as moving around, the different neighborhoods. A more detailed exploration of some neighborhoods will provide more convincing evidence of the semantic merits of the constructed embedding. For instance, Figure 10.9 provides a more detailed view of the neighborhood surrounding the word *two* (in the upper-left region of Figure 10.8). As seen from the figure, most numbers and numeric related words have been clustered together in the same region of the embedding.

In addition to word embeddings, document embeddings also constitute useful resources for text preprocessing and analysis. Although not as popular as word embeddings nowadays, several different approaches have been proposed for the construction of document embeddings. Indeed, topic models (which were presented in section 8.4) constitute a form of document embedding. More recent approaches are based on Deep Learning architectures that are trained over massive volumes of data. In general, document embeddings also surface interesting semantic properties that are otherwise hidden in the original high-dimensional sparse representations of document collections.

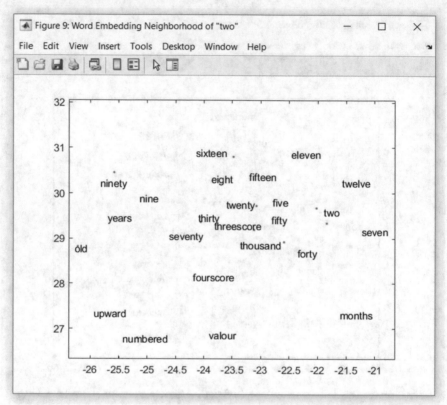

Fig. 10.9. A more detailed view of the neighborhood surrounding vocabulary word *two*

We conclude this section by building a document embedding for the Bible dataset used in this chapter. However, instead of verses, we will be using books as documents in this example. The data collection contains *66* books in total, which are available in the data file `datasets_ch10b.mat`.

```
>> clear; load datasets_ch10b                                              (10.38)
>> whos
   Name                   Size              Bytes  Class      Attributes
   books                  1x66           13503212  struct
   ranked_frequencies     1x12545          100360  double
   ranked_vocabulary      1x12545          757846  string
```

Similar to the `verses` structure array used before, the `books` structure array includes the fields `vocab`, `count` and `text`, which contain the vocabulary, word counts and raw text for each book, respectively. Additionally, the `books` structure array also includes the fields `group`, which identifies whether the book belongs to the *Old Testament* or *New Testament* subcollection, and `name`, which contains a three-letter abbreviation of the book's title.

Next, we proceed to preprocess the data. First, we create a string array with the raw text contents from `books`. Then, we tokenize, remove punctuation, and lowercase the dataset and, finally, we create a bag-of-words model and compute the corresponding TF-IDF matrix representation.

```
>> % creates a string array with the raw text from books        (10.39a)
>> for k=1:length(books), docs(k) = books(k).text; end

>> % preprocess the documents                                   (10.39b)
>> tokdocs = tokenizedDocument(docs);
>> tokdocs = erasePunctuation(tokdocs);
>> tokdocs = lower(tokdocs);

>> % creates a bag-of-words model and computes the tfidf matrix (10.39c)
>> bow = bagOfWords(tokdocs);
>> bow = removeInfrequentWords(bow,5);
>> tfidfmtx = tfidf(bow);
>> size(tfidfmtx)
ans =
          66        4806
```

Notice, from (10.39c), that because we have removed infrequent words (those appearing five or less times in total) from the bag-of-words model, the resulting TF-IDF matrix covers only *4,806* vocabulary terms. Notice also that the matrix is transposed with respect to our original definition of a TF-IDF matrix. In this case, the documents are represented by the rows of the matrix and the terms are represented by the columns of the matrix.

Next, we create a document embedding by using SVD to project the *4,806* dimensions in (10.39c) into *200* dimensions. Then, we use MDS to create a 2D map of the document embedding.

```
>> % creates a document embedding by means of SVD               (10.40a)
>> [u,s,v] = svd(full(tfidfmtx'));
>> document_embedding = u(:,1:200)'*tfidfmtx';

>> % creates a 2D map of the embedding with MDS                 (10.40b)
>> docdist = pdist(full(document_embedding'),'cosine');
>> limit = statset('MaxIter',500);
>> [map,stress] = mdscale(docdist,2,'criterion','sammon','options',limit);
```

We are finally ready to plot the 2D map of the constructed document embedding. We use the book names as markers in the map and the book groups to specify the color of the marker. This will help to identify the individual books and the two subcollections in the map. The resulting plot is presented in Figure 10.10.

```
>> for k=1:length(books) % gets information about titles and groups    (10.41a)
       titles(k) = books(k).name;
       groups(k) = books(k).group;
   end

>> hf = figure(10); % creates the plot                                  (10.41b)
>> set(hf,'Color',[1,1,1],'Name','Document Embedding Visualization');
>> textscatter(map,titles,'TextDensityPercentage',100,'ColorData',groups)
>> axis([-1,1,-1,1])
```

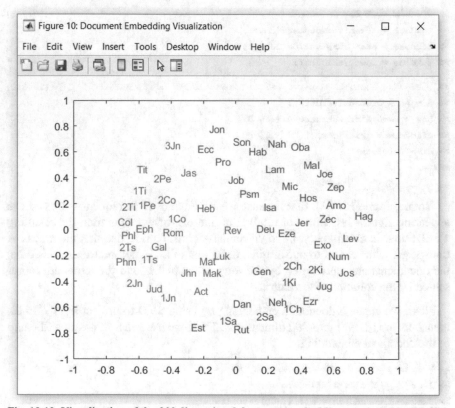

Fig. 10.10. Visualization of the *200*-dimensional document embedding computed in (10.40)

As seen from the figure, the resulting document embedding uncovers interesting properties of the collection's structure. For instance, notice the clear separation of the two subcollections, with *Old Testament* books to the right and *New Testament* books to the left. Also, notice the proximity among the gospel books: *Luk*,

Mat, Mak and *Jhn*. A more comprehensive exploration of the semantics properties of this document embedding is proposed in the second project of section 10.7.

10.5 Further Reading

Dimensionality reduction can be considered one of the core methodologies of text mining, and data mining in general. You can find a large number of materials addressing the general problem of dimensionality reduction, as well as specific dimensionality reduction methods. A good overview about dimensionality reduction can be found in (Fodor 2002).

The fundamental principles of linear projection methods can be traced back to Karl Pearson's work "On Lines and Planes of Closest Fit to Systems of Points in Space" (1901), more commonly known as Principal Component Analysis. A more recent reference for PCA would be (Jolliffe 1986). As discussed in section 10.2, this kind of analysis relies on the use of the Singular Value Decomposition (Golub and Kahan 1965). Several different but related techniques have been developed for specific analysis purposes, such as Factor Analysis (Gorsuch 1983) and Independent Component Analysis (Hyvärinen 1997), for example; as well as for improving implementation efficiency, Random Projections (Kaski 1998).

The use of linear projection methods on text processing was first proposed in the area of information retrieval (Deerwester *et al.* 1988) for document search applications. The same concepts had been applied to word models for lexical acquisition and other natural language applications (Landauer and Dumais 1997). These procedures are generally referred to as Latent Semantic Indexing (LSI) and Latent Semantic Analysis (LSA), respectively.

In an analogous way dimensionality reduction is applied to vector space models, we can also talk about dimensionality reduction within the context of statistical models. Indeed, topic models such as Probabilistic Latent Semantic Indexing (Hofmann 1999) and Latent Dirichlet Allocation (Blei *et al.* 2003) constitute dimensionality reduction mechanisms within the statistical modeling framework. In these models, documents are represented as mixtures of topics. Accordingly, topics can be considered as probabilistic analogues to dimensions of a reduced space in the vector space framework. A good introduction to probabilistic topic models, along with their main differences and similarities regarding vector space models, can be found in (Griffiths *et al.* 2007).

Regarding non-linear projection methods, MDS has been used for both generating maps of documents (Matsuda and Yamaguchi 2005) and studying semantic relationships among words (Wälchli and Cysouw 2012). A more comprehensive introduction to MDS can be found in (Cox and Cox 2002). Similarly, TSNE has been used to project high-dimensional datasets into spaces of very low dimensionality for visualization purposes. A good introduction to TSNE can be found in (van der Maaten and Hinton 2008). Both MDS and TSNE are mainly used for visualization purposes as they do not provide mapping operators to project new data samples into an existent low-dimensional representation. Other non-linear projection methods, which have the advantage of providing mapping functions that can be used to project new data samples, include neural network architectures such as, for example, Self-Organized Feature Maps (Kohonen 1990) and Auto Encoder Neural Networks (Hinton and Salakhutdinov 2006).

More specific architectures and methods to build word embeddings include skip-gram and continuous bag-of-words (Mikolov *et al.* 2013a and 2013b), GloVe (Pennington *et al.* 2014), fastText (Bojanowski *et al.* 2017) and, more recently, contextualized word embeddings such as BERT (Devlin *et al.* 2018) and ELMo (Peters *et al.* 2018), among others. Although not as popular as the work related to constructing word embeddings, some attention has been paid to the construction of sentence, paragraph, and document embeddings too. Some interesting approaches can be found in (Le and Mikolov 2014) and (Wu *et al.* 2018).

Finally, regarding vocabulary pruning and vocabulary merging techniques, a more sophisticated pruning method than the one described in this chapter can be found in (Madsen *et al.* 2004). Lemmatization and stemming algorithms are described in (Koskenniemi 1983) and (Hull 1996), respectively. Some experimental work on document pruning can be found in (Karlgren 1999).

10.6 Proposed Exercises

1. Consider the naïve stemming procedure implemented in (10.5).

   – Repeat the stemming procedure for different values of *n*. More specifically, vary *n* from *1* to *10*, and generate a plot of vocabulary size versus *n*.
   – Consider the specific case of *n* = *5*. Explore the resulting vocabulary mappings and compare those results to lemmatization and stemming produced with the Text Analytics Toolbox™ function `normalizeWords` (you might want to use `addPartOfSpeechDetails` first to improve lemmatization performance). What are your main observations?

2. In section 10.1 we have considered the case in which dimensionality reduction is applied to the column vectors of the TF-IDF matrix, i.e. the words are the variables and the documents are the data samples. In this exercise we consider applying dimensionality reduction to the row vectors of the TF-IDF matrix.

   – Can you suggest general criteria for "document pruning" and "document merging" in the case of a TF-IDF matrix?

   – Implement your proposed procedures for removing columns from a TF-IDF matrix (the "document pruning" case), as well as for combining columns (the "document merging" case).

3. Implement a function for restricting the cosine similarity scores that are computed in a reduced space to the interval *[0,1]*.

   – Consider the inclusion of an input variable that allows for selecting whether truncation or normalization should be used.
   – Generate two histograms similar to the one in Figure 10.3 by applying both restricting strategies (truncation and normalization) to the cosine scores computed in (10.14c). Evaluate the differences between the two cases.

4. Consider the distributions of similarities presented in Figures 10.1 and 10.3. Compute and plot similar histograms according to the following procedure:

   – Extract a subset of TF-IDF word vectors by randomly selecting *100* vocabulary terms (bias your sampling algorithm to avoid the selection of terms occurring less than *10* times in the collection).
   – Compute and plot histograms for word similarities at full dimensionality space as well as reduced spaces of *1000*, *500*, *100*, *50* and *10* dimensions. What are your main observations?

5. Let us explore in more detail how the relative distances between a given pair of words varies when the dimensionality of the reduced space is varied.

   – Consider the word space model defined in (10.17), which constitutes a high-dimensional representation of the vocabulary set in (10.18).
   – Use SVD to compute the *500* largest singular values and their corresponding singular vectors.
   – Compute and plot the cosine similarities between the following pairs of words: *water–drink*, *water–food* and *water–linen*, for dimensionalities varying within the range from *100* to *500*.
   – Repeat the previous step for different sets of word pairs such as *word–say*, *word–thought* and *word–desert*; *sun–moon*, *sun–light* and *sun–bread*; as well as other sets of your own selection.
   – Can you derive a general conclusion from these results?

6. Consider the linear projection method presented in section 10.2. More specifically, the procedure defined in (10.16).

   – Create a two-dimensional map for the same set of vocabulary terms in (10.28a): *bird, cloud, field, fish, flock, goat, lighting, mountain, rain, river, sea, sheep, sky, storm, thunder* and *wind*.
   – What differences do you observe between the resulting map and the one presented in Figure 10.6?

7. Consider the pretrained word embedding `fastTextWordEmbedding` that was introduced in (10.30)

- Similar to (10.33) and (10.34), test a few different types of semantic relationships among words. Consider, for instance, *cow-milk* vs. *bee-honey*, *plane-sky* vs. *boat-sea*, *doctor-patient* vs. *professor-student*, and so on.
- Consider different neighborhood sizes (from 1 to 5) and compute the success rate of the semantic association predictions for each case.
- Can you find any potential limitations of the considered type of word embeddings? (Hint: consider the case of *keyboard-typing* vs. *mouse-clicking*.)

8. Study the Text Analytics Toolbox™ functions `wordEncoding`, `word2ind` and `ind2word`, which are used to map words into numeric indexes and vice versa.

- Use `wordEncoding` to generate a word encoding for the list of *16* words defined in (10.28a), and `fastTextWordEmbedding` to load a pretrained word embedding as it was done in (10.30).
- Create a *16x300* matrix to accommodate the word vectors of the words in the list. Use `word2vec`, as in (10.34a), to retrieve each vector from the embedding along with `ind2word` to handle the words by their index.
- Use the constructed matrix to compute cosine similarities for all word pairs and build a script to retrieve the similarity score for a given pair of words w_1 and w_2. Use `word2ind` so that word strings can be passed to the script.
- Use the script to explore cosine similarities of different word pairs, such as *sheep* and *goat*, *sheep* and *fish*, *rain* and *cloud*, *rain* and *flock*, etc. Do the observed values make sense? What are your main conclusions?
- Follow the same steps in (10.28c) and (10.28d) to project the *16x300* matrix computed above into a *2*-dimensional space. Plot the resulting word map and compare it to the one in Figure 10.6. What are your observations?

10.7 Short Projects

1. Consider the complete works of William Shakespeare, available in Project Gutenberg's website http://www.gutenberg.org/ebooks/100 Accessed 6 July 2021.

- Download the plain text *UTF-8* version and manually remove all information not relevant to the works of Shakespeare that appear at the beginning and at the end of the file.
- Use the Text Analytics Toolbox function `trainWordEmbedding` to train a word embedding form this data collection.
- Use functions `tsne` and `textscatter` to create an interactive plot, like the one in Figure 10.8, to visualize the constructed embedding.

   - Explore the resulting word embedding and compare it with the one from Figure 10.8. What similarities and differences do you observe?

2. Consider the English, Spanish, and Chinese versions of the Holy Bible dataset available from the companion website http://www.textmininglab.net/ Accessed 6 July 2021.

   - Download the files and process the datasets to create a `books` structure array like the one in (10.38). The structure array must include the fields: `group`, indicating the subcollection; `name`, containing the title abbreviation; and `entext`, `estext`, and `zhtext`, containing the corresponding raw texts of the English, Spanish, and Chinese book versions, respectively.
   - Follow the procedure in (10.39) to create a string array of books for each language, preprocess the documents, create the bag-of-words models and compute the corresponding TF-IDF matrices.
   - Follow the procedures in (10.40) and (10.41) to create and visualize document embeddings for the three languages. Use the book title abbreviations as markers and the group category to determine the color of each marker.
   - Study in detail the similarities and differences observed among the three maps. What can you say about these results?
   - Repeat several times the previous steps while experimenting with different parameters and dimensions. What are your main observations?

10.8 References

Blei DM, Ng AY, Jordan MI (2003) Latent Dirichlet Allocation. Journal of Machine Learning Research 3: 993–1022

Bojanowski P, Grave E, Joulin A, Mikolov T (2017) Enriching word vectors with subword information. Transactions of the Association for Computational Linguistics 5: 135–146

Cox MF, Cox MAA (2001) Multidimensional Scaling. Chapman and Hall, Boca Raton

Deerwester S, Dumais S, Landauer T, Furnas G, Beck L (1988) Improving information retrieval with latent semantic indexing. Proceedings of the 51st Annual Meeting of the American Society for Information Science 25: 36–40, Atlanta, GA

Devlin J, Chang MW, Lee K, Toutanova K (2019) BERT: Pre-training of Deep Bidirectional Transformers for Language Understanding, https://arxiv.org/abs/1810.04805v2 Accessed 6 July 2021

Fodor IK (2002) A survey of dimension reduction techniques. U.S. Department of Energy, Lawrence Livermore National Laboratory, UCRL-ID-148494

Golub GH, Kahan W (1965) Calculating the singular values and pseudo-inverse of a matrix. Journal of the Society for Industrial and Applied Mathematics: Numerical Analysis 2(2): 205–224

Gorsuch RL (1983) Factor Analysis. Lawrence Erlbaum, Hillsdale, NJ

Griffiths T, Steyvers M, Tenenbaum JB (2007) Topics in Semantic Representation. Psychological Review 144(2): 211–244

Hinton GE, Salakhutdinov RR (2006) Reducing the dimensionality of data with neural networks. Science 313: 504–507

Hofmann T (1999) Probabilistic Latent Semantic Analysis. In Proc. of the 15th Conference on Uncertainty in Artificial Intelligence

Hull DA (1996) Stemming algorithms: a case study for detailed evaluation. Journal of the American Society of Information Sciences 47(1): 70–84

Hyvärinen A (1999) Survey on independent component analysis. Neural Computing Surveys 2: 94–128

Jolliffe IT (2002) Principal component analysis. Springer-Verlag, New York, NY

Karlgren J (1999) Stylistic experiments in information retrieval. In Strzalkowski (ed): Natural language information retrieval, pp: 147–166, Kluwer

Kaski S (1998) Dimensionality reduction by random mapping: fast similarity computation for clustering. In Proc. IEEE International Joint Conference on Neural Networks, pp: 413–418

Kohonen TK (1990) The self-organizing map. In Proc. IEEE, 78(9): 1464–1480

Koskenniemi KM (1983) Two-level morphology: a general computational model for word-form recognition and production. Technical report, University of Helsinki, Helsinki, Finland

Landauer T, Dumais S (1997) A solution to Plato's problem: the latent semantic analysis theory of acquisition, induction and representation of knowledge. Psychological Review 104(2): 211–240

Le Q, Mikolov T (2014) Distributed Representations of Sentences and Documents. Proceedings of the 31st International Conference on Machine Learning, in PMLR 32(2): 1188–1196

Madsen RE, Sigurdsson S, Hansen LK, Lansen J (2004) Pruning the vocabulary for better context recognition. In Proc. of the 7th International Conference on Pattern Recognition

Matsuda Y, Yamaguchi K (2005) An efficient MDS algorithm for the analysis of massive document collections. In Khosla, R. et al. (eds.): KES 2005, LNAI 3682: 1015–1021. Springer-Verlag Berlin Heidelberg

Mikolov T, Sutskever I, Chen K, Corrado GS, Dean J (2013a) Distributed representations of words and phrases and their compositionality. Proceedings of Advances in Neural Information Processing Systems

Mikolov T, Chen K, Corrado GS, Dean J (2013b) Efficient estimation of word representations in vector space. Proceedings of ICLR Workshop

Pearson K (1991) On lines and planes of closest fit to systems of points in space. Philosophical Magazine 2(6): 559–572

Pennington J, Socher R, Manning C (2014) GloVe: Global Vectors for word representation. In Proceedings of the Conference on Empirical Methods in Natural Language Processing, ACL

Peters ME, Neumann M, Iyyer M, Gardner M, Clark C, Lee K, Zettlemoyer L (2018) Deep contextualized word representations. Proceedings of the North American Chapter of the Association for Computational Linguistics

van der Maaten L, Hinton G (2008) Visualizing data using t-SNE. Journal of Machine Learning Research, 9: 2579–2605

Wälchli B, Cysouw M (2012) Lexical typology through similarity semantics: Toward a semantic map of motion verbs, Linguistics 50(3): 671–710, https://core.ac.uk/download/pdf/194183 08.pdf Accessed 6 July 2011

Wu L, Yen EH, Xu K, Xu F, Balakrishnan A, Chen PY, Ravikumar P, Witbrock MJ (2018) Word mover's embedding: from word2vec to document embedding. Proceedings of the Conference on Empirical Methods in Natural Language Processing (EMNLP), pp: 4524–4534

Part III: Methods and Applications

*"It is common sense to take a method and try it.
If it fails, admit it frankly and try another,
but above all, try something."*

Franklin D. Roosevelt

11 Document Categorization

This chapter opens the third part of the book, which focuses on different practical text mining applications. In this chapter we will discuss in detail the problem of document categorization. The main objective of document categorization is to assign each document in a given data collection to a class or category, according to the nature of its content. In general, document categorization, can be used to directly address different practical tasks, such as spam filtering, press clipping and document clustering, just to mention a few. Alternatively, it can be used as a component of a larger system to tackle more complex tasks, such as, for example, opinion mining and plagiarism detection.

This chapter is organized as follows. First, in section 11.1, we illustrate the preparation of the dataset to be used along the chapter. Then, in section 11.2, we focus our attention on the problem of unsupervised clustering, in which a document collection is automatically organized into categories (clusters) by only considering the text contents within it and without taking into account any additional information about the nature of the document. Finally, in sections 11.3 and 11.4, we focus our attention into the supervised learning approach to document categorization, in which the categories are known, and sample documents (training data) are used to guide the categorization process. More specifically, in section 11.3, we focus on geometrically inspired methods; and, in section 11.4, we devote our attention to statistically motivated methods.

11.1 Data Collection Preparation

Before proceeding to illustrate document categorization applications, let us first introduce the data collection we will be using in this chapter. The data collection used here has been extracted from three different books:

- Oliver Twist. A novel by English author Charles Dickens, which was published in 1838. It tells the story of an orphan boy, Oliver Twist, who escapes from his guardian and goes to London, where he meets a leader of a gang and unknowingly gets involved in their criminal activities.
- Don Quixote (English translation). A novel by Spanish writer Miguel de Cervantes, which was published in two volumes, in 1605 and 1615. It tells the adventures of a country gentleman that gets obsessed after excessively reading chivalry books and ends up thinking he is also a chevalier. Don Quixote is considered the most influential work of the Spanish literature.

R. E. Banchs, *Text Mining with MATLAB®*, https://doi.org/10.1007/978-3-030-87695-1_11

- Pride and Prejudice. A novel by English novelist Jane Austen, which was published in 1813. It tells the story of Elizabeth Bennet, the second of five sisters in a landed gentry family, who deals with matters of morality, education, and manners in her 19th century England context.

The complete books are publicly available in digital format from the Project Gutenberg website http://www.gutenberg.org/ Accessed 6 July 2021. The data collection to be used here was prepared by extracting sample paragraphs from the three books, where the only restriction imposed was for each paragraph to be approximately in the range from *60* to *300* words in length. According to this, the documents in our data collection correspond to paragraphs of the original books.

The dataset has been already formatted into a `table` object named `dataset`, which contains one sample paragraph per row and three variables (columns): `book` (the name of the book), `chapter` (the chapter number the paragraph belongs to) and `text` (the raw text of the paragraph). The `dataset` is available, along with a categorical array `books` that contains the titles of the three books under consideration, in the file `datasets_ch11.mat`. Let us now load the datasets:

```
>> clear;                                                              (11.1a)
>> load datasets_ch11
>> whos
  Name          Size              Bytes  Class          Attributes
  books         1x3                 401  categorical
  dataset       2349x3          3094265  table

>> books                                                               (11.1b)
books =
  1×3 categorical array
     OLIVER TWIST      DON QUIXOTE      PRIDE AND PREJUDICE

>> head(dataset,10)                                                    (11.1c)
ans =
  10×3 table
        book          chap                          text¹                  …
    OLIVER TWIST       1     "Among other public buildings in a certain town …
    OLIVER TWIST       1     "For a long time after it was ushered into this …
    OLIVER TWIST       1     "Although I am not disposed to maintain that th …
    OLIVER TWIST       1     "'Lor bless her dear heart, when she has lived   …
    OLIVER TWIST       1     "'You needn't mind sending up to me, if the chi …
    OLIVER TWIST       1     "What an excellent example of the power of dres …
    OLIVER TWIST       2     "For the next eight or ten months, Oliver was t …
    OLIVER TWIST       2     "Everybody knows the story of another experimen …
    OLIVER TWIST       2     "Occasionally, when there was some more than us …
    OLIVER TWIST       2     "It cannot be expected that this system of farm …
```

---

[1] `text` fields have been truncated for display purposes.

Next, we randomize[2] and preprocess the data. In the preprocessing step (11.2b), we create the string array `docs` containing a tokenized, punctuation-removed, and lowercased representation of the paragraphs, and the categorical array `labels` containing the corresponding category (book) of each paragraph. We also create the auxiliary variable `nbooks`, which indicates the number of books in the collection.

```
>> % resets the random number generator and randomizes the data       (11.2a)
>> rng('default');
>> dataset = dataset(randperm(height(dataset)),:);

>> % preprocess the data                                              (11.2b)
>> docs = tokenizedDocument(dataset.text);
>> docs = erasePunctuation(docs);
>> docs = lower(docs);
>> labels = dataset.book;
>> nbooks = length(books);
```

Before continuing, let us explore in more detail each of three subcollections contained in the dataset in terms of number of documents (paragraphs) per book, and document size (in number of words) distributions for each case. These results are illustrated in Figure 11.1, which is generated as follows:

```
>> hf=figure(1); % creates a new figure                               (11.3a)
>> set(hf,'Color',[1 1 1],'Name','Basic description of the Dataset');

>> % prints subcollection's names and number of documents             (11.3b)
>> subplot(2,2,1); axis off
>> for k=1:nbooks
      ndocs = sum(labels==books(k));
      comment = sprintf('%s\nDataset %d (%d documents)\n',books(k),k,ndocs);
      text(0,1.1-k/3,comment);
   end

>> % prints histograms for each subcollection                         (11.3c)
>> for k=1:nbooks
      index = labels==books(k);
      chapsizes = doclength(docs(index));
      [freqs,docsizes] = hist(chapsizes,min(chapsizes):max(chapsizes));
      subplot(2,2,k+1);
      bar(docsizes,freqs);
      axis([61,300,0,25]);
      text(220,23,sprintf('Dataset %d',k));
      xlabel('Document Size (in words)'); ylabel('Frequency');
   end
```

---

[2] Randomization is used here to ensure a good distribution of dataset samples across the different partitions to be created later in (11.4).

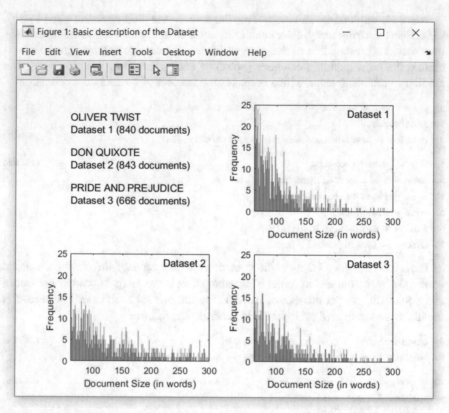

Fig. 11.1. Histograms of document sizes (in number of words) for the three subcollections composing the dataset

As seen from Figure 11.1, dataset 1 (Oliver Twist) contains *840* documents (paragraphs), which is similar in size to dataset 2 (Don Quixote) with *843* documents. On the other hand, dataset 3 (Pride and Prejudice) is the smallest of the three subsets with only *666* documents, which is approximately *20%* less documents than the ones contained in the other two subsets. Also, from Figure 11.1, it can be seen that all three subcollections exhibit a similar distribution of document lengths. However, certain predominance of shorter documents can be observed for the case of dataset 1, as well as a relatively larger number of longer documents, within the range between *200* and *300* words, for the case of dataset 2.

A more comprehensive set of statistics about the collection is summarized in Table 11.1 (the generation of the table is left to you as an exercise). Some interesting observations can be derived from the table. For instance, notice that, although exhibiting similar number of documents and vocabulary sizes, the average document size and number of running words in dataset 2 are about *30%* larger than the ones in dataset 1. This is consistent with the distribution differences previously observed in Figure 11.1 and might be explained by the more verbose nature of

Spanish as compared to English, which seems to be preserved even after translation. In the case of dataset 3 compared to dataset 1, notice that while the number of documents, the vocabulary size, and the number of running words are consistently smaller in dataset 3, the average document size is larger. This mainly responds to differences in the literary styles between these two books.

Table 11.1. Summary of main data collection statistics

Sub collection	Dataset 1	Dataset 2	Dataset 3
Book title	Oliver Twist	Don Quixote	Pride & Prejudice
Documents (paragraphs)	840	843	666
Running words	83,942	110,384	75,536
Vocabulary	8,597	8,579	5,373
Minimum doc. size	58	59	60
Maximum doc. size	284	298	298
Average doc. size	99.93	130.94	113.42

Finally, let us prepare the data collection for experimentation purposes. First, we split the dataset into a data partition of three sets: train, development, and test. The procedure is described for the test partition (11.4a) and then repeated for the development (11.4b) and train (11.4c) partitions. Later in this chapter, these partitions are used for training the models, tunning model hyper-parameters (if any), and evaluating the performance of the models, respectively.

```
>> % prepares the test set partition (200 samples)          (11.4a)
>> tstidx = 1:200; % defines the index range
>> ntst = length(tstidx); % size of the test set
>> tstdocs = docs(tstidx); % test set documents
>> tstlbls = labels(tstidx); % test set labels

>> % prepares the development set partition (200 samples)    (11.4b)
>> devidx = 201:400; ndev = length(devidx);
>> devdocs = docs(devidx); devlbls = labels(devidx);

>> % prepares the train set partition (1949 samples)         (11.4c)
>> trnidx = 401:length(docs); ntrn = length(trnidx);
>> trndocs = docs(trnidx); trnlbls = labels(trnidx);
```

Next, we proceed to compute a bag-of-words model for the dataset. Notice this is done with the train set partition. For the computed bag-of-words we remove infrequent words (those appearing 5 or less times in the train set) and frequent words (the 40 most frequent words). We also create two auxiliary variables containing the vocabulary and the vocabulary size, which will be used later.

```
>> trnbow = bagOfWords(trndocs); % computes bag-of-words model   (11.5a)
```

```
>> % removes infrequent words (counts <= 5)                          (11.5b)
>> trnbow = removeInfrequentWords(trnbow,5);

>> % removes 40 most frequent words                                  (11.5c)
>> frqwds = topkwords(trnbow,40).Word;
>> trnbow = removeWords(trnbow,frqwds);

>> vocab = trnbow.Vocabulary; % stores the vocabulary               (11.5d)
>> vsize = length(vocab); % vocabulary size
```

Now, we use the bag-of-words model from (11.5) to compute TF-IDF matrices for all three partitions in the dataset:

```
>> % computes tfidf matrices for train, development, and test       (11.6)
>> trntfidf = tfidf(trnbow);
>> tsttfidf = tfidf(trnbow,tstdocs);
>> devtfidf = tfidf(trnbow,devdocs);
```

and we save all variables related to the generated partitions, which are used in the rest of the chapter, into the file `datasets.mat`.

```
>> save datasets books nbooks vocab vsize ntst tstdocs tstlbls ...   (11.7)
   tsttfidf ndev devtfidf devdocs devlbls ntrn trndocs trnlbls trnbow trntfidf
```

Table 11.2 summarizes and describes all variables saved to `datasets.mat`.

Table 11.2. All variables in the dataset partition to be used for document categorization

Variable	Size	Class	Description
books	1x3	categorical	Categories (books) in the dataset
nbooks	1x1	double	Number of categories (books)
vocab	1x3168	string	Dataset vocabulary
vsize	1x1	double	Vocabulary size
ntst	1x1	double	Number of documents in the test set
tstdocs	200x1	tokenizedDocument	The documents in the test set
tstlbls	200x1	categorical	The labels in the test set
tsttfidf	200x3168	double (sparse)	TF-IDF matrix of the test set
ndev	1x1	double	Number of documents in the development set
devdocs	200x1	tokenizedDocument	The documents in the development set
devlbls	200x1	categorical	The labels in the development set
devtfidf	200x3168	double (sparse)	TF-IDF matrix of the development set
ntrn	1x1	double	Number of documents in the train set
trndocs	1949x1	tokenizedDocument	The documents in the train set
trnlbls	1949x1	categorical	The labels in the train set
trnbow	1x1	bagOfWords	The bag-of-words model
trntfidf	1949x3168	double (sparse)	TF-IDF matrix of the train set

11.2 Unsupervised Clustering

Being able to organize a collection of objects into groups or categories is the basis for the abstraction and generalization processes of human thinking. Such a categorization process is based on the observed similarities and differences among the most salient attributes of the objects under consideration, which allow for describing the subjacent structure of the given data collection. In the case of text, such as in a collection of documents, the salient features of the documents are the words they contain and the semantic relationships they define across documents within the collection. In this sense, the problem of organizing a collection of documents into groups or categories is a problem of semantic nature.

More specifically, unsupervised clustering, or simply clustering, is the process of automatically grouping the objects in a given collection according to the similarities and differences of their salient features. While elements within each group or cluster are expected to exhibit large similarities among them, elements across different groups or clusters are expected to exhibit large differences among them. The process is referred to as unsupervised, as far as no information about the categories coexisting in the collection is previously know or used during the process. Unsupervised clustering is a powerful procedure for discovering the hidden structure, if any, of a given data collection. In this section we will focus our attention into a specific method of cluster analysis denominated k-means clustering.

The k-means clustering algorithm is an iterative process in which a collection of n observations is partitioned into k clusters, which are updated iteratively until they converge into a stable partition. Each algorithm's iteration is performed in two steps: assignment and updating. In the assignment step, each element in the collection is assigned to the closest cluster by means of a distance metric. The distance between a cluster and a given element is computed in a vector space representation by measuring the distance between the given element representation and the centroid of the cluster. In the updating step, all cluster centroids are updated by taking into consideration the new partition generated during the assignment step. In k-means clustering, centroid vectors are represented by the mean vector of all elements belonging to the corresponding clusters. The algorithm starts from a predefined set of centroids (which can be generated either randomly or by means of any other criterion) and performs consecutive iterations of the assignment and updating steps until no more changes to the partition are observed over consecutive iterations. Figure 11.2 illustrates the main idea behind k-means clustering.

The MATLAB® function `kmeans` implements the k-means clustering algorithm. The function syntax is `[class,centroids] = kmeans(dataset,k)`, where the input variable `dataset` is a matrix representation of the object collection, the rows being the observations (documents) and the columns being the variables (terms), and `k` is the desired number of clusters. The output variable `class` is a numeric array containing the index of the cluster each element in the collection has been assigned to, and `centroids` is a matrix containing the corresponding cluster centroid locations. The function `kmeans` also admits additional input parameters as well as it can return additional output variables. For a more detailed description of this function, you can either refer to the function's online documentation or just enter `help kmeans` in the command window.

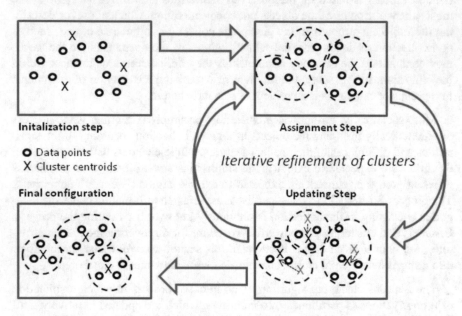

Fig. 11.2. Illustrative example on the operation of the *k*-means clustering algorithm

Notice that *k* (the number of clusters) is an input variable for the *k*-means algorithm. This means that before applying *k*-means to a data collection, the number of clusters must be defined. In the particular case of the data collection we are using in this chapter, we already know it is composed of paragraphs extracted from three different books. Accordingly, a value of *k=3* seems to be an appropriate choice in this case. However, in general, this type of knowledge is not available, and selecting a value of *k* for a given dataset is a non-trivial and difficult problem.

The selection of *k* is typically done by comparing clustering qualities for different values of *k*. Different metrics of clustering quality are available, which are conceptually based on similar principles. We focus our attention here on the Dunn Index, which assesses cluster quality by looking at the ratio between inter-cluster distances (i.e. between elements from different clusters) and intra-cluster distances (i.e. between elements within the same cluster). More specifically, the Dunn Index is computed as the ratio between the minimum inter-cluster distance and the maximum intra-cluster distance:

$$dunn(k) = min_{1 \leq i < j \leq k} \, d(c_i, c_j) \, / \, max_{1 \leq m \leq k} \, d(c_m)$$

(11.8)

where $d(c_i, c_j)$ represents the inter-cluster distance between clusters i and j, and $d(c_m)$ represents the intra-cluster distance within cluster m.

As good clusters are expected to be concentrated (i.e. all elements within the same cluster must be close to each other) and well separated (i.e. clusters must be far away from each other), they are expected to exhibit small intra-cluster distances and large inter-cluster distances. Accordingly, a higher Dunn Index will be an indication of better clustering. Notice from (11.8) that the index definition does not impose any specific distance formulation. Indeed, different formulations can be used, resulting in different versions of the index. In the specific implementation used here, we compute inter-cluster distances $d(c_i, c_j)$ by considering the distances between the cluster centroids, and intra-cluster distances $d(c_m)$ by considering the maximum distance between any element in the cluster and its centroid.

Let us now prepare for the experimental work in this section by clearing the workspace, resetting the random variable generator, and loading the dataset previously saved to **datasets.mat** in (11.7).

```
>> clear; rng('default'); load datasets                          (11.9)
```

Next, let us compute the Dunn Index for different values of k over the train set of the data collection. As resulting Dunn Index values typically vary due to different random initializations of the k-means algorithm, we run 5 different simulations for each different value of k and report Dunn Index averages.

```
>> % computes Dunn Index for numbers of clusters between 2 and 10      (11.10a)
>> for ncls=2:10, n = ncls-1;
       for m=1:5 % computes 5 different simulations for each case
           % applies the k-means clustering algorithm
           [idxs,centroids] = kmeans(trntfidf,ncls,'Distance','cosine');
           % initializes variable for intra-cluster distances
           indist = zeros(1,ncls);
           for k=1:ncls % computes intra-cluster distances
               temp = 1-cosineSimilarity(trntfidf(idxs==k,:),centroids(k,:));
               indist(k) = max(temp);
           end
           indistmax = max(indist); % maximum intra-cluster distance
           % computes the minimum inter-cluster distance
           outdistmtx = 1-cosineSimilarity(centroids);
           outdistmin = min(outdistmtx(triu(outdistmtx,1)>0));
           % computes the Dunn Index
           dunn(n,m) = outdistmin/indistmax;
           nclusts(n) = ncls;
       end
   end

>> averagedunn = mean(dunn'); % computes Dunn Index averages          (11.10b)
```

The resulting Dunn Index averages for each considered number of clusters k are depicted in Figure 11.3 (the generation of the bar plot is left to you as an exercise). As seen from the figure, the maximum Dunn Index value is obtained for $k=3$, which confirms our original intuition about the best number of clusters to be aligned with the number of books (**nbooks**) in the data collection.

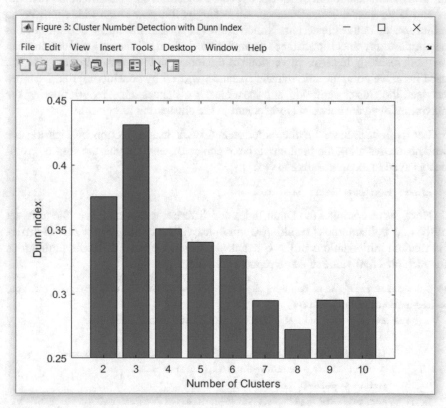

Fig. 11.3. Dunn Index averages for different numbers of clusters *k*

To evaluate the quality of the proposed clustering mechanism in terms of document categorization, we use the test set for evaluation purposes. In order to do that, we need to compute the cluster assignments for each sample in the test set. Accordingly, we first apply k-means to the train set (with $k=3$) to obtain the cluster centroids and, afterwards, we assign each test set sample to one of the three clusters based on its proximity to the cluster centroids:

```
>> % applies k-means (with k=nbooks) to the train set              (11.11a)
>> [~,centroids] = kmeans(trntfidf,nbooks,'Distance','cosine');

>> % finds the closest centroid for each sample in the test set    (11.11b)
>> [~,idxs] = max(cosineSimilarity(centroids,tsttfidf));
```

At this point, we should realize about the problem of mapping the cluster indexes `idxs` into the three books in the collection. As the clustering in (11.11a) is an unsupervised process, there is no knowledge about the three books associated to the resulting clusters that relates the cluster indexes to the books. So, in order to evaluate classification accuracy, we need to infer this mapping.

We can visually infer the mapping between cluster indexes and books by creating a cross-plot of cluster indexes vs. books for the samples in the test set. As we expect the resulting clusters to be mainly aligned with the three book categories, we should be able to observe a larger concentration of samples in those cases in which the cluster index and the book correspond to each other. We create the cross-plot by using the following procedure (notice that a small amount of noise is added for visualization purposes).

```
>> hf = figure(4);                                                          (11.12)
>> figtitle = 'Cross-plot between cluster indexes and category labels';
>> set(hf,'Color',[1 1 1],'Name',figtitle);
>> for n=1:length(tstlbls) % gets book labels of samples in the test set
        nlbl(n,1) = find(tstlbls(n)==books);
   end
>> plot(nlbl+randn(size(nlbl))/10,idxs+randn(size(idxs))/10,'.');
>> xlabel('Actual Category Labels'); ylabel('Cluster Indexes');
>> xticks([1,2,3]); xticklabels(books);
```

The resulting cross-plot is depicted in Figure 11.4. From the figure, it becomes evident that cluster #1 mainly corresponds to Pride and Prejudice, cluster #2 to Oliver Twist, and cluster #3 to Don Quixote.

Such a mapping between clusters and books can also be inferred automatically by a brute-force exploration of the amount of overlap between clusters and books across all possible mappings. This is illustrated by the following procedure:

```
>> % considers all possible mappings                                        (11.13a)
>> permutations = perms([1,2,3]);

>> % computes cluster-book overlaps for all possible mappings               (11.13b)
>> for n=1:size(permutations,1), temp = 0;
        for k = 1:nbooks
            clusters = idxs'==k;
            permutedbooks = tstlbls==books(permutations(n,k));
            temp = temp + sum(clusters & permutedbooks);
        end
        overlaps(n) = temp;
   end

>> % gets the best mapping (i.e. maximum overlap)                           (11.13c)
>> [~,best] = max(overlaps);
```

```
>> books(permutations(best,:))
ans =
  1×3 categorical array
     PRIDE AND PREJUDICE        OLIVER TWIST        DON QUIXOTE
```

where, again, we confirm that clusters *#1*, *#2* and *#3* correspond to Pride and Prejudice, Oliver Twist, and Don Quixote, respectively.

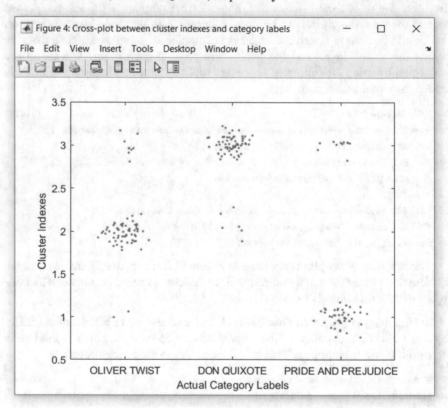

Fig. 11.4. Cross-plot of cluster assignments versus book categories (some noise has been added to the indexes and categories to facilitate visualization)

Once we have established the appropriate mapping between the generated cluster indexes and the actual categories in the collection, we are ready to assess the quality of the performed categorization over the test set. We do this by means of classification accuracy, which is defined as the percentage of successful categorizations with respect to the total amount of elements being categorized:

$$accuracy = correct\_cases \,/\, all\_cases \times 100\% \qquad \text{(11.14)}$$

where *correct_cases* refer to the total amount of documents assigned to the correct category, and *all_cases* refer to the number of documents being categorized.

Before computing the accuracy, we apply the mapping inferred in (11.13) to all cluster indexes `idxs` assigned to the samples in the test set:

```
>> predictions = books(permutations(best,idxs))';                    (11.15)
```

such that they can be directly compared to the array of test set labels `tstlbls`

```
>> accuracy = sum(predictions==tstlbls)/ntst*100                     (11.16)
accuracy =
    86
```

As seen from (11.16) the resulting accuracy is *86%*. This result happens to be far above random selection (i.e. assigning each document to one of the three clusters at random), which would result in an accuracy of about *33%*. Indeed, the *k*-means clustering algorithm has exploited the implicit structure of the dataset for generating a partition that approximates to a good extent the actual categories in the dataset. Notice that, except for the number of categories coexisting in the dataset, no previous knowledge about the nature of dataset has been used.

For better understanding the differences between the resulting partition and the original document categories, a confusion matrix can be computed. Like the cross-plot depicted in Figure 11.4, the confusion matrix provides useful information about the degree of overlap among different clusters and categories.

```
>> % computes and displays the confusion matrix                      (11.17)
>> confusion_mtx = confusionmat(predictions,tstlbls,'ORDER',books);3
>> cmtx = array2table(confusion_mtx,'VariableNames',string(books));
>> cmtx.('Classified as') = string(books)';
>> disp(cmtx)
```

OLIVER TWIST	DON QUIXOTE	PRIDE AND PREJUDICE	Classified as
62	6	0	"OLIVER TWIST"
9	65	11	"DON QUIXOTE"
2	0	45	"PRIDE AND PREJUDICE"

The result in (11.17) is basically the same one from Figure 11.4. The only difference is that, by using the permuted cluster indexes obtained in (11.13), the largest cluster-to-category overlaps lay along the main diagonal of the matrix. The information provided by the confusion matrix is very useful to understand how well the resulting clusters approximate or represent the actual document categories composing the dataset. If we think about the clustering conducted in (11.11) as an unsupervised classification task, the confusion matrix provides the required information to evaluate the performance of such a task.

---

[3] Notice we have swapped predictions and reference labels in the function call (see `help confusionmat`). Accordingly, in the confusion matrix presented in (11.17), actual labels correspond to columns and predicted labels to rows, which is consistent with the representation in Figure 11.4.

Indeed, we can directly compute the same classification accuracy score, as defined in (11.14), from the confusion matrix by noticing that the elements along its main diagonal are precisely the *correct_cases* from (11.14). Then, the classification accuracy can be also derived from the confusion matrix as follows:

```
>> accuracy = sum(diag(confusion_mtx))/sum(sum(confusion_mtx))*100          (11.18)
accuracy =
   86
```

Next, we shift our attention to supervised methods for document categorization.

11.3 Supervised Classification in Vector Space

In the previous section, we conducted unsupervised clustering, in which no previous knowledge about the data collection was supposed to be available. In this section we will focus our attention on the supervised classification approach. In the supervised classification setting, we have access to information about the different categories in the data collection. This information is usually available in the form of data samples for which the categories are know. These data samples are used by the algorithms to learn the main properties and characteristics of the different data categories, which are then exploited to categorize new data samples.

In this section, more specifically, we start by considering a very common supervised approach known as k nearest neighbors or *knn*. This method constitutes a remarkably simple but robust classification algorithm that, similarly to k-means clustering, operates over a vector space model. The basic idea of the *knn* algorithm is to assign a new data sample to a category based on the categories of the closest samples for which the categories are known. In other words, given a new data sample x, its k closest samples are extracted from the train set, which are then referred to as the nearest neighbors of x. Finally, x is assigned to the most common category that is observed among its neighbors.

As we did in the previous section, we prepare for the experimental work in this section by clearing the workspace, resetting the random variable generator, and loading the dataset in `datasets.mat`.

```
>> clear; rng('default'); load datasets                                     (11.19)
```

In the first exercise illustrated here, we aim at training a *knn* model using the train set and evaluating it over the test set. However, before training the model, we must select an optimal value for k. We do this by exploring the resulting classification accuracy for different *knn* models using different values of k. This optimization process if carried over the development set. Accordingly, we use the train set to train *knn* models and the development set to evaluate them for a range of k values. We use functions `fitcknn` and `predict` for the training and inference tasks, respectively (you might want to use the `help` command to get more detailed information about these two functions).

```
>> kvals = 3:10; % considers k values from 3 to 10                    (11.20a)
>> for k=1:length(kvals)
       knn_model = fitcknn(trntfidf,trnlbls,...
                          'Distance','cosine','NumNeighbors',kvals(k));
       predictions = predict(knn_model,devtfidf);
       accuracy(k) = sum(devlbls==predictions)/ndev*100;
   end

>> [maxaccuracy,idxoptim] = max(accuracy);                            (11.20b)
>> koptim = kvals(idxoptim) % optimum value of k
>> fprintf('koptim = %d, maxacc = %5.2f\n',koptim,maxaccuracy)
koptim = 6, maxacc = 95.50
```

As seen from (11.20b), the optimum value of k is 6, for which the accuracy value is *95.5%*. Resulting accuracies for all the different values of k considered are depicted in Figure 11.5 (the generation of the plot is left to you as an exercise).

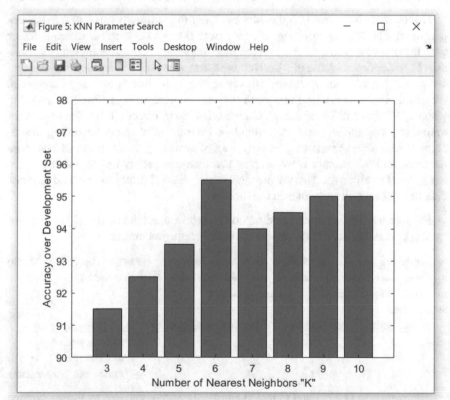

Fig. 11.5. Accuracy values computed over the development set for different values of k

Notice from the figure that the accuracy values obtained for other values of k are always lower that the one at $k=6$, but they are still much higher than the accuracy achieved by means of unsupervised clustering in (11.16), in the previous section, where the obtained accuracy was *86%*.

Having determined an optimal value for the parameter k, we can now train the *knn* algorithm over the train set with k = `koptim` and evaluate its performance over the test set. Again, we use functions `fitcknn` and `predict` for the training and inference tasks:

```
>> % trains a new knn model over the train set with k = koptim          (11.21a)
>> knn_model = fitcknn(trntfidf,trnlbls,...
                       'Distance','cosine','NumNeighbors',koptim);

>> % computes accuracy of the generated model over the test set          (11.21b)
>> predictions = predict(knn_model,tsttfidf);
>> accuracy = sum(predictions==tstlbls)/ntst*100
accuracy =
    91
```

Notice that after training the model by using the optimum value of k = `koptim` derived in (11.20), the resulting accuracy over the test set is *91%*, which is lower than the one observed over the development set *95.5%*. This is completely reasonable as the optimum value of k derived with one subset of the dataset is not necessarily optimal for another subset. However, the logic here is that any knowledge we can derive from a subset in a partition will be better than just random guessing when applied to a different subset. On the other hand, directly using the test set for optimizing any model parameter would be considered cheating according to machine learning best practices. The only way of accessing model performance in an unbiased and fair manner is by keeping test data completely unknow to the learning process at all times. This ensures test data is a good proxy to new and unknow data the system will encounter in the future.

For a more detailed understanding on how the *knn* model in (11.21) is discriminating across categories, we can compute the confusion matrix:

```
>> confusion_mtx = confusionmat(predictions,tstlbls,'ORDER',books);      (11.22)
>> cmtx = array2table(confusion_mtx,'VariableNames',string(books));
>> cmtx.('Classified as') = string(books)';
>> disp(cmtx)
```

OLIVER TWIST	DON QUIXOTE	PRIDE AND PREJUDICE	Classified as
60	2	1	"OLIVER TWIST"
7	68	1	"DON QUIXOTE"
6	1	54	"PRIDE AND PREJUDICE"

The results in (11.22) and (11.21b) can be directly compared to those obtained in (11.17) and (11.18), as they have been computed over exactly the same test data partition. Notice the accuracy improvement from *86%* in (11.18), for the case of unsupervised clustering, to *91%* in (11.21b), for the case of supervised classification. More detailed evidence on how each specific category assignments have improved can be derived from directly comparing both confusion matrices. It is interesting to notice how in the case of Pride and Prejudice and Don Quixote the number of erroneous category assignments have been significantly reduced from *11* to *2* and from *6* to *3*, respectively. However, in the case of Oliver Twist erroneous category assignments have slightly increased from *11* to *13* (mainly explained by a significant increase on confusion between the Pride and Prejudice and Oliver Twist categories). Indeed, there are *9+3–2=10* less errors in (11.22) than in (11.18), which accounts for the *10/200*100% = 5%* boost in accuracy observed in the supervised classification results.

Next, let us consider another type of supervised classification algorithm, the multilayer perceptron (MLP), which is typically expected to perform better than the simple *knn* algorithm. First, let us provide a quick introduction to this type of systems,[4] after which we will train and evaluate an MLP classifier and compare its performance with the *k*-means and *knn* classifiers previously used.

Artificial neural networks, or simply neural networks, refer to a huge family of computational models that were originally inspired in biological neural systems. Although there are significant differences between artificial neural networks and their biological counterparts, they all are based on the same fundamental concepts. A neural network is a computational structure composed of an interconnection of multiple basic units or neurons, which are also referred to as perceptrons. Figure 11.6 illustrates the basic architecture of the perceptron (top half of the figure) and the general architecture of a multilayer perceptron (bottom half of the figure).

As seen from the figure, the perceptron performs two basic operations to generate an output from a set of input variables. First, a linear combination of the input values is computed and, second, a function is applied to the result of the linear combination. The function applied in the latter step, which is referred to as the activation function, is typically non-linear in nature. Training a perceptron implies finding an optimal set of weights w_k in the linear combination step, such that desired target outputs are generated for the corresponding sets of inputs. Notice the perceptron also includes a bias term b, which is an additional parameter that must be optimized during the learning process. Common implementations of the perceptron treat the bias term b as an additional input with a fixed value, such as $b=x_0=1$, and include an additional weight parameter w_0, which must be learned along with the rest of the weights in the linear combination. Accordingly, the output of the perceptron can be express in terms of an activation function that is applied to the inner product between an input vector and a vector of weights.

---

[4] For a more comprehensive introduction to the topic of neural networks, you must refer to the references provided in the *Further Reading* section. Other neural network architectures, as well as other supervised learning methods, are explored in more detail in exercises 11.6-5 and 11.6-6.

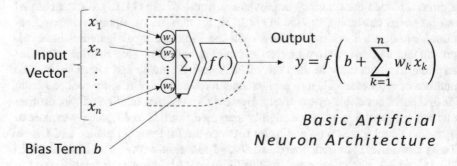

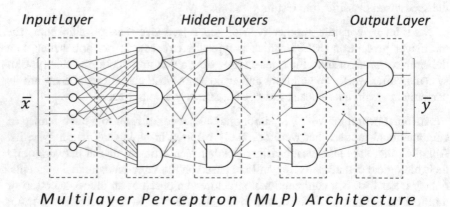

Fig. 11.6. Artificial neuron (top) and multilayer perceptron (bottom) architectures

The MLP architecture, which is illustrated in the bottom half of Figure 11.6, is a type of neural network in which the individual neurons are arranged in layers. Typically, for all neurons in a given layer, the inputs are fully connected to the outputs in the previous layer, and the outputs are fully connected to the inputs in the following layer. More specifically, only forward connections between consecutive layers are allowed, reason for which this type of architecture is also commonly referred to as a feed-forward network.

Three different types of layers must be distinguished in an MLP architecture: input, output, and hidden layers. The input layer, which is responsible for handling the inputs to the network, is composed of input nodes (no actual neurons are used in the input layer). Its size must be equal to the number of variables in the problem space. For instance, in the case of the document categorization problem considered here, the number of input variables is the size of the vocabulary. The output layer, on the other hand, contains the neurons that are responsible for producing the outputs of the system. Accordingly, its size must be equal to the number of outputs in the problem under consideration, which, in this case, corresponds to the number of categories (books) in the data collection. Finally, hidden layers, which might vary in number and sizes, are all the layers in between the input and the output layers. They behave as feature extractors in the sense that they transform and project the input variables into alternative representation spaces that facilitate the resolution of the problem at hand. MLPs with one or two hidden layers are consider shallow networks, while MLPs with three or more hidden layers are typically considered deep neural networks (DNNs).

Training a single perceptron is a relatively simple task. The basic learning algorithm uses an error signal, which is computed as the difference between a given input and its desired output, to update the weights. This is recursively done for all the weights in the perceptron and all data samples in the training set, until a convergence criterion is achieved. Training an MLP network, on the other hand, is a much more complex task. Indeed, it was not practically possible until the backpropagation algorithm[5] was introduced. As its name suggests it, this learning algorithm adjusts the weights of the neurons in an MLP by propagating the error signal backwards from the output layer to the rest of the layers.[6]

The MATLAB™ function `trainNetwork` allows for training neural network architectures of different types, including MLPs. With it, we can train an MLP for the three-book categorization problem we have been considering in this chapter. However, we need to define the MLP model architecture and set some training options first. Let us proceed to define the architecture by using some of the predefined layer objects:

```
>> layers = [                                                              (11.23)
        featureInputLayer(vsize)
        fullyConnectedLayer(10)
        tanhLayer
        fullyConnectedLayer(nbooks)
        softmaxLayer
        classificationLayer];
```

As seen from (11.23), the model architecture is defined by using six different layer definitions. Notice the concept of a layer definition here is not the same as the conceptual layers previously depicted in Figure 11.6. Indeed (11.23) is defining a three-layer MLP architecture. The first (input) layer is required to have the same number of units as the number of variables in our data representation. This is given by the vocabulary size, previously saved into the auxiliary variable `vsize` in (11.5d). The second (hidden) layer has been set to have *10* units with activation functions of type hyperbolic tangent (`tanhLayer`). Finally, the third (output) layer must have as many units as the number of categories in the dataset, previously

---

[5] Although originally derived in the context of control systems, backpropagation was formally introduced and popularized in the context of neural networks by Rumelhart *et al.* (1986).

[6] A detailed description of backpropagation is beyond the scope of this book. For more details about it and other important concepts related to neural networks, you must refer to the relevant works cited in the *Further Reading* section.

saved into in the auxiliary variable `nbooks` in (11.2b). Additionally, a SoftMax normalization is applied to the output layer (`softmaxLayer`) and, in the case of classification networks, a `classificationLayer` must be added after normalization, which assigns the normalized outputs to their corresponding categories.

Next, we can proceed to set some of the training options manually. We do this by using the function `trainingOptions`:

```
>> options = trainingOptions('sgdm', ...                                   (11.24)
        'MaxEpochs',500,...
        'MiniBatchSize',64, ...
        'Shuffle','every-epoch', ...
        'ValidationData',{full(devtfidf),devlbls},...
        'ValidationFrequency',10,...
        'ValidationPatience',5,...
        'Verbose',false);
```

As seen from (11.24), we passed the value `'sgdm'` as first argument to the function. This is a required argument, and it refers to the solver to be used for training the network and the specific one we are using here `'sgdm'` stands for Stochastic Gradient Descent with Momentum.[7] Other optional arguments include the following key value pairs, which define some of the available training options:

- `MaxEpochs`: defines the maximum number of times the train set is presented to the network during training. A value of *500* is much more than needed and training should stop before that based on the `ValidationPatience`.
- `MinibatchSize`: controls the number of train samples used by the learning algorithm at each iteration. Here, we are defining mini batches of size *64*.
- `Shuffle`: determines whether and how data randomization is used. `'every-epoch'` enforces shuffling the train set before each training epoch.
- `ValidationData`: this is where we pass the development set to the training algorithm. In the context of neural networks, which can easily overfit the train set, validation data is used to monitor the system performance. This information is used to inform some decisions such as early stopping.[8]
- `ValidationFrequency`: determines how often the prediction error over the validation set is computed. This frequency is given in number of iterations.
- `ValidationPatience`: determines how many times the validation error can be larger or equal to the ones in previous iterations before training stops.

---

[7] This method, which belongs to the family of Stochastic Gradient Descent methods, adds a momentum term to the weight updating step aiming at reducing oscillations. The default value for the momentum parameter, as used here, is *0.9*.

[8] Early stopping is a training strategy used to avoid overfitting. By monitoring the prediction error of a neural network over an independent validation set, the training process can be stopped when not reduction of such error is observed after few consecutive iterations.

- **Verbose**: controls the amount of training progress information displayed in the command window. We have set it to **false**, which omits displaying any information at all. You can set it to **true** to see training progress information in real time. Additionally, you can use the key-value pair **'Plots'**, **'training-progress'** to display training progress graphs in real time.

More training options can be set with function **trainingOptions**. We have just used a few of them. To see all options available along with their corresponding set or default values, you can display the **options** object created in (11.24) by executing the command **disp(options)**.

Finally, we are ready to train and evaluate the MLP model. First, we use function **trainNetwork** to train the model. The training is conducted over the train set, while the development set is used for deciding when to stop the training. Then, the function **classify** is used to generated predictions for the test set; and, finally, the accuracy and confusion matrix are computed.

```
>> % trains with train set and uses dev set for validation          (11.25a)
>> mlp_model = trainNetwork(full(trntfidf),trnlbls,layers,options);

>> % computes accuracy of the generated model over the test set      (11.25b)
>> predictions = classify(mlp_model,full(tsttfidf));
>> acc = sum(tstlbls==predictions)/ntst*100
acc =
    96

>> % computes the confusion matrix                                    (11.25c)
>> confusion_mtx = confusionmat(predictions,tstlbls,'ORDER',books);
>> cmtx = array2table(confusion_mtx,'VariableNames',string(books));
>> cmtx.('Classified as') = string(books)';
>> disp(cmtx)
```

OLIVER TWIST	DON QUIXOTE	PRIDE AND PREJUDICE	Classified as
71	5	0	"OLIVER TWIST"
1	65	0	"DON QUIXOTE"
1	1	56	"PRIDE AND PREJUDICE"

Notice from (11.25b) a classification accuracy of *96%* has been achieved with the MLP model, which is significantly better than the *91%* and *86%* previously obtained with *knn* and *k*-means, respectively. This is consistent with what should be expected given the more powerful mathematical and computational framework offered by neural networks. The actual benefit comes from the fact that the MLP is able to build a much better representation of the problem space under consideration. This is actually happening in the hidden layer of the network, where the original dimensionality of the data is reduced to *10* dimensions, facilitating the final classification process occurring at the output layer.

Notice from (11.25c) how errors across the different categories have been reduced. In the case of the Pride and Prejudice, all samples have been correctly assigned to the category. In the case of Oliver Twist and Don Quixote, on the other hand, 2 and 6 samples have been assigned to wrong categories, respectively. In particular, notice how one sample of each of the latter two categories have been wrongly assigned to the Pride and Prejudice. This last observation helps to introduce here a very important concept that we will discuss in more detail later in chapter 12. The fact that all Pride and Prejudice data samples have been assigned correctly to the Pride and Prejudice category does not necessarily mean that the classifier is perfectly discriminating this category from the others. In fact, as mentioned, there are two samples from the other categories that were wrongly assigned to the Pride and Prejudice category.

Accordingly, we must distinguish two types of errors. Samples of a referential category wrongly assigned to other categories are referred to as false negatives. As seen from (11.25c), there are no false negatives in the Pride and Prejudice category. On the other hand, samples of any other category that are wrongly assigned to referential category are referred to as false positives. As seen from (11.25c), there are two false positives in the Pride and Prejudice category. Notice these two error types are relative to a referential category, which means false negatives and false positives of a given category become false positives and false positives for other categories. (These concepts will be discussed in more detail in chapter 12.)

As verified with the experiments conducted in this section, in general, supervised methods achieve better accuracy values than unsupervised methods. However, we must not disregard the value of unsupervised methods, as they constitute excellent exploration tools in those situations in which no relevant information about the data (i.e. training data) is available. On the other hand, supervised methods become useful analysis tools only when good-quality training data is at hand.

Up to this point we have only focused our attention on documents, but we can also think in the document categorization problem from the perspective of words. In fact, in vector space, the dimensions of document vectors correspond to the vocabulary terms in the document collection, and the distance scores used to assess similarities among document vectors rely on word co-ocurrences and distributions across documents. By using the train set that is available for our collection of documents, some interesting analyses can be performed over the vocabulary set.

For instance, we might be interested in identifying the most representative vocabulary words in each of the three categories in our data collection. To this end, let us first compute the average document vector for each of the three categories in the collection. We do this over the train set:

```
>> for k=1:nbooks                                                              (11.26)
       % computes the average document vector for the kth category
       kindexes =  trnlbls==books(k);
```

```
      meanvect(k,:) = sum(trntfidf(kindexes,:),1)/sum(kindexes);
      % normalizes the computed average vector
      meanvect(k,:) = meanvect(k,:)/norm(meanvect(k,:));
  end
```

From a geometrical point of view, each of the vectors computed in (11.26) constitutes the centroid of its corresponding category's set of vectors and, accordingly, it provides a vector-based representation for the whole category. We can think about these vectors as "average documents", which are the most representative documents of each category. Consequently, vocabulary terms exhibiting the largest TF-IDF scores within these average vectors should be the most representative words in their corresponding categories. In this way, we can extract the most relevant terminology associated to each of the categories in the dataset.

Let us extract, for instance, the *10* most representative vocabulary terms (i.e. keywords) in each category. As we are not interest in words that are commonly relevant to all categories, we subtract the two other-category vectors from each considered vector. The keyword extraction process is implemented as follows:

```
>> % gets the 10 most relevant words in category 1 (Oliver Twist)      (11.27a)
>> [~,idx1] = sort(meanvect(1,:)-meanvect(2,:)-meanvect(3,:),'descend');
>> disp(vocab(idx1(1:10)))
   Columns 1 through 7
     "oliver"    "old"    "boy"    "bumble"    "jew"    "sikes"    "fagin"
   Columns 8 through 10
     "gentleman"    "brownlow"    "upon"

>> % gets the 10 most relevant words in category 2 (Don Quixote)       (11.27b)
>> [~,idx2] = sort(meanvect(2,:)-meanvect(1,:)-meanvect(3,:),'descend');
>> disp(vocab(idx2(1:10)))
   Columns 1 through 6
     "don"    "sancho"    "quixote"    "thou"    "thee"    "worship"
   Columns 7 through 10
     "senor"    "thy"    "knight"    "dulcinea"

>> % gets the 10 most relevant words in cat 3 (Pride & Prejudice)      (11.27c)
>> [~,idx3] = sort(meanvect(3,:)-meanvect(1,:)-meanvect(2,:),'descend');
>> disp(vocab(idx3(1:10)))
   Columns 1 through 6
     "elizabeth"    "darcy"    "jane"    "bennet"    "bingley"    "miss"
   Columns 7 through 10
     "mrs"    "wickham"    "collins"    "lydia"
```

As seen from (11.27), the extracted sets of words are characteristic of their corresponding categories. Indeed, any person who is familiar with the three books used to generate the data collection would be able to tell what the books are, just by looking at the three sets of words in (11.27).

To better illustrate the discriminative power of these sets of words with respect to other words in the vocabulary, let us construct some dendrograms for different sets of words. In particular, we will be considering three groups of words: the set of most discriminative words (ranks *1* to *7*) for each category, and two sets of less discriminative words (ranks *101* to *107* and ranks *1001* to *1007*) for each category. In each case, a total amount of *21* words (*7* from each category ranking) are to be considered. For these computations, we operate over the vector representations of words rather than documents:

```
>> % selects the 7 most representative words from each category         (11.28a)
>> words = [idx1(1:7),idx2(1:7),idx3(1:7)];
>> wordmtx = trntfidf(:,words)';
>> % computes and plots the corresponding dendrogram
>> hf = figure(7); ha = subplot(1,3,1);
>> set(hf,'Color',[1 1 1],'Name','Word dendrograms for different word sets');
>> y = pdist(wordmtx,'cosine'); z = linkage(y,'average');
>> [h,t] = dendrogram(z,0,'labels',vocab(words),'orientation','right');
>> temp = axis; axis([0,1,temp(3:4)]); set(ha,'Xcolor',[1 1 1]);
>> ylabel('Words in ranks 1 to 7 for each category');

>> % selects words in ranks 101 to 107 from each category               (11.28b)
>> words = [idx1(101:107),idx2(101:107),idx3(101:107)];
>> wordmtx = trntfidf(:,words)';
>> % computes and plots the corresponding dendrogram
>> ha = subplot(1,3,2);
>> y = pdist(wordmtx,'cosine'); z = linkage(y,'average');
>> [h,t] = dendrogram(z,0,'labels',vocab(words),'orientation','right');
>> temp = axis; axis([0,1,temp(3:4)]); set(ha,'Xcolor',[1 1 1]);
>> ylabel('Words in ranks 101 to 107 for each category');

>> % selects words in ranks 1001 to 1007 from each category             (11.28c)
>> words = [idx1(1001:1007),idx2(1001:1007),idx3(1001:1007)];
>> wordmtx = trntfidf(:,words)';
>> % computes and plots the corresponding dendrogram
>> ha = subplot(1,3,3);
>> y = pdist(wordmtx,'cosine'); z = linkage(y,'average');
>> [h,t] = dendrogram(z,0,'labels',vocab(words),'orientation','right');
>> temp = axis; axis([0,1,temp(3:4)]); set(ha,'Xcolor',[1 1 1]);
>> ylabel('Words in ranks 1001 to 1007 for each category');
```

The resulting dendrograms are presented in Figure 11.7. As seen from the figure, when considering the *7* most relevant words for each category, a clear distinction among the three categories can be observed (left-most plot in Figure 11.7). In fact, the words from Pride and Prejudice appear grouped together in the upper section of the dendrogram, similarly, the words from Don Quixote appear together in the lower section and the words from Oliver Twist in the central section. It is important to recall here that the `linkage` function, which implements a hierarchical clustering algorithm, did not use any information about the categories. It just grouped the words based on the cosine distances among their corresponding vector representations.

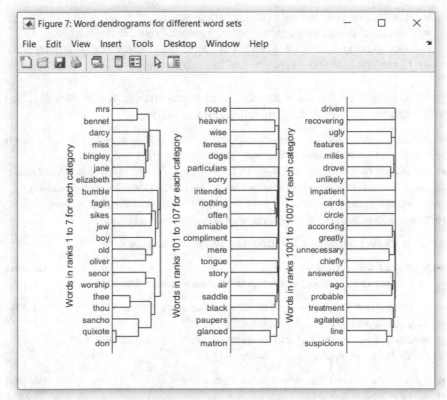

Fig. 11.7. Dendrograms for different sets of words within ranks *1* to *7* (left), ranks *101* to *107* (center), and ranks *1001* to *1007* (right)

In the case of words in ranks *101* to *107* from each of the three categories (central plot in Figure 11.7), no clear distinction among the categories can be inferred. However, as seen from the corresponding dendrogram, some word groupings can still be identified. In fact, a detailed inspection of the words reveals that some of the words relevant to Oliver Twist are located towards the lower section of the dendrogram, some words relevant to Don Quixote appear towards the upper section, and most of the words relevant to Pride and Prejudice appear in the central section. Finally, when considering words in ranks *1001* to *1007* (right-most plot in Figure 11.7), no distinction at all among the categories is possible. In this case, words from the three document categories can be found at any position within the dendrogram. Indeed, in most word pair associations observed in this dendrogram, the words come from different categories.

Continuing our exploratory analysis of words, and as a final exercise for this section, let us construct a word cloud with the set of most representative words from each document category. The word cloud will be constructed according to the following specifications:

- Words must be projected over a spherical surface.
- Font sizes should be varied according to words' relative importance.
- Font colors should be assigned according to the corresponding category.
- The word cloud should spin around the sphere center.

First, we project the same set of 7 most representative words from (11.28a) over a sphere's surface by using multidimensional scaling:

```
>> % selects the 7 most representative words from each category           (11.29)
>> words = [idx1(1:7),idx2(1:7),idx3(1:7)];
>> % applies multidimensional scaling to construct a 3-dimensional embedding
>> map = mdscale(pdist(trntfidf(:,words)','cosine'),3,'criterion','sammon');
>> % projects the 3-diemnsional word representations on the unit sphere
>> for k=1:size(map,1), normap(k,:) = map(k,:)/norm(map(k,:)); end
```

then, we display the word cloud into a three-dimensional plot:

```
>> % creates a new figure                                                 (11.30a)
>> hf = figure(8); set(hf,'Color',[1 1 1],'Name','Word Cloud');

>> % creates and plots a 3-D white semi-transparent sphere                (11.30b)
>> [x,y,z] =sphere(20); c = ones(21,21);
>> surf(x,y,z,c,'EdgeAlpha',0.0,'FaceColor',[1 1 1],'FaceAlpha',0.3);
>> axis off; axis square; axis vis3d;

>> % makes font sizes proportional to term-frequencies                    (11.30c)
>> tfmtx = trnbow.Counts;
>> fsize = sum(tfmtx,1); fsize = 3*ceil(log(full(fsize(words)))');

>> % assigns a color to each document category                            (11.30d)
>> fcolor = char(['r'*ones(7,1);'b'*ones(7,1);'k'*ones(7,1)]);

>> hold on; % plots the word could                                        (11.30e)
>> for k=1:length(words)
        x = 1.5*normap(k,1); y = 1.5*normap(k,2); z = 1.5*normap(k,3);
        w = vocab(words(k)); fc = fcolor(k); fs = fsize(k);
        text(x,y,z,w,'HorizontalAlignment','center','Color',fc,'Fontsize',fs);
    end; hold off;
>> axis(1.1*[-1,1,-1,1,-1,1]);
```

and, finally, we spin the word cloud around...

```
>> % captures current view's azimuth and elevation                        (11.31a)
>> [ai,ei] = view;
```

```
>> % generates a trajectory for the azimuth
>> ai = ai/180*pi; af = ai+2*pi; az = linspace(ai,af,700);
>> % generates a trajectory for the elevation
>> ei = ei/180*pi; ef = ei-pi/2; el = linspace(ei,ef,700);
>> % converts azimuths and elevations into cartesian coordinates
>> x = cos(el).*sin(az); y =-cos(el).*cos(az); z = sin(el);

>> % rotates the view of the word cloud                    (11.31b)
>> for k=1:length(x), view([x(k),y(k),z(k)]); pause(0.01); end
```

The final rotated view of the word cloud is shown in Figure 11.8. Words relevant to Oliver Twist and Pride and Prejudice can be seen towards the left lower and upper regions of the sphere, respectively, and words relevant to Don Quixote can be seen towards the right side of the sphere.

Fig. 11.8. Rotated view of a three-dimensional word cloud containing the 7 most discriminative words from each category in the data collection

11.4 Supervised Classification in Probability Space

In this section, we continue studying the supervised classification approach to document categorization. However, different from the methods discussed in the previous section, here we focus our attention on statistical methods. More specifically, we first consider the likelihood ratio approach, in which the probability of a given document is estimated by means of different category-dependent statistical models, and the ratios among such probabilities is used to determine the most probable category for the given document.

Before proceeding with the experimental work, let us introduce in more detail the likelihood ratio framework. Consider, for instance, a binary partition of a given dataset into two mutually exclusive categories A and B. By applying the Bayes rule, we can express the probability of occurrence for each category given an observed document D as follows:

$$p(A|D) = p(D|A) \times p(A)/p(D) \qquad\qquad (11.32a)$$

$$p(B|D) = p(D|B) \times p(B)/p(D) \qquad\qquad (11.32b)$$

and by taking the ratio between (11.32a) and (11.32b), we get:

$$p(A|D)/p(B|D) = p(D|A)/p(D|B) \times p(A)/p(B) \qquad\qquad (11.33)$$

The result in (11.33) provides a very simple criterion for binary classification: if $p(A|D)/p(B|D)>1$, we can say that document D most likely belongs to category A; otherwise, if $p(A|D)/p(B|D)<1$, we can say that D most likely belongs to category B. For evaluating whether this probability ratio is either greater or less than one, we consider the logarithm of the right-hand side of (11.33):

$$log(p(D|A)/p(D|B)) + log(p(A)/p(B)) \qquad\qquad (11.34)$$

where the first term is the logarithm of the ratio of likelihoods and the second term is the logarithm of the ratio of priors. In this implementation (because of the logarithm) D will be said to belong to category A or B, depending on whether the expression in (11.34) is greater or less than zero, respectively.

Further elaboration of (11.34), allows for restating the likelihood ratio classification criterion derived from (11.33) as follows:

$$D \leftarrow A \quad if \quad [log(p(D|A))-log(p(D|B))] > \zeta \qquad\qquad (11.35a)$$

$$D \leftarrow B \quad if \quad [log(p(D|A))-log(p(D|B))] < \zeta \qquad\qquad (11.35b)$$

where ζ is the logarithm of the inverted prior ratio $p(B)/p(A)$, and the likelihoods $p(D|A)$ and $p(D|B)$ represent probability estimates for the document D given statistical models derived from categories A and B, respectively. In the practical implementation of the method, the train data samples belonging to category A and the train data samples belonging to category B are used to independently train the parameters of a predefined class of statistical model. The two resulting models are then used to estimate $p(D|A)$ and $p(D|B)$, respectively. In the case of ζ, although $p(A)$ and $p(B)$ can be also estimated from the train data, the common practice is to search for an optimal value of ζ over a development set.

As seen from (11.35), the likelihood ratio framework is indeed very general, as it does not impose any restriction on the class of statistical model to be used. Indeed, it can be implemented by means of any model as far as good model parameters can be estimated from the available train data and good likelihood estimates can be derived from the model. In the experimental work presented next, we will use the LDA topic models that we had previously described in chapter 8. However, as suggested later in the exercise section, you can repeat the experiments by using other types of statistical models.

First, as in the previous subsections, we start by clearing the workspace, resetting the random variable generator, and loading the dataset in `datasets.mat`.

```
>> clear; rng('default'); load datasets                              (11.36)
```

Next, we proceed to compute three LDA models, one for each book category, for which we use the documents in the train set. First, we start by setting the number of topics to 25. Then, we use function `encode` and the bag-of-words model `trnbow` previously computed in (11.5a) to obtain the word counts for the documents in the train set, one category at a time. Then we use function `fitlda` to compute the LDA models.

```
>> ntopics=25; % sets the number of topics                           (11.37a)

>> for k = 1:nbooks % computes an LDA model for each book category   (11.37b)
       counts = encode(trnbow,trndocs(trnlbls==books(k)));
       models(k) = fitlda(counts,ntopics,'Verbose',0);
   end
```

Notice that in our experimental setting we are dealing with three categories. So, different from the binary classification problem stated in (11.35), we need to handle a multi-category scenario. A simple approach is to consider three binary classification problems: Oliver Twist versus non Oliver Twist, Don Quixote versus non Don Quixote, and Pride and Prejudice versus non Pride and Prejudice. This would require training six different models, one for each category and one for each category complement. However, as seen from (11.37), we have computed three LDA models only, one for each category. So, in this case, we need to compute probability estimates for the three category complements by using a simple linear combination of models. Finally, regarding category assignment decisions, instead of making independent binary decisions for each category, we implement a simple multi-category decision process, which selects the maximum log likelihood ratio out of the three being computed.

We can now proceed to apply likelihood ratio classification to our data collection. Before applying the procedure to the test data, we apply it to the development data. This will help us to select appropriate values for ζ_1, ζ_2 and ζ_3.

First, we compute probability estimates for each document in the development set by using each of the three LDA models from (11.37b). We use function `logp` to estimate the log probabilities:

```
>> logprobs = zeros(length(devdocs),nbooks);                          (11.38)
>> for n=1:ndev % iterates over the samples in the development set
        counts = encode(trnbow,devdocs(n));
        for k=1:nbooks % iterates over the LDA models
            logprobs(n,k) = logp(models(k),counts);
        end
    end
```

Next, we compute the logarithm of the likelihood ratios for each document with respect to each category. As log probability estimates for each complement category, we just average the log probabilities for the two complement categories:

```
>> for n=1:size(logprobs,1)                                           (11.39)
        loglirat(n,1) = logprobs(n,1)-(logprobs(n,2)+logprobs(n,3))/2;
        loglirat(n,2) = logprobs(n,2)-(logprobs(n,1)+logprobs(n,3))/2;
        loglirat(n,3) = logprobs(n,3)-(logprobs(n,1)+logprobs(n,2))/2;
    end
```

Now, we are ready to search for optimal values for ζ_1, ζ_2 and ζ_3. Instead of implementing a formal procedure, let us conduct an *ad hoc* exploration of the solution space. For this, we generate accuracy curves by varying one ζ value at a time (from *-100* to *100*), while maintaining the other two values equal to zero.

```
>> zeta = -100:100;                                                    (11.40)
>> acc = zeros(length(zeta),nbooks);
>> for k=1:length(zeta)
        % assigns categories to samples in the development set
        [~,tmp1] = max((loglirat+repmat([zeta(k) 0 0],ndev,1))');
        [~,tmp2] = max((loglirat+repmat([0 zeta(k) 0],ndev,1))');
        [~,tmp3] = max((loglirat+repmat([0 0 zeta(k)],ndev,1))');
        % computes the classification accuracies over the development set
        correct = [sum(devlbls==books(tmp1)'),...
                    sum(devlbls==books(tmp2)'),sum(devlbls==books(tmp3)')];
        acc(k,:) = correct/ndev*100;
    end
```

The resulting accuracy curves are presented in Figure 11.9 (the generation of the plot is left to you as an exercise).

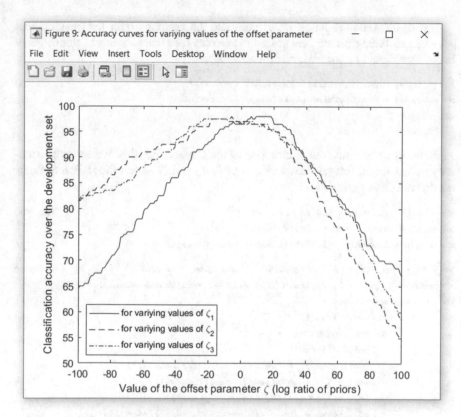

Fig. 11.9. Classification accuracy, measured over the development set, for varying values of log prior ratios ζ_1, ζ_2 and ζ_3

As seen from the figure, the highest classification accuracies are observed for the intervals $0<\zeta_1<20$, $-20<\zeta_2<20$, and $-30<\zeta_3<0$ when each of these three parameters is varied independently from the others. Although there is not any guaranty that a high accuracy value will be obtained when concentrating on the region defined by those three intervals, we propose to explore that region in more detail. However, before doing that, let us recapitulate a moment on how the accuracies have been computed in (11.40).

The decision criterion for category assignment implemented in (11.40) follows the form of (11.34) rather than (11.35). First, we added a matrix of ζ values (one different value per column) to the matrix of log likelihood ratios `loglirat`. Then, we extracted the indexes of the maximum values along each row. Notice that this is actually a multi-category selection criterion, as we are forcing the selection of one category from the three possible ones. This selection is based on the highest binary classification score, of the form of (11.34), from the three computed scores. Finally, the accuracies were computed by counting and normalizing the total amount of correct category assignments, in the same way it was done in (11.16).

Just to illustrate the latter, let us compute the accuracy over the development set when all three ζ parameters are set to zero (as seen from Figure 11.9, this accuracy value should be somewhere around *95%*):

```
>> [~,temp] = max((loglirat+zeros(ndev,nbooks))');                    (11.41)
>> accuracy = sum(devlbls==books(temp)')/ndev*100
accuracy =
    96.5000
```

Now, let us continue our exploration of the ζ space. For this, we compute accuracy values for all integer combinations of ζ_1, ζ_2 and ζ_3 within the region of interest derived from Figure 11.9:

```
>> % specifies the region of interest                                 (11.42a)
>> zeta1 = 0:20; zeta2 = -20:20; zeta3 = -30:0;
>> ncases = length(zeta1)*length(zeta2)*length(zeta3);

>> % computes accuracy values within the region of interest           (11.42b)
>> index = 0; acc = zeros(ncases,1); zeta = zeros(ncases,nbooks);
>> for k=1:length(zeta1)
        for n=1:length(zeta2)
            for j=1:length(zeta3)
                index = index+1;
                zeta(index,:) = [zeta1(k),zeta2(n),zeta3(j)];
                zeta_mtx = repmat(zeta(index,:),ndev,1);
                [~,temp] = max((loglirat+zeta_mtx)');
                acc(index) = sum(devlbls==books(temp)')/ndev*100;
            end
        end
    end

>> % finds the case with maximum accuracy                             (11.42c)
>> [maxacc,index] = max(acc)
maxacc =
    99
index =
    440

>> % gets the optimum values of ζ1, ζ2 and ζ3                         (11.42d)
>> optzeta = zeta(index,:)
optzeta =
    0    -6    -25
```

The values of ζ_1, ζ_2 and ζ_3 in (11.42d) produce the highest accuracy over the development set within the considered region of the ζ space. Notice the improvement of accuracy, from *96.5%* in (11.41) to *99%* in (11.42c), that has been achieved by adjusting the ζ parameters. However, it can be verified that this optimal set of parameters is not unique! Indeed, there are many more value combinations for these parameters that produce the same maximum accuracy of *99%*:

```
>> sum(acc==maxacc)                                                  (11.43)
ans =
   291
```

Now, we are finally ready to apply the likelihood ratio classification algorithm to the test data. First, we follow the same procedures used in (11.38) and (11.39) to compute the document log probabilities and log likelihood ratios:

```
>> logprobs = zeros(ntst,nbooks);                                    (11.44a)
>> for n=1:ntst % iterates over the samples in the test set
       counts = encode(trnbow,tstdocs(n));
       for k=1:nbooks % iterates over the LDA models
           logprobs(n,k) = logp(models(k),counts);
       end
   end

>> for n=1:size(logprobs,1)                                          (11.44b)
       loglirat(n,1) = logprobs(n,1)-(logprobs(n,2)+logprobs(n,3))/2;
       loglirat(n,2) = logprobs(n,2)-(logprobs(n,1)+logprobs(n,3))/2;
       loglirat(n,3) = logprobs(n,3)-(logprobs(n,1)+logprobs(n,2))/2;
   end
```

and, finally, we compute the resulting accuracy and confusion matrix:

```
>> zeta_mtx = repmat(optzeta,ntst,1);                                (11.45a)
>> [~,idxs] = max((loglirat+zeta_mtx)');
>> predictions = books(idxs)';
>> accuracy = sum(predictions==tstlbls)/ntst*100
accuracy =
   95.5000

>> confusion_mtx = confusionmat(predictions,tstlbls,'ORDER',books);  (11.45b)
>> cmtx = array2table(confusion_mtx,'VariableNames',string(books));
>> cmtx.('Classified as') = string(books)';
>> disp(cmtx)
```

OLIVER TWIST	DON QUIXOTE	PRIDE AND PREJUDICE	Classified as
71	2	3	"OLIVER TWIST"
2	69	2	"DON QUIXOTE"
0	0	51	"PRIDE AND PREJUDICE"

The values obtained in (11.45a) and (11.45b) can be directly compared to those obtained in (11.16) and (11.17) for the case of *k*-means clustering, (11.21b) and (11.22) for the case of *k* nearest neighbors, and (11.25b) and (11.25c) for the case of the MLP classifier. Although we can safely say that both, likelihood ratio (with *95.5%* accuracy) and MLP (with 96% accuracy), outperform *k* nearest neighbors (with *91%* accuracy) and *k*-means clustering (with *86%* accuracy); we cannot so easily compare performances between likelihood ratio and the MLP classifier. To determine whether the apparent better performance of the latter with respect to the likelihood ratio approach is significant or not, we need to conduct a statistical significance test (more about this in exercise 11.6-8).

As a final classification exercise in this chapter, we will use an MLP classifier on the LDA probability space. Similar to what we did in the previous section, the MLP will be used as the classification mechanism but, instead of using a TF-IDF vector space, we use the LDA topic probability space as our feature space for representing the documents. In this case, we use the complete train set to train a single LDA model rather than training one model per category. For this, we use function `fitlda`, as we did in (11.37b),[9] and consider the same number of *25* topics.

```
>> rng('default'); % restarts the random number generator          (11.46)
>> ldamodel = fitlda(trnbow,ntopics,'Verbose',0); % computes the LDA model
```

Now we can use the computed model to project all documents in the collection (train, development, and test) into the LDA topic probability space. We do this by means of function `transform`:

```
>> trnfeatmtx = transform(ldamodel,trndocs);                        (11.47)
>> tstfeatmtx = transform(ldamodel,tstdocs);
>> devfeatmtx = transform(ldamodel,devdocs);
```

The feature matrices computed in (11.47) constitute low-dimensional representations of the documents in the data collection. The documents have been projected into an LDA space of *25* topics. This space is a probability space, which can be easily verified by looking at the sum of all topic scores for each document. Indeed, each entry in the matrices represents the conditional probability of a topic given the corresponding document.

```
>> whos trnfeatmtx tstfeatmtx devfeatmtx                            (11.48a)
  Name            Size           Bytes  Class      Attributes
  devfeatmtx      200x25          40000  double
  trnfeatmtx      1949x25        389800  double
  tstfeatmtx      200x25          40000  double

>> sum(trnfeatmtx(1,:)) % sum of topic scores for train document 1  (11.48b)
ans =
     1
```

Next, we proceed to train and evaluate the MLP classifier. We follow exactly the same procedure used in the previous section (11.23), (11.24) and (11.25) to define the architecture, set the training options, train the MLP and evaluate its per-

---

[9] Notice that, in this case, we pass as main argument to `fitlda` the bag-of-words model `trnbow`, rather than a matrix of frequency counts. Function `fitlda` admits both argument types.

formance. The only difference is that, in this case, we use the probabilistic feature
space constructed in (11.47) as input.

```
>> % defines the MLP architecture                                    (11.49a)
>> layers = [
        featureInputLayer(ntopics)
        fullyConnectedLayer(10)
        tanhLayer
        fullyConnectedLayer(nbooks)
        softmaxLayer
        classificationLayer];
```

```
>> % sets the training options                                       (11.49b)
>> options = trainingOptions('sgdm', ...
        'MaxEpochs',500,...
        'MiniBatchSize',64, ...
        'Shuffle','every-epoch', ...
        'ValidationData',{devfeatmtx,devlbls},...
        'ValidationFrequency',10,...
        'ValidationPatience',5,...
        'Verbose',false);
```

```
>> % trains the MLP classifier                                       (11.49c)
>> mlpnet = trainNetwork(trnfeatmtx,trnlbls,layers,options);
```

```
>> % computes predictions and accuracy over the test set            (11.49d)
>> predictions = classify(mlpnet,tstfeatmtx);
>> acc = sum(predictions==tstlbls)/ntst*100
acc =
    95
```

```
>> % computes and displays the confusion matrix                      (11.49e)
>> confusion_mtx = confusionmat(predictions,tstlbls,'ORDER',books);
>> cmtx = array2table(confusion_mtx,'VariableNames',string(books));
>> cmtx.('Classified as') = string(books)';
>> disp(cmtx)
```

OLIVER TWIST	DON QUIXOTE	PRIDE AND PREJUDICE	Classified as
68	2	0	"OLIVER TWIST"
4	68	2	"DON QUIXOTE"
1	1	54	"PRIDE AND PREJUDICE"

As seen from (11.49d) and (11.49e), the performance of the MLP classifier
over the LDA topic probability space is comparable to both the log likelihood ratio approach presented before (11.45a) and (11.45b), and the MLP classifier operating over the TF-IDF space from the previous section (11.25b) and (11.25c). To
determine whether the small differences observed among these three systems are
significant, a statistical significance test is needed (see exercise 11.6-8).

We conclude this section by exploring in more detail the distribution of topics across the three different categories in the data collection, as well as the top vocabulary words associated to some of these topics. For this, we use the function **predict**, which returns the most probable LDA topic for each document, and we plot histograms of most probable topics for documents in each category. The resulting histograms are depicted in Figure 11.10.

```
>> name = 'Distribution of Topics over Document Categories';                    (11.50)
>> hf = figure(10);
>> set(hf,'Color',[1 1 1],'Name',name);
>> for k=1:nbooks
        x = predict(ldamodel,trndocs(trnlbls==books(k)));
        subplot(nbooks,1,k)
        histogram(x,'BinMethod','integers','Normalization','probability')
        xlabel(books(k));
        axis([0,ntopics+1,0,0.4]);
    end
```

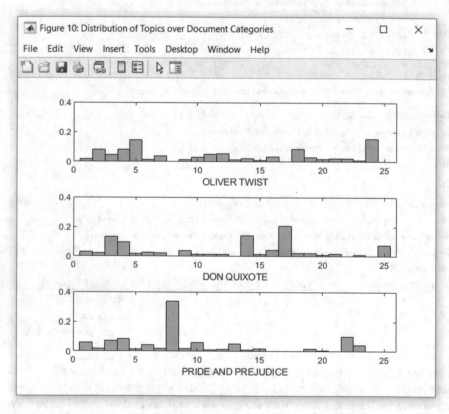

Fig. 11.10. Distribution of the *25* LDA topics over the three book categories in the collection

As seen from the figure, different distributions of topics are obtained for each of the three book categories. A visual inspection and a quick comparative analysis reveal that there are some topics that are exclusive to some of the categories and there are other topics that are shared across the categories. Accordingly, the most relevant words within each of those topics will be either exclusive to or shared across the categories. Consider, for instance, topics 24, 17 and 8, which seem to be the most important topics exclusively associated to Oliver Twist, Don Quixote and Pride and Prejudice, respectively. Also consider topics 1 and 4, which seem to be similarly important across the three categories. We can use function **topkwords** to extract and list the top vocabulary words associated with each of those topics:

```
>> topics = [24,17,8,1,4];                                                   (11.51)
>> tbl = table();
>> for k=1:length(topics)
       colname = sprintf('Topic %d',topics(k));
       tbl.(colname) = topkwords(ldamodel,10,topics(k)).Word;
   end
>> disp(tbl)
```

Topic 24	Topic 17	Topic 8	Topic 1	Topic 4
"oliver"	"don"	"mr"	"could"	"we"
"old"	"quixote"	"elizabeth"	"would"	"if"
"mr"	"sancho"	"much"	"only"	"will"
"gentleman"	"said"	"darcy"	"no"	"us"
"boy"	"one"	"herself"	"its"	"are"
"young"	"these"	"jane"	"own"	"do"
"these"	"knight"	"most"	"such"	"can"
"upon"	"here"	"miss"	"being"	"our"
"when"	"take"	"did"	"though"	"think"
"rose"	"thy"	"bingley"	"between"	"dear"

By looking at the top *10* words in the five considered topics (11.51), it can be confirmed that while topics *24*, *17* and *8* are exclusively associated to the different categories; topics *1* and *4*, on the other hand, are much more generic in nature.

11.5 Further Reading

The concept and the algorithm of *k*-means clustering are older than the term itself, which was first adopted by (James MacQueen 1967). In addition to centroid-based clustering, other clustering methods are also available such as, for example, hierarchical clustering (Defays 1977) and density-based clustering (Ester *et al.* 1996). For a good reference book on unsupervised clustering in general, you can refer to (Gan 2007). For a more specific treatment on clustering of text data, you should refer to a text mining book such as, for instance, (Berry 2003). For a more detailed discussion about the Dunn Index and other algorithms for clustering quality evaluation, you can refer to chapter 17 in (Gan 2007),

Regarding supervised text categorization, good surveys are available from (Sebastiani 2002) and (Ikonomakis *et al.* 2005). With respect to specific methods for supervised classification, you will find that there is a huge amount of literature available. For a detailed description of the likelihood ratio approach described in section 11.4, along with other similar methods, refer to (Severini 2000). When the likelihood ratio method is used with word unigram models, it is commonly referred to as naïve Bayes classification (Rish 2001). For a good example on text classification by means of Neural Networks see (Li and Park 2007). The original introduction to backpropagation within the context of neural network architectures was presented in (Rumelhart *et al.* 1986) and more details on the Stochastic Gradient Descent optimization method can be found in (Bottou 1998). For an excellent and comprehensive introduction to neural networks see (Haykin 1999).

The area of artificial neural networks has experimented a significant boost in recent years due to the increased availability of computational power and the introduction of deep learning techniques (Bengio *et al.* 2015). A more recent and comprehensive review on text classification by means of deep neural networks (DNNs) is provided in (Minaee *et al.* 2021).

Another method not mentioned in this chapter worth of further attention is Support Vector Machines (SVMs). For a good example on text classification by means of SVMs refer to (Joachims 1998). An excellent introduction to text classification by using the MATLAB's Text Analytics Toolbox™ can be found in (The MathWorks 2021).

Document categorization is actually a core technology that can be found behind many different text mining services such as, just to mention a few, spam filtering (Sahami 1998), e-mail routing (Busemann *et al.* 2000), topic detection and tracking (Allan 2002), press-clipping (Gründel *et al.* 2001) and language identification (Cavnar and Trenkle 1994).

11.6 Proposed Exercises

1. Consider the unsupervised clustering performed in section 11.2, for which an accuracy value of *86%* was measured over the test set:

    – Randomly generate *50* different train, development, and test partitions as the one generated in (11.4).
    – Randomly generate *50* different initial sets of centroids for starting the k-means clustering algorithm (`kmeans` can receive via the additional parameter `'start'`, a matrix containing the initial centroids for each cluster).

- Conduct the corresponding *2,500* independent simulations of k-means clustering and compute the resulting accuracy for each case (use $k=3$).
- What are the mean and variance of the accuracies measured over the complete set of *2,500* experiments? What are the means and variances for each of the elements in the confusion matrix?
- Plot distributions of means and variances observed over sets of experiments when only varying the initial set of centroids. What do you observe?
- Plot distributions of means and variances observed over sets of experiments when only varying the considered test set. What do you observe?

2. Consider the dimensionality reduction techniques described in chapter 10. Here you will conduct different k-means clustering experiments (with $k=3$) to evaluate the effect of different dimensionality reduction techniques. With the same data partition created in (11.4), the *50* different initial set of centroids generated in the previous exercise and $k=3$, consider each of the following cases:

- No vocabulary pruning used. Note we already applied vocabulary pruning to the dataset used in this chapter, more specifically in (11.5b) and (11.5c).
- Other vocabulary pruning. Explore other pruning options, such as removing a standard list of stop-words and/or using different pruning thresholds.
- Vocabulary merging. Consider two cases: lemmatization and stemming.
- Projection methods. Also consider two cases: linear and non-linear projections. Explore different dimensionality values.
- Compute the resulting accuracy for each of the considered cases. How these results compare to the *86%* accuracy observed in section 11.2?

3. Consider the k nearest neighbors method and the MATLAB® function `fitcknn` we used in section 11.3.

- Study function `fitcknn` in more detail, specifically look at input parameters `'NSMethod'`, `'DistanceWeight'`, `'Distance'` and `'OptimizeHyperparameters'`.
- Consider the optimal value of $k=6$ derived in (11.20) and experiment with different combinations of options for parameters `'NSMethod'`, `'DistanceWeight'`, and `'Distance'` (set `'OptimizeHyperparameters'` to `'none'` for these experiments). Can you identify an optimal configuration?
- Consider the optimal configuration identified in the previous step. Explore its performance for different values of k. Is $k=6$ still an optimum value?
- Set `'OptimizeHyperparameters'` to `'auto'`, which optimizes parameters `'Distance'` and `'NumNeighbors'` simultaneously. What is the resulting optimum combination of parameters in this case?

4. Let us propose and evaluate a very simple classification method that operates in vector space and is similar to the *knn* algorithm. Use the same dataset partition (for train, development and test) used along the chapter.

- First, compute the average vector (centroid) for each document category over the train dataset, in the same way they were computed in (11.26).
- Then, for each document in the test set, compute its cosine distance to each of the three average vectors computed in the previous step.
- Finally, assign to each document in the test set, the same category of the average vector that is closest to it.
- How the resulting accuracy compares to the one achieved by means of *knn* classification in section 11.3?
- Repeat the experiment by using different partitions of the data set. Compute mean and average values for the observed accuracies. What are your conclusions?

5. Consider MATLAB® function `fitcecoc`, which can be used to train a Support Vector Machine multiclass model.

- Repeat the supervised classification in vector space from section 11.3 by using Support Vector Machines to implement the classifier.
- Compute the accuracy and compare this result with those obtained in section 11.3 for the *k* nearest neighbors and MLP classifiers.

6. Consider the Convolutional Neural Network (CNN) architecture. You can find detailed instructions on how to implement a CNN on page 2-73 of the Text Analytics Toolbox™ User's Guide (The MathWorks 2021).

- Use function `fastTextWordEmbedding` to load the *fastText* word embedding model and apply it to the input text (if the support package is not installed, the function will provide a link to download and install it).
- Define the network architecture by using a sequence length of *10* and convolutional layer blocks of *50* filters for *n*-gram lengths of *2, 3,* and *4.*
- Repeat the supervised classification from the previous exercise by using the implemented Convolutional Neural Network as classifier.
- Compute the accuracy and compare this result with those obtained in the previous exercise and the different sections of the chapter.

7. Consider the LDA topic models computed in section 11.4. In this exercise you will be computing perplexities, which can be done by means of function `logp`.

- First, consider the three LDA topic models computed in (11.37), one for each of the three categories in the data collection.
- Split the test set into three subsets according to each of the three document categories, and compute the perplexity for each of the three LDA models over each of the three subsets, i.e. generate a *3×3* perplexity matrix.
- Look at the resulting perplexity matrix. What do you observe? Are these results in accordance with what you would expect?

- Repeat the previous steps with the development set. What main differences and/or similarities do you observe?
- Next, consider the single LDA model computed in (11.46). Compute multiple similar models by varying the number of topics (about *10* models for each different number of topics) in the range between *5* and *30*.
- Compute the average perplexities for each number of topics over the development set and generate a plot of average perplexities vs number of topics. Can you identify an optimum number of topics?

8. Let us evaluate the statistical significance of the different results: *96%* for the MLP with TF-IDF in (11.25b), *95.5%* for the log likelihood ratio in (11.45a), and *95%* for the MLP with LDA in (11.49d). To do this, you need to conduct pairwise comparisons by using a procedure denominated bootstraping:

- Randomly resample the test set by allowing repetitions, i.e. perform consecutive random selections of documents over the complete test set until you end up with a sampled set as large as the original test set.
- Compute the accuracies for a pair of methods over the sampled test set. Notice that you need to keep the indexes of the selected documents during the sampling phase, so you can match the corresponding estimated and actual categories when computing the accuracies.
- Repeat the previous two steps several times (at least *1,000* times) and compute the difference between both accuracies at each time.
- Generate and plot a histogram for the resulting accuracy differences and observe their distribution.
- You can say that the difference between two accuracies $A_1\%$ and $A_2\%$ is statistically significant with a confidence of $X\%$, if $X\%$ of the differences in the distribution are above the observed accuracy difference $|A_1\%-A_2\%|$. Conventionally, accepted values for $X\%$ should be above *95%* or *99%*.
- Alternatively, you can use MATLAB® functions **bootstrp** and **bootci** for computing the distribution and the confidence.

9. The procedure we used in section 11.4 for selecting the values of parameters of ζ_1, ζ_2 and ζ_3 was simple but inefficient. We just explored the solution space and retained the best set of parameters we observed. A more appropriate approach than simple exploration would be to use an optimization procedure:

- Consider the MATLAB® function **fminsearch** and study its operation.
- Create an objective function **fobj** with variables ζ_1, ζ_2 and ζ_3 such that attains a minimum value when classification accuracy reaches its maximum.
- Use **fminsearch** to find a set of values for ζ_1, ζ_2 and ζ_3 that minimizes the objective function **fobj** (you will probably need to use scaled versions of ζ_1, ζ_2 and ζ_3 in your implementation for **fminsearch** to work properly).

10. Consider the likelihood ratio method presented in section 11.4, where LDA topic models were used for computing likelihood estimates. If word unigram models are used instead, the procedure is reduced to what is commonly known as a Naïve Bayes classifier.

   – Repeat the likelihood ratio classification experiment presented in section 11.4 by using word unigram models instead.
   – Repeat once more the likelihood ratio classification experiment presented in section 11.4 by using word bigram models instead.
   – How do the resulting accuracies compare between them and with the one obtained by using LDA topic models?

11. Recompute the LDA topic model from (11.46) without restarting the random number generator. Next, compute the new feature matrices (11.47) and retrain and evaluate the MLP classifier (11.49). What accuracy value do you get? Did you expect this result? Repeat the previous procedure *100* times and generate a histogram of accuracies? What can you conclude?

12. Consider all classification methods used in this chapter and create an ensemble of classifiers by combining all different results.

   – Consider a simple majority voting approach, which assigns to a given data sample the category that most of the classifiers are assigning to it.
   – Consider a more elaborated weighted approach, which weighs the decision of each classifier before performing the voting. You can use the development set to determine an optimal set of weights.
   – Compute accuracies over the test set for each of the two ensemble methods described above. How they compare to each other and with the individual classifier accuracies?

11.7 Short Projects

1. Select one of the books used in this chapter and download it from the Project Gutenberg website http://www.gutenberg.org/ Accessed 6 July 2021.

   – Create a dataset like the one used in this chapter, in which each paragraph from the book should be considered a different document. Keep the information about the corresponding chapter for each document (paragraph) by means of a field in the table or, alternatively, in a separated array.
   – Compute the TF-IDF matrix for the data collection. Extract the *10* most important words of the complete data collection by computing the average vector of the whole data collection and considering those vocabulary terms

with largest average TF-IDF weight. Compare the extracted terms with the ones extracted for the same book in section 11.3.

- How can you explain the observed differences? What can you do to improve the selection of words in the previous step?
- Compute average vectors for each chapter and extract the *10* most representative words for each chapter by following a similar procedure to the one used in section 11.3.
- Select only two to three representative words per chapter. Consider their corresponding word vectors and use a non-linear projection method to generate a two-dimensional word cloud for the selected set of words. Use different colors for words in different chapters.
- Recompute the word cloud by adding an additional dimension to the word vectors. This additional dimension should contain the number of the chapter from which the corresponding word was extracted. Experiment with different relative weights for this additional dimension.
- Construct a summary of the book by extracting the most representative paragraph from each chapter (select the paragraph exhibiting the shortest cosine distance to the corresponding chapter's average vector).
- Establish a desired range for the number of words to be permitted in each selected paragraph. Therefore, if for a given chapter the extracted paragraph is too short, you can decide to combine two or more paragraphs; and, on the other hand, if the extracted paragraph is too long, you can consider using the second or third most representative paragraph instead.
- Perform a subjective evaluation of the constructed summary.

2. Consider the *Reuters-21578* dataset, which contains a categorized collection of Reuter's newswires. The dataset is freely available at the UCI KDD Archive http://kdd.ics.uci.edu/databases/reuters21578/reuters21578.html Accessed 6 July 2021.

- Read the related documentation, download, and preprocess the complete collection. Select only those documents for which topic categories are available and maintain those topic categories that have been assigned to at least *20* or more documents (you should find *57* categories).
- Generate a dataset similar to the one used in this chapter. Notice that you will need to create several category indexes as *5* different category sets are used in the data collection. Also create a variable to preserve the document ID numbers as they are in chronological order.
- Conduct an unsupervised clustering exercise and evaluate the possible correspondences and differences between the resulting clusters and the actual topic categories. Conduct several independent clustering exercises and experiment with different numbers of clusters. Try to determine which pairs of categories are more likely to be mixed in clusters and which are more likely to be separated.

- Generate average vectors for each topic category (notice that one document can belong to more than one category). Construct dendrograms, as the ones generated in section 11.3, including relevant words from different sets of topics. Are results consistent with your previous observations?
- Extract the most representative terms for each topic. Select some few topics and construct word clouds for each of the selected ones.
- Randomize the data collection and partition it into train, development, and test sets. Ensure that a significant number of samples from all categories appear in all three subsets.
- Implement and train an MLP classifier for topic categories. Evaluate classification results over the test set by computing the classification accuracy for each individual category. (Note this is a multilabel classification problem, i.e. documents can be assigned to more than one category. The output layer of the MLP and the label targets must be defined accordingly).
- Generate a bar plot displaying the accuracies for all topic categories.
- Implement classifiers for the other four category sets: exchanges, organizations, people, and places (consider only those categories with more than 20 occurrences). Evaluate the classifiers and generate bar plots for the results.
- Consider again the complete data collection (just before partitioning it into train, development, and test sets). Use the ID numbers of the documents for obtaining their chronological order.
- Use a window of size $k=N/10$ documents (where N is the total number of documents) to extract the oldest k documents in the collection and compute the 10 most representative words for such a subset.
- Move the window $k/2$ documents ahead in time and compute the 10 most representative words for the new subset. Repeat this process recursively until reaching the newest subset of k documents in the collection.
- Plot a timeline placing each group of most representative words in the appropriate position along the timeline. Experiment with different values of k and compare the resulting timelines. What are your main observations?

11.8 References

Allan J (2002) Topic detection and tracking: event-based information organzation. Kluwer international series on information retrieval, Springer

Bengio Y, LeCun Y, Hinton GE (2015). Deep Learning. Nature. 521 (7553): 436–444

Berry MW (ed.) (2003) Survey of Text Mining I. Springer

Bottou, L (1998) Online Algorithms and Stochastic Approximations. Online Learning and Neural Networks. Cambridge University Press.

Busemann S, Schmeier S, Arens RG (2000) Message classification in the call center. In Proceedings of the 6th Applied Natural Language Processing Conference, pp. 158–165

Cavnar WB, Trenkle JM (1994) N-gram-based text categorization. In Proceedings of the Third Annual Symposium on Document Analysis and Information Retrieval, pp. 161–175

Defays D (1977) An efficient algorithm for a complete link method. The Computer Journal of the British Computer Society 20(4): 364–366

Ester M, Kriegel HP, Sander J, Xu X (1996) A density-based algorithm for discovering clusters in large spatial databases with noise. In Proceedings of the Second International Conference on Knowledge Discovery and Data Mining, pp. 226–231

Gan G (2007) Data clustering: theory, algorithms and applications. ASASIAM Series on Statistics and Applied Probability 20, Society for Industrial and Applied Mathematics

Gründel H, Naphtali T, Wiech C, Gluba JM, Rohdenburg M, Scheffer T (2001) Clipping and analysing news using machine learning techniques. Lecture Notes in Computer Science 2226: 87–99

Haykin SS (1999). Neural Networks: A Comprehensive Foundation. Prentice Hall.

Ikonomakis M, Kotsiantis S, Tampakas V (2005) Text classification using machine learning techniques. WSEAS Transactions on Computers 4(8): 966–974

Joachimns T (1998) Text categorization with support vector machines: learning with many relevant features. In Proceedings of the 10th European Conference on Machine Learning, pp. 137–142

Li CH, Park SC (2007) Neural network for text classification based on singular value decomposition. In Proceedings of the Seventh International Conference on Computer and Information Technology, pp. 47–52

MacQueen JB (1967) Some methods for classification and analysis of multivariate observation. In proceedings of the Fifth Berkelet Symposium on Mathematical Statistics and Probability, pp. 281–297

Minaee S, Kalchbrenner N, Cambria E, Nikzad N, Chenaghlu M, Gao JF (2021) Deep Learning Based Text Classification: A Comprehensive Review, https://arxiv.org/abs/2004.03705 Accessed 6 July 2021

Rish I (2001) An empirical study of the naïve Bayes classifier. In Proceedings of the IJCAI 2001 Workshop on Empirical Methods in Artificial Intelligence

Rumelhart DE, Hinton GE, Williams RJ (1986) Learning representations by back-propagating errors. Nature 323 (6088): 533–536

Sahami M, Dumais S, Heckerman D, Horvitz E (1998) A Bayesian approach to filtering junk e-mail. In Proceedings of the AAAI Workshop on Learning for Text Categorization

Sebastiani F (2002) Machine learning in automated text categorization. ACM Computing Surveys 34(1): 1–47

Severini TA (2000) Likelihood Methods in Statistics. New York: Oxford University Press.

The MathWorks (2021) Text Analytics ToolBox™ User's Guide, https://www.mathworks.com/help/pdf_doc/textanalytics/textanalytics_ug.pdf Accessed 6 July 2021

12 Document Search

This chapter focuses on a fundamental problem in information management, which is also relevant for text mining applications: document search. The area of study that deals with this specific problem in detail is known as information retrieval. In this chapter, we present and discuss several methods and applications that are closely related to the field of information retrieval. However, it is important to mention that such field is indeed much broader and more extensive than what we actually explore here.

This chapter is organized as follows. First, in section 12.1, we introduce the basic evaluation metrics of precision and recall, as well as present some examples of binary search. Then, in section 12.2, we focus our attention on vector search, which is based on the vector space model presented in chapter 9. In this section, we also discuss the problems of query expansion, relevance estimation and relevance feedback. In section 12.3, we introduce and describe the BM25 family of ranking methods. Finally, in section 12.4, we focus our attention on the problem of cross-language document search, for which we introduce some basic concepts and present some related examples.

12.1 Binary Search

Before proceeding to illustrate document search methods and applications, let us first describe the data collection to be used in this chapter. The data collection used here is basically the same we used in the previous chapter, which consists of a collection of paragraphs extracted from three books: Oliver Twist, by Charles Dickens, Don Quixote by Miguel de Cervantes, and Pride and Prejudice by Jane Austen. Three basic differences with respect to the way this data collection is preprocessed and used in this chapter with respect to the previous chapter are to be noticed:

- Here, the full collection is considered as a single partition, without making any distinctions among train, development and test subsets.
- Vocabulary pruning of infrequent words includes only singletons (i.e. words occurring once in the collection) and, in the cases of frequent words, the standard list of English stopwords defined in `stopWords` are removed.
- The category index has been modified to account for both the book and the chapter of origin for each individual document (paragraph), rather than only accounting for the book as it was done in chapter 11.

© The Author(s), under exclusive license to Springer Nature Switzerland AG 2021
R. E. Banchs, *Text Mining with MATLAB®*, https://doi.org/10.1007/978-3-030-87695-1_12

Let us load the new version of the dataset, which is already preprocessed and available in the data file `datasets_ch12a.mat`:

```
>> clear; load datasets_ch12a                                              (12.1)
>> whos
  Name            Size                Bytes  Class             Attributes
  bowmodel        1x1               3038586  bagOfWords
  docs            2349x1           20362531  tokenizedDocument
  labels          2349x1              18792  double
  ndocs           1x1                     8  double
  tfidfmtx        2349x8379         1753232  double               sparse
  vocab           1x8379             508658  string
```

where `bowmodel` is a bag-of-words model created with all documents `docs` available in the collection, `tfidfmtx` is the corresponding TF-IDF matrix, `ndocs` provides the total number of documents, `vocab` contains the vocabulary, and `labels` is the new category index vector that encodes information about both the book and the chapter each document belongs to. The convention used in this new index is as follows: paragraphs belonging to Oliver Twist are identified with indexes ranging from *1000* to *1053*, paragraphs in Don Quixote with indexes from *2000* to *2126*, and paragraphs in Pride and Prejudice with indexes from *3000* to *3061*. The complete distribution of documents (paragraphs) along books and chapters is depicted in Figure 12.1 (the generation of the plot is left to you as an exercise).

Let us now proceed to illustrate, with a very simple example, a basic procedure known as binary search. For this, consider chapter XVII of the Second Part of Don Quixote, which corresponds to category *2069* in our book-chapter composite index. Such chapter is entitled *Chapter XVII: Wherein is shown the furthest and highest point which the unexampled courage of Don Quixote reached or could reach; together with the happily achieved adventure of the lions*.

Suppose now that we are not able to recall either such a long title or the chapter number, but we are interest in finding all the documents (i.e. paragraphs in our data collection) that belong to that chapter. We should be able to think about some keywords that are relevant to the chapter and construct a query for conducting the search. Let us consider the following two keywords: *courage* and *lions*. Then, we look for all the documents in the data collection that contain both keywords. With this basic strategy (and some luck) we will be retrieving some of the paragraphs within this chapter. We implement this search strategy as follows:

```
>> % creates indexes for the query keywords                                (12.2a)
>> idx1 = vocab=="courage";
>> idx2 = vocab=="lions";

>> % performs the search                                                    (12.2b)
>> tfmtx = bowmodel.Counts;
>> bsearch_and = find((tfmtx(:,idx1)>0)&(tfmtx(:,idx2)>0));
```

```
>> % displays the categories of the retrieved documents          (12.2c)
>> labels(bsearch_and)'
ans =
        2069          2069
```

The basic search strategy implemented in (12.2b) is referred to as a binary search, as we are not concerned about the frequencies of occurrence of the keywords but on whether they occur or not within the documents. Notice also that keyword occurrences are combined by using the logical *AND* (&) operator, aiming at recovering those documents that contain both keywords. Hence, we have named the variable containing the search result as **bsearch_and**.

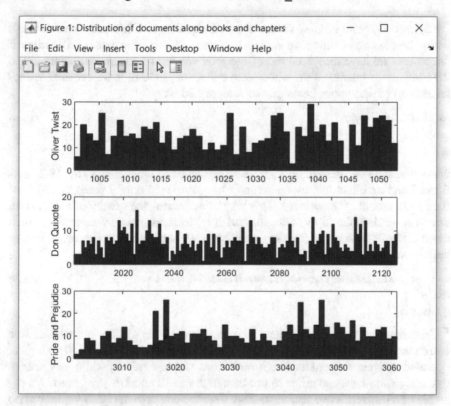

Fig. 12.1. Distribution of documents in the data collection along the book-chapter composite categories

A natural question arising from (12.2c) would be: how good is this result? At a first glance, it appears to be an excellent result as we have obtained paragraphs from the chapter we were aiming for. However, the answer to this question can be approached from two different, but related, standpoints. In the first case, we would be interested on the percentage of correct answers that have been selected, i.e.

from all the retrieved documents, what is the proportion of correct ones? This measure is referred to as precision, and it is actually what most common commercial search engines take care about (because it is precisely what you, as final user, can easily notice).

From (12.2c), we can already deduce that the precision resulting from our search exercise in (12.2) is *100%*. In a more general case, this value can be computed as follows:

```
>> sum(labels(bsearch_and)==2069)/length(bsearch_and)*100          (12.3)
ans =
    100
```

The second point of view for assessing the quality of a document search result would be the one in which we are more concerned about the actual proportion of documents we have recovered out of the total amount of documents belonging to the category of interest in the whole data collection. Indeed, if we consider the total amount of paragraphs belonging to category *2069*:

```
>> sum(labels==2069)                                                (12.4)
ans =
    13
```

we will change our perception that the search strategy implemented in (12.2) performed and excellent job, as we actually have retrieved only *2* paragraphs out of the *13* we should have retrieved. This way of measuring the quality of a document search result is referred to as recall; and it is, in most practical scenarios, much more difficult to account for than precision. We can compute the recall of our search exercise in (12.2) as follows:

```
>> sum(labels(bsearch_and)==2069)/sum(labels==2069)*100            (12.5)
ans =
    15.3846
```

In general, as we will see along the remaining of this section, more restrictive search strategies, as the one implemented in (12.2) by means of the *AND* logical operator, are prone to produce high precisions and low recalls, while less restrictive search strategies are prone to produce high recalls and low precisions. An example of a less restrictive search strategy than the one used in (12.2) would be one using the *OR* logical operator instead of *AND*. But, before proceeding with such an example, let us provide a formal introduction to precision and recall.

Consider the diagram in Figure 12.2, where both the actual category and the retrieved document subsets for a hypothetical document search are presented. As seen from the figure, the target category set (the class of documents we are interested in retrieving) and the selected document set (the group of documents that the search strategy is able to find) partition the overall document collection into four subsets. These four subsets are typically referred to as:

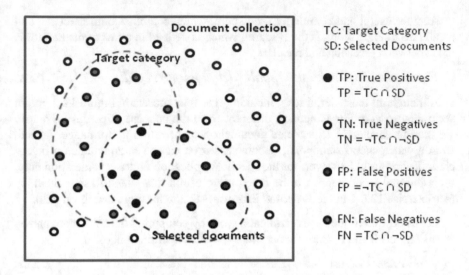

Fig. 12.2. Target category and selected document subsets in a generic document search

- True Positives (TP): which are the documents correctly identified as belonging to the target category. This subset corresponds to the intersection between the target category and the selected documents subsets.
- True Negatives (TN): which are the documents correctly identified as not belonging to the target category. This subset corresponds to the intersection between the complements of the target category and the selected documents subsets.
- False Positives (FP): which are the documents mistakenly identified as belonging to the target category. This subset corresponds to the intersection between the selected documents and the complement of the target category subsets. False positives are also often referred to as wrong selections.
- False Negatives (FN): which are the documents mistakenly identified as not belonging to the target category. This subset corresponds to the intersection between the target category and the complement of the selected documents subsets. False negatives are also often referred to as missed selections.

We can now formally present definitions for precision and recall in terms of the cardinalities of the TP, TN, FP and FN subsets as follows:

$$precision = TP / (TP+FP) \times 100\%$$
(12.6a)

$$recall = TP / (TP+FN) \times 100\%$$
(12.6b)

where the notations *TP*, *TN*, *FP* and *FN*, here refer to the cardinalities of their corresponding subsets.

Another useful and commonly used score, which combines both precision and recall into a single score, is the F-score[1], which is defined in terms of the harmonic mean between precision and recall:

$$F\text{-}score = 2 \times precision \times recall \,/\, (precision + recall)$$ (12.7)

Although all these definitions, including the four subsets in Figure 12.2, might seem new to you, this is nothing different from the concept of confusion matrix we used in chapter 11 to evaluate document categorization performance. Indeed, from a very general standpoint, document search can be interpreted as a binary classification task. However, for the specific application under consideration here, precision and recall happen to be much more useful than classification accuracy (see exercise 12.6-1 in the *Proposed Exercises* section for more details on this).

Let us now continue the experimental work by computing the confusion matrix, precision, recall and F-score for our binary search experiment in (12.2):

```
>> % generates an indicator vector for the selected documents       (12.8a)
>> selected = zeros(ndocs,1);
>> selected(bsearch_and) = 1;

>> % generates an indicator vector for the target category          (12.8b)
>> targets = double(labels==2069);²

>> % computes the cardinality of the TP, TN, FP, FN subsets         (12.8c)
>> true_positives = sum((selected==1)&(targets==1));
>> true_negatives = sum((selected==0)&(targets==0));
>> false_positives = sum((selected==1)&(targets==0));
>> false_negatives = sum((selected==0)&(targets==1));

>> % constructs and displays the confusion matrix                   (12.8d)
>> temp = confusionmat(selected,targets,'ORDER',[1,0]);
>> cmtx = array2table(temp,'VariableNames',["Category","Non-Category"]);
>> cmtx.Selections = ["Selected";"Non-Selected"];
>> disp(cmtx)
      Category    Non-Category     Selections
         2             0          "Selected"
        11           2336         "Non-Selected"

>> % computes and displays the precision, recall and F-score        (12.8e)
>> precision = true_positives/(true_positives+false_positives)*100;
>> recall = true_positives/(true_positives+false_negatives)*100;
```

---

[1] F-score actually refers to a family of F_x-scores, each of which apply a different weight to precision and recall, overemphasizing either one or the other. The unweighted harmonic mean, known as F_1-score, is the most used one. Here, we use F-score to refer to the F_1-score.

[2] We cast to **double** here because **confusionmat** in (12.8d) requires both inputs **selected** and **targets** to be of the same type. This casting is not needed for the computations in (12.8c).

```
>> fscore = 2*precision*recall/(precision+recall);
>> scores = [precision,recall,fscore];
>> fprintf('Precision = %6.2f   Recall = %6.2f   F-score = %6.2f\n',scores)
Precision = 100.00   Recall =  15.38   F-score =  26.67
```

As we will use steps (12.8c) and (12.8e) very often along the rest of the chapter, the function **getscores** has been prepared aiming at avoiding the continuous repetition of these steps. The syntax for this function is as follows:

$$\text{scores = getscores(selected,targets);} \tag{12.9}$$

where **selected** and **targets** are indicator vectors for the selected documents and target category subsets, as computed in (12.8a) and (12.8b), respectively; and the output variable **scores** is an array containing the corresponding precision, recall and F-score, as computed in (12.8e).

Let us now illustrate the case of a less restrictive binary search strategy than the one used in example (12.2). In this case, instead of using the *AND* operator for combining the occurrences of both keywords, we will use the *OR* (|) operator, i.e. the search strategy consists of finding all documents in the dataset that contains at least one of the keywords. Notice that with this search strategy, we expect to recover more documents than in (12.2).

```
>> % searches by using the OR operator with keyword matches          (12.10a)
>> bsearch_or = find((tfmtx(:,idx1)>0)|(tfmtx(:,idx2)>0));
```

```
>> % counts the number of documents retrieved                        (12.10b)
>> disp(length(bsearch_or))
    33
```

```
>> % counts the number of documents in the correct category          (12.10c)
>> disp(sum(labels(bsearch_or)==2069))
    5
```

As seen from (12.10b), a total of *33* documents have been retrieved this time. However, notice from (12.10c) that only *5* out of the *33* retrieved documents belong to the category of interest, which means precision has been severely compromised. We can compute precision, recall and F-score for this result by using the function **getscores** as follows:

```
>> % generates an indicator vector for the selected documents        (12.11a)
>> selected = zeros(ndocs,1);
>> selected(bsearch_or) = 1;
```

```
>> % computes and displays precision, recall and F-score             (12.11b)
>> scores = getscores(selected,targets);
>> fprintf('Precision = %6.2f   Recall = %6.2f   F-score = %6.2f\n',scores)
Precision =  15.15   Recall =  38.46   F-score =  21.74
```

Notice from (12.11b), that the new search strategy has provided a significant increase in recall, but at the expense of a dramatic reduction in accuracy. Overall, as suggested by both F-score values reported in (12.8e) and (12.11b), the *AND* strategy seems to be better than the *OR* strategy. However, in general, the prevalence of one over the other will always depend on the specific application at hand.

In general, and as confirmed by the scores presented in (12.8e) and (12.11b), for a given search model or technology, precision and recall tend to antagonize each other. This means that efforts to increase precision will generally compromise recall, and efforts to increase recall will generally compromise precision. Consider, for instance, the naïve search strategy of selecting all the documents in the collection. This will guarantee a *100%* of recall, but precision will be dropped to a very low value. For the case of collection at hand, which contains *2,349* documents, and the category under consideration, which contains *13* documents, such a naïve search strategy precision would be *0.55%* and the F-score *1.1%*. Improving both precision and recall together is, in general, a very difficult task!

As a single experiment can be certainly misleading when studying or evaluating the performance of a search method, in practice, several queries are typically used and evaluated. Accordingly, average precision, recall and F-scores are reported. With this idea in mind, our last experiment in this section reconsiders the binary search strategies used in (12.2) and (12.10) and evaluates them over a small subset of search examples.

To this end, a set of specific queries have been defined for eight different categories within the data collection. Table 12.1 presents the keywords composing the queries and the corresponding target categories for each case.

Table 12.1. Keywords composing the queries and their corresponding target categories

Qry	Book	Chap	Category	Keywords
1	Oliver Twist	2	1002	oliver, twist, board
2	Oliver Twist	8	1008	london, road
3	Oliver Twist	14	1014	brownlow, grimwig, oliver
4	Don Quixote	53	2053	curate, barber, niece
5	Don Quixote	69	2069	courage, lions
6	Don Quixote	93	2093	arrival, clavileno, adventure
7	Pride & Prejudice	18	3018	darcy, dance
8	Pride & Prejudice	43	3043	gardiner, housekeeper, bingley

The keywords and categories of the queries depicted in Table 12.1 are available in the file `datasets_ch12b.mat`. This file contains three variables: `queries` (the tokenized documents containing the keywords for each query), `qrylbls` (the corresponding book-chapter composite indexes), and `nqrys` (the number of queries).

We start by loading the file containing the keywords and their categories, and constructing a TF matrix for the eight queries under consideration:

```
>> load datasets_ch12b                                            (12.12a)
>> whos queries qrylbls nqrys
   Name          Size         Bytes  Class               Attributes
   nqrys         1x1              8  double
   qrylbls       8x1             64  double
   queries       8x1           5299  tokenizedDocument

>> % gets the TF matrix (word counts) for the queries            (12.12b)
>> qrytfmtx = encode(bowmodel,queries);
```

Next, we proceed to perform binary searches for each query and each of the two considered search strategies. We implement this by means of a double loop in which each query is tested against each document, as follows:

```
>> bsearch_or_selected = zeros(ndocs,nqrys);                     (12.13)
>> bsearch_and_selected = zeros(ndocs,nqrys);
>> for q=1:nqrys
       keywordmask = full(qrytfmtx(q,:)>0);
       for d=1:ndocs
           keywordsindoc = full(tfmtx(d,keywordmask)>0);
           bsearch_and_selected(d,q) = prod(keywordsindoc+0)>0;
           bsearch_or_selected(d,q) = sum(keywordsindoc)>0;
       end
   end
```

Before continuing, let us discuss in more detail the algorithm in (12.13). For each query (outer loop), a binary vector `keywordmask` is constructed, which indicates the specific keywords contained in the query. Then, for each document (inner loop), `keywordmask` is applied to the document counts to generate the binary vector `keywordsindoc`, which indicates whether each of the keywords is present or not in the document at hand.

In the case of the *OR* search strategy, the summation is used to identify whether at least one of the keywords is present. In the case of the *AND* search strategy, the product is used to identify whether all the keywords are present. Notice than in the latter case, zero has been added to the binary vector `keywordsindoc`. This is needed because (differently from `sum`) `prod` does not operate with binary variables. So, adding zero automatically casts the binary vector into a real valued vector that can be used as input for `prod`.

Notice that each entry of the two binary matrices derived from (12.13) indicates whether the corresponding document (row) satisfies or not the given search criterion for the corresponding query (column). According to this, each of the columns in these matrices are equivalents to the `selected` vectors in (12.8a) and (12.11a).

Having computed the selected document subsets for the queries under consideration, we can now compute precision, recall and F-score for each case (notice we also compute the number of selected documents for each case):

```
>> % computes the scores for each query and search strategy          (12.14a)
>> for k=1:nqrys
        targets = labels==qrylbls(k); % sets the target category vector
        % AND-based search strategy:
        selected = bsearch_and_selected(:,k);
        scores_and(k,:) = getscores(selected,targets);
        ndocs_and(k) = sum(selected);
        % OR-based search strategy:
        selected = bsearch_or_selected(:,k);
        scores_or(k,:) = getscores(selected,targets);
        ndocs_or(k) = sum(selected);
    end
```

```
>> % prints the report for the AND search strategy                   (12.14b)
>> variables = ["Selections","Precision","Recall","F-score"];
>> querynumber = "Q"+string(1:nqrys)';
>> tbl_and = array2table([ndocs_and',scores_and],'VariableNames',variables);
>> disp(addvars(tbl_and,querynumber,'NewVariableNames',"Query",'Before',1))
```

Query	Selections	Precision	Recall	F-score
"Q1"	3	33.333	5	8.6957
"Q2"	5	20	4.5455	7.4074
"Q3"	3	33.333	4.7619	8.3333
"Q4"	3	33.333	11.111	16.667
"Q5"	2	100	15.385	26.667
"Q6"	1	100	7.1429	13.333
"Q7"	4	25	3.8462	6.6667
"Q8"	1	100	4	7.6923

```
>> % prints the report for the OR search strategy                    (12.14c)
>> tbl_or = array2table([ndocs_or',scores_or],'VariableNames',variables);
>> disp(addvars(tbl_or,querynumber,'NewVariableNames',"Query",'Before',1))
```

Query	Selections	Precision	Recall	F-score
"Q1"	272	6.6176	90	12.329
"Q2"	100	4	18.182	6.5574
"Q3"	283	5.3004	71.429	9.8684
"Q4"	92	5.4348	55.556	9.901
"Q5"	33	15.152	38.462	21.739
"Q6"	70	12.857	64.286	21.429
"Q7"	185	10.27	73.077	18.009
"Q8"	178	7.3034	52	12.808

Notice, from (12.14b) and (12.14c), that our previous observation is actually in agreement with the general trend. While the *AND* search strategy tends to retrieve reduced sets of documents resulting in high precision and low recall, on the other hand, the *OR* search strategy tends to retrieve extensive sets of documents resulting in low precision and high recall. However, with regards to their overall performance, which can be judged by means of the F-score, it is not clear whether one of the strategies performs better than the other.

Finally, we can compute average precision, recall and F-score for both binary search strategies in order to better appreciate the global differences between them. Next, we compute average scores, which are presented in a bar plot, for comparative purposes. The bar plot is depicted in Figure 12.3 (the generation of the plot is left to you as an exercise).

```
>> % computes score averages                                    (12.15)
>> or_averages = mean(scores_or);
>> and_averages = mean(scores_and);
```

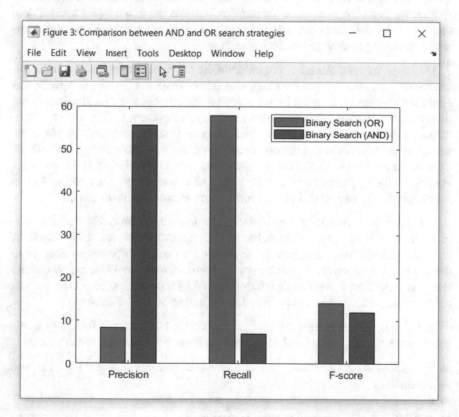

Fig. 12.3. Average performance, over the eight queries in Table 12.1, for both *OR* and *AND* binary search strategies

12.2 Vector-based Search

In this section we introduce the vector-based search paradigm that, as implied by its name, is based on the vector space model. This paradigm exploits the concepts of distance and similarity to define a search criterion that also provides a natural ranking mechanism for the selected documents. So, in this section we also introduce the notions of relevance and ranking, along with query expansion.

Before presenting and illustrating vector-based search, we need to generate a TF-IDF representation for the queries in Table 12.1. We do this by using the bag-of-words model **bowmodel** available for the data collection:

```
>> qrytfidfmtx = encode(bowmodel,queries);                                (12.16)
```

Next, we illustrate how to conduct vector search with the same sample query we used in (12.2) and (12.10), which happens to be query number 5 in the set of queries introduced in Table 12.1. The basic intuition behind vector search is very simple: we just compute cosine similarities between the given query and all documents in the collection and use the resulting similarities for ranking the documents according to their proximity to the query.

Different from the examples presented in the previous section, in which the relevance of a document given the query was binary in nature (i.e. the document was either selected or not selected by the search criterion), the use of a similarity metric allows for introducing a continuous notion of relevance. In this scenario, documents are not just relevant or not with respect to a given query, on the other hand, we can talk about a given document being more or less relevant than others. Consequently, search results can be presented as a ranked list of documents according to their relative relevance with respect to the query. This notion should sound familiar to you, as this is what most commercial search engines do!

A natural question arising here is about how to compute precision, recall and F-score if there is not precise definition of the selected document subset at all. A common practice is to compute the evaluation scores at different ranking positions. Then, for example, we can compute precision, recall and F-score by considering only the first n documents within the ranked output. In such a case we will refer to the resulting scores as precision@n, recall@n and F-score@n.

In the following example we conduct the vector-based search for query 5 and compute the three scores at each rank position from the first to the last document:

```
>> % computes cosine similarities and ranks with respect to query 5     (12.17a)
>> qrynum = 5;
>> vsearch = full(cosineSimilarity(qrytfidfmtx,tfidfmtx));
>> [~,ranks] = sort(vsearch(qrynum,:),'descend');

>> % computes the scores at each rank position                          (12.17b)
```

```
>> for k=1:ndocs
      selected = zeros(ndocs,1);
      selected(ranks(1:k)) = 1;
      targets = labels==qrylbls(qrynum);
      scores(k,:) = getscores(selected,targets);
   end
```

Now, let us present these results in a plot of scores versus rank. The curves are depicted in Figure 12.4 (the generation of the plot is left to you as an exercise).

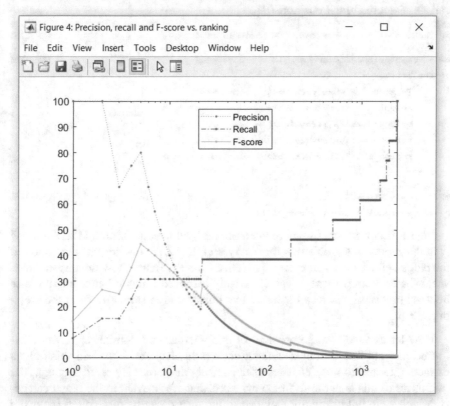

Fig. 12.4. Curves of precision, recall and F-score versus rank position for a sample vector-based search using query 5 from Table 12.1

Two interesting observations can be derived from Figure 12.4. First, notice the clear opposite trends of both precision and recall. While high precision and low recall values are exhibited at top ranks (left-hand side of the graph), low precisions and high recalls are exhibited at large rank positions (right-hand side of the graph). This confirms, again, our intuition about the antagonistic behavior between precision and recall. The second observation, on the other hand, has to do with the behavior of the F-score. Notice how F-score exhibits a maximum in an intermediate region of the rank scale. In this particular example, the optimal region for F-scores appears to be in the rank interval from *4* to *10*. This behavior is indeed a general trend. In general, the F-score tends to achieve its maximum value in the region of ranks close to where precision and recall intercept with each other.

For better appreciating the advantages of vector-based search with respect to binary search, let us proceed to conduct a more complete evaluation (over the set of eight queries in Table 12.1) of the procedure illustrated in (12.17). We compare this result with those from (12.8) and (12.10). For this, we compute precision, recall and F-score at rank position *10*.

```
>> rankloc = 10; % sets the rank position at 10                        (12.18a)

>> % computes the scores for all the queries                          (12.18b)
>> for k=1:nqrys
       [~,ranks] = sort(vsearch(k,:),'descend');
       selected = zeros(ndocs,1);
       selected(ranks(1:rankloc)) = 1;
       targets = labels==qrylbls(k);
       scores_v10(k,:) = getscores(selected,targets);
   end

>> % computes score averages                                          (12.18c)
>> v10_averages = mean(scores_v10);
```

Figure 12.5 displays the score averages for vector-based search (12.18c) along with those previously obtained for binary search (12.15) (again, the generation of the plot is left to you as an exercise). Notice from the figure, how the implemented vector-based search strategy has effectively achieved a much better compromise between precision and recall than the two binary search strategies from the previous section.

Now let us focus on the problem of query expansion. Similar to how we extracted relevant terms from document clusters in chapter *11*, we can think about extracting terms from a given set of documents that are relevant to a query. The usefulness of this is twofold: first, new terms that are relevant to the target category, which were not originally included in the query, can be discovered (keyword extraction); second, a new query can be constructed by adding these new terms to the original query (query expansion). The new query can be used to conduct a new search, which in most of the cases allows for improving the quality of the selected document set.

To illustrate these points, let us consider again query *5* from Table 12.1, for which the corresponding keywords *courage* and *lions* can be directly derived from its TF-IDF vector representation as follows:

```
>> qrynum = 5;                                                        (12.19)
>> [weight,idxs] = sort(qrytfidfmtx(qrynum,:),'descend');
```

```
>> vocab(idxs(weight>0))
ans =
  1×2 string array
    "courage"    "lions"
```

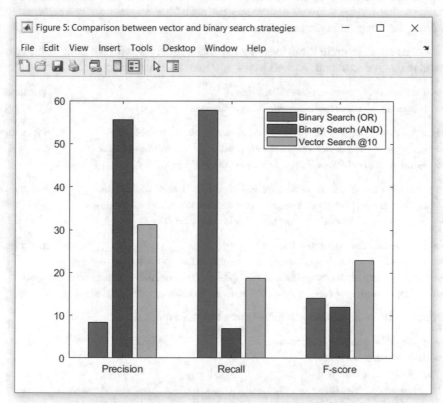

Fig. 12.5. **Average performance, over the eight queries in Table 12.1, for vector-based search _@10_ and both, _OR_-based and _AND_-based, binary search strategies**

As seen from Figure 12.4, the first two ranked documents for this query belong to the corresponding target category, so these two documents can be used to infer additional terms that are relevant to the query. What we do in practice is to use the vectors associated to these relevant documents to "pull" the original query vector towards a "good" region in the model space.

Notice that, in most common practical situations, we do not know how many of the top rank documents are actually relevant to the target category. However, in general, we can assume that there will be more relevant documents than non-relevant ones if we just restrict ourselves to a small set of documents within the top rank region. (In our example here, we consider only the first two documents as we already know that the third document does not belong to the target category.)

Another useful source of information for both keyword extraction and query expansion is the set of least relevant documents with respect to the query. These are actually the documents located in the other extreme of the ranking scale, and represent the region of space we are not interested in. So, we use the vectors associated to these least relevant documents to "push" the original query vector away from a "bad" region in the model space.

Let us now illustrate these ideas by expanding query *5* with the help of the two most relevant documents, as well as the *100* least relevant ones:

```
>> % recomputes document ranks for query 5                              (12.20a)
>> [~,ranks] = sort(vsearch(qrynum,:),'descend');

>> % computes the contribution of the most relevant documents           (12.20b)
>> mrv = 2; % number of most relevant samples to be considered
>> vplus = sum(tfidfmtx(ranks(1:mrv),:))/mrv;

>> % computes the contribution of the least relevant documents          (12.20c)
>> lrv = 100; % number of least relevant samples to be considered
>> vminus = sum(tfidfmtx(ranks(end-lrv+1:end),:))/lrv;

>> % computes the expanded the query                                    (12.20d)
>> a = 0.90; % sets the value of the combination parameter
>> newqry = qrytfidfmtx(qrynum,:) +a*vplus -(1-a)*vminus; % expands the query
>> newqry = newqry.*(newqry>0); % discards resulting negative entries (if any)
```

from which the most representative terms can be inspected:

```
>> [~,idxs] = sort(newqry,'descend');                                   (12.21)
>> disp(vocab(idxs(1:15)))
  Columns 1 through 6
    "lions"      "courage"     "delay"     "opening"    "foot"     "keepers"
  Columns 7 through 11
    "cages"      "withhold"    "resolute"    "plant"    "temerity"
  Columns 12 through 15
    "encourage"    "fearing"    "presents"    "fright"
```

As seen from (12.21), in addition to the original keywords *courage* and *lions*, the most representative terms in the expanded query also include words that are directly related to *courage* (such as *resolute* and *temerity*) and *lions* (such as *keeper* and *cages*), along with other terms that are representative, to a lesser degree, of the context under consideration. In this manner, the procedure in (12.20) has proven useful to expand the original query with new terms that we had not thought about in the first place, which seem to be relevant to the target category.

The next question to be addressed is about whether the expanded query computed in (12.20) is actually better for selecting documents in the target category under consideration than the original query *5* from Table 12.1. To answer this

question, we conduct vector-based search experiments for both, the original and
the expanded, queries and compute the corresponding evaluation scores at several
different rank positions.

```
>> % creates the target category vector                             (12.22a)
>> targets = labels==qrylbls(qrynum);

>> % vector-based search for the original query                     (12.22b)
>> vsearch = full(cosineSimilarity(qrytfidfmtx(qrynum,:),tfidfmtx));
>> [~,ranks] = sort(vsearch,'descend');
>> for k=1:50
       selected = zeros(ndocs,1);
       selected(ranks(1:k)) = 1;
       original(k,:) = getscores(selected,targets);
   end

>> % vector-based search for the expanded query                     (12.22c)
>> vsearch = full(cosineSimilarity(newqry,tfidfmtx));
>> [~,ranks] = sort(vsearch,'descend');
>> for k=1:50
       selected = zeros(ndocs,1);
       selected(ranks(1:k)) = 1;
       expanded(k,:) = getscores(selected,targets);
   end
```

Finally, we can visually compare results in (12.22b) and (12.22c) by generating
bar plots for both cases at some selected rank positions.

```
>> % defines the rank positions to be displayed and their tags      (12.23a)
>> ranks = 1:2:30;
>> for k=1:length(ranks), xlbls(k)= sprintf("@%02d",ranks(k)); end

>> % initializes the figure                                         (12.23b)
>> figname = 'Comparison between original and expanded queries';
>> hf = figure(6); set(hf,'Color',[1 1 1],'Name',figname);

>> hs = subplot(3,1,1); % displays the precision bar plot           (12.23c)
>> bar([original(ranks,1),expanded(ranks,1)]);
>> legend('original','expanded');
>> set(hs,'XTickLabel',xlbls); ylabel('Precision');

>> hs = subplot(3,1,2); % displays the recall bar plot             (12.23d)
>> bar([original(ranks,2),expanded(ranks,2)]);
>> set(hs,'XTickLabel',xlbls); ylabel('Recall');

>> hs = subplot(3,1,3); % displays the f-score bar plot            (12.23e)
>> bar([original(ranks,3),expanded(ranks,3)]);
>> set(hs,'XTickLabel',xlbls); ylabel('F-score');
```

The three resulting bar plots are presented in Figure 12.6. Notice from the figure how all three scores benefit from the query expansion strategy, mainly as the rank positions are incremented and more documents are selected.

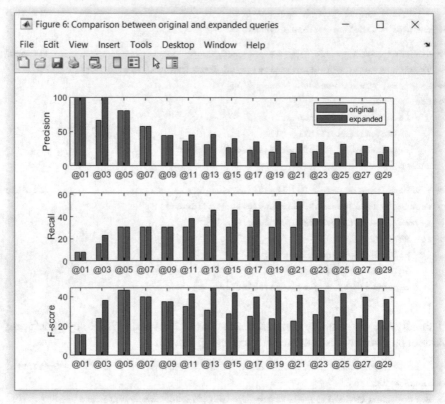

Fig. 12.6. Comparative evaluation between the performances of the original query 5 from Table 12.1 and the expanded query 5 computed in (12.20)

12.3 The BM25 Ranking Function

In this section, we focus our attention on the BM25 family of ranking functions, which derives from the probabilistic relevance framework and constitutes the most popular TF-IDF based retrieval approach used in document search. BM25 is implemented in MATLAB® function `bm25Similarity`.

Similar to the vector-based search presented in the previous section, BM25 aims at scoring the documents in a collection according to their relevance to a given query, based on the query terms that occur in each document. Different from the vector-based search, BM25 pays attention to the IDF component while controlling the TF component with a set of parameters that allow for compensating for different term frequency and document length effects. The generic BM25 score can be formulated as follows:

$$BM(q,d) = \sum_i IDF_{BM}(q_i) \times TF_{BM}(q_i,d) \qquad (12.24)$$

where q and d refer to the query and the document, and q_i to the ith keyword within the query. The IDF_{BM} component is typically computed as:

$$IDF_{BM}(q_i) = ln(\ (N-n(q_i)+0.5)\ /\ (n(q_i)+0.5) + 1\ ) \qquad (12.25)$$

where N is the total number of documents in the collection and $n(q_i)$ refers to the document frequency associated to keyword q_i, i.e. the number of documents in the collection containing q_i. The TF_{BM} component, on the other hand, is computed as:

$$TF_{BM}(q_i,d) = TF(q_i,d) \times (k+1)\ /\ (\ TF(q_i,d) + k\ (1-b+b|d|/|d|_{av})\ ) + \delta \qquad (12.26)$$

where $TF(q_i,d)$ is the term frequency of keyword q_i in document d, $|d|$ is the length of document d in number of words, $|d|_{av}$ is the average length of the documents in the collection, and k, b and δ are adjustable parameters, referred to as term frequency scaling factor (k), document length scaling factor (b) and document length correction factor (δ), which can be optimized for the task at hand.

In the standard BM25 implementation, the parameters are set as follows: $k=1.2$, $b=0.75$ and $\delta=0$. Different value settings of these parameters give raise to different variants of the algorithm. For instance, when setting $b=1$ and $b=0$ the resulting variants are referred to as BM11 and BM15, respectively. In the case of BM11, it is when the document length has the most influence on the score, while in the case of BM15, the effect of the document length is completely ignored. When $\delta>0$, the resulting ranking function is known as BM25+, which aims at preventing over-penalizing long documents.

Next, we use the BM25 ranking function **bm25Similarity** to repeat the search exercise with all queries from Table 12.1. Similar to what we did in (12.18), we retrieve documents up to rank position *10*.

```
>> % computes BM25 relevance scores                              (12.27a)
>> bmsearch = bm25Similarity(docs,queries);

>> % gets document rankings and evaluates performance            (12.27b)
>> for k=1:nqrys
        targets = labels==qrylbls(k);
        [~,ranks] = sort(bmsearch(:,k),'descend');
        selected = zeros(ndocs,1);
        selected(ranks(1:rankloc)) = 1;
        scores_bm25(k,:) = getscores(selected,targets);
   end
```

```
>> % computes score averages                                          (12.27c)
>> bm25_averages = mean(scores_bm25);
```

Now, let us compare the results in (12.27) with the ones from (12.18) previously obtained with the vector-based search strategy.

```
>> % displays results from BM25 (12.27) and vector search (12.18)      (12.28)
>> datavalues = [v10_averages;bm25_averages];
>> tbl = array2table(datavalues,'VariableNames',variables(2:end));
>> methodnames = ["Vector @10";"BM25 @10"];
>> disp(addvars(tbl,methodnames,'NewVariableNames',"Method",'Before',1))
```

Method	Precision	Recall	F-score
"Vector @10"	31.25	18.777	22.897
"BM25 @10"	33.75	20.124	24.608

As seen from (12.28), BM25 consistently overperforms the vector-based search for the task under consideration in all three evaluation metrics. In general, BM25 is considered state-of-the-art with regards to TF-IDF based approaches (see exercise 12.6-6 for a more detailed comparison between BM25 and vector-based search performances).

Finally, let us compare the standard BM25 function, used in (12.27), with some of its variants. Consider, for instance, the BM11 variant, for which we set parameter `'DocumentLengthScaling'` to *1* and follow the procedure used in (12.27):

```
>> % implements BM11 search                                           (12.29)
>> bm11search = bm25Similarity(docs,queries,'DocumentLengthScaling',1);
>> for k=1:nqrys
        targets = labels==qrylbls(k);
        [~,ranks] = sort(bm11search(:,k),'descend');
        selected = zeros(ndocs,1); selected(ranks(1:rankloc)) = 1;
        scores_bm11(k,:) = getscores(selected,targets);
   end
>> bm11_averages = mean(scores_bm11);
```

and, similarly, by setting `'DocumentLengthScaling'` to *0*, we implement BM15:

```
>> % implements BM15 search                                           (12.30)
>> bm15search = bm25Similarity(docs,queries,'DocumentLengthScaling',0);
>> for k=1:nqrys
        targets = labels==qrylbls(k);
        [~,ranks] = sort(bm15search(:,k),'descend');
        selected = zeros(ndocs,1); selected(ranks(1:rankloc)) = 1;
        scores_bm15(k,:) = getscores(selected,targets);
   end
>> bm15_averages = mean(scores_bm15);
```

Figure 12.7 (left to you as an exercise) presents a bar plot comparing the performances among the different BM25 search variants from (12.27), (12.29) and (12.30). As seen from the figure, for the task under consideration, BM15 seems to provide the best performance across all metrics, followed by BM25 and BM11.

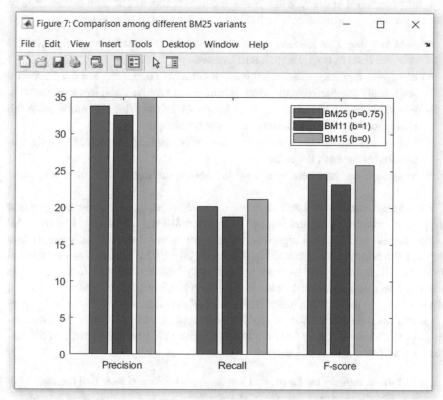

Fig. 12.7. Average performances, over the eight queries in Table 12.1, for different variants of the BM25 ranking function: BM25 (12.27), BM11 (12.29) and BM15 (12.30)

12.4 Cross-language Search

In the previous sections we have deal with the problem of searching documents in a monolingual context. However, it is very common in some practical applications to face the problem of document search in a cross-language scenario (the query and the document collection are in different languages), a multilingual scenario (the document collection contains documents in different languages), or both. In this section we focus on the problem of cross-language search. More specifically, we explore some general properties of parallel data collections and illustrate how to exploit them to mine word associations across languages, as well as to conduct cross-language search.

The parallel dataset we are going to use in this section is a bilingual collection of short paragraphs and sentences in English and Spanish. The data is available in the file `datasets_ch12c.mat`, which contains the following variables:

- `rawtext_eng` and `rawtext_spa`: string arrays containing each document's original texts in both, English and Spanish.
- `docs_eng` and `docs_spa`: tokenized documents containing the preprocessed texts (only standard tokenization, punctuation removal, and lowercasing).
- `bowmodel_eng` and `bowmodel_spa`: bag-of-words models computed from the corresponding tokenized document representations.
- `vocab_eng` and `vocab_spa`: string arrays containing the corresponding vocabularies for each language.
- `tfidfmtx_eng` and `tfidfmtx_spa`: the corresponding TF-IDF matrices.

The dataset consists of a collection of *1,304* short paragraphs and sentences that have been extracted from the English and Spanish versions of Don Quixote. All *1,304* English and Spanish segments of text are fully parallel, which means that they are the corresponding translations of each other. The dataset was constructed such that all segment lengths are approximately between *10* and *50* words. This was done in a semiautomatic manner, by using the English and Spanish versions of Don Quixote that are available from the Project Gutenberg's website and applying some basic alignment heuristics. Some amount of manual intervention was required to correct and verify the alignments. Table 12.2 presents some basic statistics for this parallel bilingual dataset.

Table 12.2. Basic statistics for the parallel English-Spanish dataset from Don Quixote

	English	Spanish
Text segments	1,304	1,304
Minimum segment length	9	10
Maximum segment length	49	50
Average segment length	26.58	24.90
Running words	34,657	32,476
Vocabulary terms	3,908	4,866
Singleton vocabulary terms	2,070	2,940
Common vocabulary terms	208	208

Notice from the table that this dataset, although containing more text segments, is actually smaller than the Don Quixote subset used in the previous sections. Indeed, the total amount of running words in the English portion of this dataset is *34,657* while the total amount of running words in the *843* paragraphs of the Don

Quixote subset used in the previous sections is a little bit more than three times larger: *110,384*. In general, and because of obvious reasons, parallel data will be always scarcer than monolingual data.

Another interesting observation from Table 12.2 is that the Spanish vocabulary is significantly larger than the English one. In fact, the English vocabulary size is about *80%* of the Spanish vocabulary size. This is an unavoidable consequence of Spanish having a richer morphology than English. Also notice the huge proportion of singletons in both cases, being around *53%* and *60%* of the total vocabulary sizes for English and Spanish, respectively.

Regarding segment lengths, it can be seen from Table 12.2 that, on average, English segments tend to be two words longer than Spanish segments. However, an interesting fact about this disproportion is that it is not evenly distributed over all segment lengths. To explore this issue in more detail, let us build a cross-plot between Spanish and English segment lengths and compute a regression curve for it. We do this as follows:

```
>> load datasets_ch12c % loads the bilingual collection                    (12.31a)

>> % gets the total number of words in each text segment                   (12.31b)
>> nw_spa = doclength(docs_spa);
>> nw_eng = doclength(docs_eng);

>> % computes a linear regression between nw_spa and nw_eng                 (12.31c)
>> coeff = regress(nw_spa',[nw_eng',ones(length(nw_eng),1)]);
>> x_reg = [10,50]; y_reg = coeff(1)*x_reg + coeff(2);

>> % generates the cross-plot                                              (12.31d)
>> hf = figure(8);
>> figname ='Cross-plot between English and Spanish segment lengths';
>> set(hf,'Color',[1 1 1],'Name',figname);
>> plot(nw_eng,nw_spa,'.',x_reg,y_reg,'-k',x_reg,x_reg,':k');
>> legend('data samples','regression line','symmetry line',...
          'Location','NorthWest');
>> axis([10,50,10,50]);
>> xlabel('Length of English text segments');
>> ylabel('Length of Spanish text segments');
```

The resulting cross-plot and regression line are depicted in Figure 12.8. Notice from the figure, in which the identity (or symmetry) line has been also plotted for reference purposes, how the regression line between English and Spanish segment lengths departs from the symmetry line as the segment lengths increase. Indeed, according to the regression line, a Spanish segment of *10* words in length has, on average, an English translation of *10* words; while a Spanish segment of *45* words in length has, on average, and English translation of *50* words.

This specific pattern between segment lengths is not actually a general trend. In some other bilingual datasets, Spanish tends to be more verbose than English. What is observed in Figure 12.8 seems to be an effect related to Spanish being the original language of the documents and the English versions being a translation.

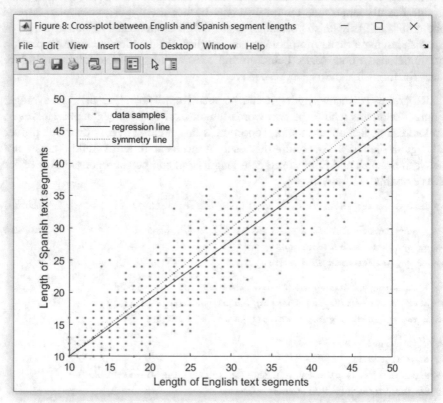

Fig. 12.8. Cross-plot and regression line between English and Spanish segment lengths for the Don Quixote parallel dataset described in Table 12.2

A final observation with respect to Table 12.2 that is worth to be notice is related to common vocabulary terms. Although English and Spanish are different languages, we can see that there is a small proportion of common terms in both languages. The total number of such terms is *208*, which represents about *5%* and *4%* of the total vocabularies for English and Spanish, respectively.

A more detailed exploration of this subset of terms reveals that there are different categories of, as well as reasons for, such commonalities. The largest group corresponds to Spanish proper names such as *Dulcinea*, *Sancho* and *La Mancha*, which are written in English exactly in the same way they are written in Spanish. Other groups of common terms include words, different from proper names, that are the same in both languages (lexical cognates) such as *doctor*, *singular*, *univer-*

sal and *no*; and words that are spelled in the same way in both languages but actually have a different meaning, such as *he* (an auxiliary verbal from in Spanish), *quite* (a verbal form in Spanish) and *a* (a preposition in Spanish).

Next, let us illustrate some of the useful things that can be done with a bilingual parallel dataset like the one described in Table 12.2. In our first example, we attempt at automatically inferring the corresponding Spanish translations of some given English terms. However, before proceeding, let us conduct some pruning to the English and Spanish vocabularies in the parallel dataset. More specifically, we will eliminate all vocabulary terms occurring only in a single document (think a little bit about this and convince yourself that this is slightly different from removing singletons):

```
>> % applies vocabulary pruning to the English subset          (12.32a)
>> mask = full(sum(bowmodel_eng.Counts>0)>1);
>> pvocab_eng = vocab_eng(mask);
>> ptfidfmtx_eng = tfidfmtx_eng(:,mask);
```

```
>> % applies vocabulary pruning to the Spanish subset          (12.32b)
>> mask = full(sum(bowmodel_spa.Counts>0)>1);
>> pvocab_spa = vocab_spa(mask);
>> ptfidfmtx_spa = tfidfmtx_spa(:,mask);
```

Now, we are ready to proceed with the translation inference example. Consider again query 5 from Table 12.1, which was composed of two terms: *lions* and *courage*. We can extract, from the bilingual dataset, the closest Spanish terms that are associated to these English words by using a procedure similar to the one we used in (12.20) to expand a query. The basic idea is as follows: first, we identify all segments of text containing the English term of interest; then, we use the corresponding Spanish segments to compute the most-relevant-document vector contribution and the rest of the segments to compute the least-relevant-document vector contribution; finally, from the resulting vector, we extract the indexes for the most representative Spanish words.

Let us first illustrate the proposed procedure by extracting the top five Spanish translation candidates for the English term *lions*:

```
>> % creates and index for the English word "lions"           (12.33a)
>> widx_eng1 = pvocab_eng=="lions";
>> % gets pointers to the documents in which the word occurs
>> docs_eng1 = full(ptfidfmtx_eng(:,widx_eng1)>0);
```

```
>> % derives the Spanish vector of translation candidates      (12.33b)
>> % 1. computes the contribution of the most relevant documents
>> vplus = sum(ptfidfmtx_spa(docs_eng1,:))/sum(docs_eng1);
>> % 2. computes the contribution of the least relevant documents
>> vminus = sum(ptfidfmtx_spa(not(docs_eng1),:))/sum(not(docs_eng1));
```

```
>> % 3. computes the complete vector of translation candidates
>> spav_eng1 = full(vplus-vminus);

>> % gets the 5 most representative Spanish terms for "lions"          (12.33c)
>> [vals_eng1,ranks] = sort(spav_eng1,'descend');
>> spaw_eng1 = pvocab_spa(ranks(1:5))
spaw_eng1 =
  1×5 string array
    "leones"     "han"     "estremo"     "toma"     "llegar"
```

Next, let us find the five most relevant Spanish words to the English term *courage*. We do it by following the same procedure used in (12.33) (this time, we omit the explanations):

```
>> widx_eng2 = pvocab_eng=="courage";                                 (12.34a)
>> docs_eng2 = full(ptfidfmtx_eng(:,widx_eng2)>0);

>> vplus = sum(ptfidfmtx_spa(docs_eng2,:))/sum(docs_eng2);            (12.34b)
>> vminus = sum(ptfidfmtx_spa(not(docs_eng2),:))/sum(not(docs_eng2));
>> spav_eng2 = full(vplus-vminus);

>> [vals_eng2,ranks] = sort(spav_eng2,'descend');                     (12.34c)
>> spaw_eng2 = pvocab_spa(ranks(1:5))
spaw_eng2 =
  1×5 string array
    "ánimo"     "estremo"     "valgan"     "llegar"     "imposible"
```

Effectively, the first options derived in both (12.33c) and (12.34c) constitute the corresponding Spanish words for *lions* and *courage* that are used in the Spanish version of Don Quixote. Notice that, in a more general context, the most common translations for *courage* would probably be *coraje* or *valor*; and the most common English translations for *ánimo* would be *mood* or *disposition*. However, in the context under consideration, the English word that was actually used for translating *ánimo* was *courage*. This should be a reminder of the fact that data-driven approaches will always be biased and affected by the specific domain and other particularities of the corpus under consideration.

Now, let us extract the five most representative English words for Spanish words *leones* and *ánimo*. We do this by repeating the procedure already illustrated in (12.33) and (12.34), but in the opposite direction (from Spanish to English):

```
>> widx_spa1 = pvocab_spa=="leones";                                  (12.35a)
>> docs_spa1 = full(ptfidfmtx_spa(:,widx_spa1)>0);
>> vplus = sum(ptfidfmtx_eng(docs_spa1,:))/sum(docs_spa1);
>> vminus = sum(ptfidfmtx_eng(not(docs_spa1),:))/sum(not(docs_spa1));
>> engv_spa1 = full(vplus-vminus);
>> [vals_spa1,ranks] = sort(engv_spa1,'descend');
>> engw_spa1 = pvocab_eng(ranks(1:5))
```

```
engw_spa1 =
  1×5 string array
    "lions"     "character"     "reached"     "happily"     "courage"
```

```
>> widx_spa2 = pvocab_spa=="ánimo";                                      (12.35b)
>> docs_spa2 = full(ptfidfmtx_spa(:,widx_spa2)>0);
>> vplus = sum(ptfidfmtx_eng(docs_spa2,:))/sum(docs_spa2);
>> vminus = sum(ptfidfmtx_eng(not(docs_spa2),:))/sum(not(docs_spa2));
>> engv_spa2 = full(vplus-vminus);
>> [vals_spa2,ranks] = sort(engv_spa2,'descend');
>> engw_spa2 = pvocab_eng(ranks(1:5))
engw_spa2 =
  1×5 string array
    "heart"     "courage"     "hast"     "province"     "delay"
```

Notice from (12.35) that the first recovered word for *leones* is actually *lions*. However, for the case of *ánimo*, *courage* happens to be the second recovered word, after *heart*. This is because the word *ánimo* is used in the text collection under consideration with two different senses, which are precisely *heart* and *courage*; being the former the most commonly used one. This can be easily verified by looking at the occurrences of *courage* and *ánimo* in the corresponding English and Spanish segments in which they occur:

```
>> % English segments in which the word "courage" occurs                (12.36a)
>> find(docs_eng2)'
ans =
   744    753
```

```
>> % Spanish segments in which the word "ánimo" occurs                   (12.36b)
>> find(docs_spa2)'
ans =
    60    172    393    744    753    969    978
```

where text segment *753* is a good example of the *courage* sense of *ánimo*:

```
>> disp(rawtext_eng(753))                                                (12.37a)
"What dost thou think of this, Sancho?" said Don Quixote. "Are there any
enchantments that can prevail against true valour? The enchanters may be
able to rob me of good fortune, but of fortitude and courage they
cannot."
```

```
>> disp(rawtext_spa(753))                                                (12.37b)
-¿Qué te parece desto, Sancho? -dijo don Quijote-. ¿Hay encantos que valgan
contra la verdadera valentía? Bien podrán los encantadores quitarme la
ventura, pero el esfuerzo y el ánimo, será imposible.
```

and text segment *978* is a good example of the *heart* sense of *ánimo*:

```
>> disp(rawtext_eng(978))                                                    (12.38a)
Here Sancho exclaimed, "I don't mount, for neither have I the heart nor
am I a knight."
```

```
>> disp(rawtext_spa(978))                                                    (12.38b)
-Aquí -dijo Sancho- yo no subo, porque ni tengo ánimo ni soy caballero.
```

Except for (12.35b), if you take a look at the second, third, fourth and fifth ranked words in (12.33c), (12.34c) and (12.35a), you will find that most of them are not related at all to the word we are looking a translation for. A natural question that arises here is whether there is, or not, a way to discriminate those words that might actually be a translation from those that are not. The answer is yes! The way to perform this discrimination is by looking at the scores associated to the ranked words, which we have stored already in `vals_eng1`, `vals_eng2`, `vals_spa1` and `vals_spa2`, in (12.33c), (12.34c), (12.35a) and (12.35b), respectively.

The rationale for using the scores to discriminate valid translations from other unrelated words occurring within the top ranks is based on the assumption that the actual translations would exhibit larger scores than other words that are unrelated. Then, we could expect a sudden drop in the scores when crossing the rank boundary between the set of valid and non-valid translations. The performance quality of this procedure depends, of course, on the quality and the extension of the parallel dataset being used.

To illustrate this procedure, we generate bar plots for all four-translation inference conducted in examples (12.33c), (12.34c), (12.35a) and (12.35b). The generated plots are presented in Figure 12.9.

```
>> % performs the basic figure settings                                      (12.39a)
>> figname = 'Bar plots from translation inference examples';
>> hf = figure(9); set(hf,'Color',[1 1 1],'Name',figname);
```

```
>> % creates the bar plot for the "lions" example (12.33)                    (12.39b)
>> hs = subplot(2,2,1); barh(vals_eng1(1:5)); axis ij;
>> set(hs,'YTickLabel',spaw_eng1); title('Spanish words for "\it{lions}"');
```

```
>> % creates the bar plot for the "courage" example (12.34)                  (12.39c)
>> hs = subplot(2,2,3); barh(vals_eng2(1:5)); axis ij;
>> set(hs,'YTickLabel',spaw_eng2); title('Spanish words for "\it{courage}"');
```

```
>> % creates the bar plot for the "leones" example (12.35a)                  (12.39d)
>> hs = subplot(2,2,2); barh(vals_spa1(1:5)); axis ij;
>> set(hs,'YTickLabel',engw_spa1); title('English words for "\it{leones}"');
```

```
>> % creates the bar plot for the "ánimo" example (12.35b)                   (12.39e)
>> hs = subplot(2,2,4); barh(vals_spa2(1:5)); axis ij;
>> set(hs,'YTickLabel',engw_spa2); title('English words for "\it{ánimo}"');
```

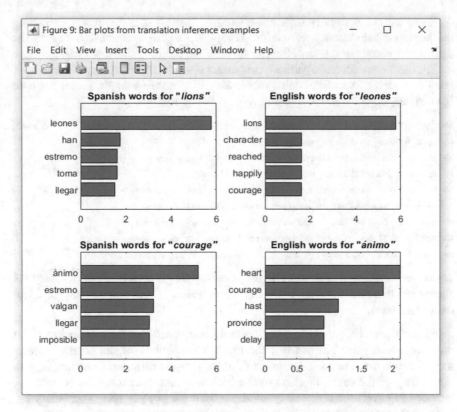

Fig. 12.9. Bar plots for the five most representative Spanish (English) words for the English (Spanish) terms composing query 5 from Table 12.1

Notice from the figure, how in the case of *lions* and *leones* the score gaps between the actual translations and the unrelated terms are very clear. However, in the cases of *courage* and *ánimo*, the gaps are significantly smaller. If we set an absolute threshold to retain *ánimo* in the former case, we will probably miss *courage* in the latter one.

The last example we are going to illustrate in this section focus on the problem of cross-language search. Suppose we are given the Spanish keywords in query 5 from Table 12.1 (i.e. *leones* and *ánimo*) but we are required to find relevant paragraphs in the English version of the book (indeed, from the English dataset we used in sections 12.1 and 12.2, which also contains texts from Oliver Twist and Pride and Prejudice). As you might be thinking already, this is a more challenging task than just retrieving relevant documents from the parallel collection used in this section. One simple way of addressing this problem is by using the inferred translations from (12.35a) and (12.35b) to construct a new English query and conduct the document search just as we did in sections 12.1 and 12.2 (we leave this method for you to work on later in exercise 12.6-9).

Here, we use a slightly different approach. Instead of individually extracting the keyword translations, we use the complete English vector associated to the Spanish keywords as a query. Similar to what we did in (12.35), we will generate an English vector for the Spanish terms under consideration but, instead of generating one vector for each Spanish word, we will generate a single vector for both Spanish keywords *leones* and *ánimo*. We do this as follows:

```
>> % combines documents that are relevant to both Spanish keywords        (12.40)
>> mask = docs_spa1|docs_spa2;
>> % computes the contribution of the relevant documents
>> vplus = sum(ptfidfmtx_eng(mask,:))/sum(mask);
>> % computes the contribution of the non-relevant documents
>> vminus = sum(ptfidfmtx_eng(not(mask),:))/sum(not(mask));
>> a = 0.2; % sets the value of the combination parameter
>> xqry = full(a*vplus-(1-a)*vminus); % computes the complete vector
>> xqry = xqry.*(xqry>0); % discards negative entries (if any)
```

where the resulting vector `xqry` is the cross-language derived English query we should use to search for documents that are relevant to both Spanish keywords *leones* and *ánimo*.

Nevertheless, before continuing, it should be noticed that there is an important problem we still need to address: the English vocabulary of the parallel dataset used for inferring the cross-language English query is different from the English vocabulary of the data collection over which we want to conduct the search! So, we need to covert the obtained query `xqry` from the `pvocab_eng` "vocabulary coordinates" into the corresponding `vocab` "vocabulary coordinates":

```
>> % intersects the vocabularies and gets indexes for common terms        (12.41)
>> [~,index1,index2] = intersect(vocab,pvocab_eng);
>> converted_qry = zeros(1,length(vocab)); % initializes the new query
>> converted_qry(index1) = xqry(index2); % performs the conversion
```

Now, we are ready to conduct the search and compute the associated scores (precision, recall and F-score) by following the same procedure used in (12.22). The resulting scores are plotted in Figure 12.10 along with those obtained for the original vector-based search in (12.22b) (the generation of the plot is left to you as an exercise).

```
>> targets = labels==qrylbls(qrynum);                                     (12.42)
>> vsearch = full(cosineSimilarity(converted_qry,tfidfmtx));
>> [~,ranks] = sort(vsearch,'descend');
>> for k=1:50
        selected = zeros(ndocs,1);
        selected(ranks(1:k)) = 1;
        crosslang(k,:) = getscores(selected,targets);
    end
```

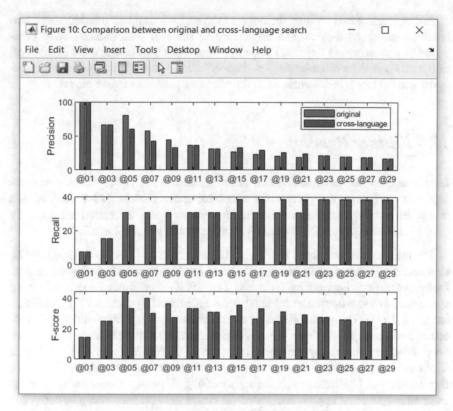

Fig. 12.10. Comparative evaluation between the performances of the original vector-based search conducted in (12.22b) and the cross-language search in (12.42)

As seen from the figure, the performance of the implemented cross-language search is comparable to the original vector-based search in (12.22b). At several rank positions its performance is either below or above the original system performance but, on average, they actually perform the same:

```
>> % computes and displays original and cross-language score averages    (12.43)
>> datavalues = [mean(original);mean(crosslang)];
>> tbl = array2table(datavalues,'VariableNames',variables(2:end));
>> methodnames = ["Original";"Cross-Language"];
>> disp(addvars(tbl,methodnames,'NewVariableNames',"Method",'Before',1))
```

Method	Precision	Recall	F-score
"Original"	27.868	34	25.085
"Cross-Language"	27.307	34.462	25.135

In general, cross-language document search is a harder problem than monolingual search, and then its performance is typically worse. The similar performance observed here is due to the fact that the bilingual dataset used to implement the cross-language search strategy belongs to the same domain as the document collection over which the search was performed (indeed, both datasets were composed of text segments extracted from the same book). In general, the domain affinity between the available parallel dataset and the specific cross-language task under consideration is determinant to the resulting performance on the task.

12.5 Further Reading

In this chapter we have introduced some fundamentals concepts on document search, mostly from the perspective of vector space models. More detailed and comprehensive introductions to information retrieval and its related areas can be found in (Manning *et al.* 2008) and (Baeza-Yates and Ribeiro-Neto 2011).

For more details on binary search, you should refer to (Fox *et al.* 1992). Precision and recall biased behaviors of *AND* and *OR* binary search strategies, respectively, were first reported by (Lee and Fox 1988). The fundamentals of vector space models for information retrieval were established by (Jones 1972) and (Salton *et al.* 1975). The probabilistic relevance framework was introduced by (Robertson and Jones 1976), from which other probabilistic models have been developed, including the binary independence model (Yu and Salton 1976) and BM25 (Robertson *et al.* 1994). For a more recent revision on the probabilistic relevance framework, see (Robertson and Zaragoza 2009). A good introduction to current information retrieval approaches based on neural networks and deep learning can be found in (Mitra and Craswell 2018).

The basic idea behind the method illustrated in (12.20) for redefining a query by using the most and least relevant documents in a ranked list is due to (Rocchio 1971). For more information on the specific problem of query expansion refer to (Efthimiadis 1996). An interesting discussion about the concept of relevance is provided by (Saracevic 1975) and a good review on ranking is given in (Belkin and Croft 1987). Another important topic we did not address in this chapter is the one related to the construction of indexes. A good book on this topic can be found in (Witten *et al.* 1999). A more recent machine learning based approach to the ranking problem, is the one known as learning to rank (Liu 2009), which also has found applications in areas different from information retrieval, such as fault detection (Xuan and Monperrus 2014) and news recommendation (Lv *et al.* 2011).

Finally, for more information on the topic of cross-language information retrieval (CLIR), see (Grefenstette 1998) and (Wang and Oard 2021). The specific method you are requested to implement in short project 12.7-2 is described in detail in (Littman *et al.* 1998). Other subfields of information retrieval not discussed here include multimedia information retrieval (MIR) (Nordbotten 2008) and human-computer information retrieval (HCIR) (White *et al.* 2013).

12.6 Proposed Exercises

1. Consider the definition of accuracy presented in chapter 11 for assessing the quality of a document categorization result:

    – Show that such definition is completely equivalent to the following definition, which uses the cardinalities of the subsets presented in Figure 12.2, $accuracy = (TP+TN)/(TP+TN+FP+FN) \times 100\%$.
    – Use this formula to compute the accuracy of all experimental results presented in section 12.1. What do you observe? Can you give an explanation to these results?
    – What can you conclude about the utility of accuracy to evaluate the quality of a given set of selected documents?

2. Consider the five queries from Table 12.1 that contain three keywords.

    – Implement a binary search script able to combine both, *AND* and *OR*, operators into a single search.
    – Conduct search experiments with different keyword operations. For example, use (*oliverANDtwist)ORboard*, *oliverAND(twistORboard)*, (*oliverOR twist)ANDboard*, etc. for the first query, and do the same with the rest.
    – How do the resulting evaluation scores (precision, recall and F-score) compare to the ones reported in (12.14b) and (12.14c)?
    – Which specific examples benefited from the new binary search strategy?

3. Reproduce example (12.17) for all queries in Table 12.1.

    – Plot the curves for precision, recall and F-score versus the number of selected documents for each case, as it was done for query *5* in Figure 12.4.
    – What similarities and differences do you observe among the eight plots?
    – Could you infer from the plots which is the most appropriate rank cutoff, for each case, such that the best compromise between precision and recall is achieved? What are your conclusions?

4. The bar plot presented in Figure 12.5 considered the average evaluation scores for vector-based search at rank position *10*.

    – Reproduce the average score computations for the vector-based search strategy at rank positions varying from *1* to *20*.
    – What specific problem do you encounter when computing F-scores at the first rank positions? Can you find the cause?
    – Use individual query's score-versus-rank curves you generated in the previous exercise to explain the problem detected in the previous step.
    – Propose a strategy to overcome the detected problem and implement it.

– Can you determine the most appropriate rank cutoff, for the overall set of eight queries, such that the best compromise between precision and recall is achieved?
– How does it compare to the individual query cutoffs derived in the previous exercise? What are your conclusions?

5. Consider de query expansion procedure described in (12.20).

– Repeat the experiment for different values of the combination parameter and try to determine an optimum value for it.
– Repeat the experiment by considering different amounts of most and least relevant documents. Can you get optimal values for these quantities too?
– Repeat the exercise to determine the most appropriate values of combination parameter for each of the other queries in Table 12.1.
– What are your main conclusions?

6. Consider de vector-based ranking approach described in section 12.2 and the BM25 ranking function described in section 12.3.

– Consider the range of ranking positions from 1 up to the total number of documents `ndocs`.
– Compute average scores (precision, recall and F-scores), at each rank position, for searches conducted over the eight queries in Table 12.1.
– Plot, on a semilogarithmic scale, the average values of precision, recall and F-score vs. the ranking position for both vector-based and BM25 searches.
– How do vector-based search compare to BM25?

7. Consider the bilingual parallel corpus used in section 12.4.

– Implement a script for computing all entries in Table 12.2.
– Consider now the parallel Corpora of the United Nations for Research Purposes https://opus.nlpl.eu/UN.php Accessed 6 July 2021.
– Select a language pair from these corpora and use the script from the first step to compute the same statistics for this corpus.
– Generate a cross-plot, similar to the one in Figure 12.8, for your selected language pair. What are your main conclusions?

8. Consider the translation inference procedure illustrated in Figure 12.9.

– Define and implement a heuristic to determine the number of valid translations that can be inferred for a given term.
– Test your heuristic on the same examples illustrated in Figure 12.9. What are the main problems you encounter?
– Reconsider the translation inference procedure from (12.33) an incorporate a combination parameter into the vector calculation.

-   Vary the combination parameter for the translation inference. Can you find
    any value such that your heuristic provides better results?
-   Repeat the exercise with some other English and Spanish words.
-   Repeat the whole exercise but, instead of using the pruned vocabularies
    from (12.32), consider the complete English and Spanish vocabularies.
-   What differences do you observe when using the pruned or the full vo-
    cabularies?

9. Use your best translation inference procedure from the previous exercise to
   conduct cross-language search:

-   Given the Spanish words for query 5 (*ánimo* and *leones*), infer the most
    appropriate English translations.
-   Construct different English queries by considering different numbers and
    combinations of the inferred translations.
-   Conduct the corresponding document searches.
-   How do these results compare to the one presented in section 12.4? What
    are your main conclusions?

10. Consider the procedure used in (12.40) for constructing an English query given
    some Spanish keywords.

-   How many English keywords are actually being used in this case?
-   Repeat the experiment by using different values for the combination pa-
    rameter. Can you find an optimal value for this parameter?
-   Perform similar document searches for the cases of queries *4* and *6* from
    Table 12.1. What are your conclusions?

12.7 Short Projects

1. Consider the *Reuters-21578* dataset, you already used in short project 11.7-2,
   http://kdd.ics.uci.edu/databases/reuters21578/reuters21578.html Accessed 6 July 2021. Create a script for implementing a vector-based search with the following specifications:

-   The user must be able to define a query by specifying one or more key-
    words or, alternative, a complete phrase or sentence.
-   The algorithm must return a ranked list with the first *n* documents that are
    relevant to the query.
-   The results should be displayed in a two-dimensional map where the fol-
    lowing information is to be presented: the query, the centroids of the *10*
    closest topic categories (out of *57* categories), and the selected documents.

- The user should be able to identify the different types of elements in the map, as well as to see some keywords related to the contents in the different regions of the map.
- The user should be able to click on the documents in the map to select and open them. (Hint: consider using the `ButtonDownFun` property of the plot object to define a callback function for displaying the content of the document being clicked on.)

2. Consider the English-Spanish parallel dataset used in section 12.4:

- Concatenate both TF-IDF matrices `tfidfmtx_eng` and `tfidfmtx_spa` into a single matrix of size $M \times (N_E + N_S - N_c)$, where M is the total number of text segments, N_E is the English vocabulary size, N_S is the Spanish vocabulary size and N_c is the number of common terms in both vocabularies.
- Apply SVD to the concatenated TF-IDF matrix, and retain the corresponding projection matrix. (Consider a reduced dimensionality of *100*.)
- Construct a vector space representation for the English and Spanish collections of paragraphs from the book of Don Quixote. (The English and Spanish paragraph collections are available in the files `DonQuixote.mat` and `DonQuijote.mat`, respectively.)
- Notice that vector space representations for English and Spanish documents should use the extended vocabulary of size $N_E + N_S - N_c$.
- Project both the English and Spanish datasets into a *100*-dimensional space by using the linear projection matrix derived from the SVD.
- Consider queries *4*, *5* and *6* from Table 12.1, as well as some other queries you might create or derive from Don Quixote's chapter titles.
- Construct a vector space representation for the queries and project them into the reduced space. Conduct document search in the reduced space for both English and Spanish documents indifferently.
- Evaluate your results and conduct different experiments by using different dimensionalities.

12.8 References

Baeza-Yates R, Ribeiro-Neto B (2011) Modern Information Retrieval: The Concepts and Technology behind Search (2nd Edition), Assison-Wesley Professional

Belkin NJ, Croft WB (1987) Retrieval Techniques. In Williams ME (Ed.) Annual Review of Information Science and Technology, pp. 109–154

Efthimiadis EN (1996) Query Expansion. In Williams ME (Ed.), Annual Review of Information Systems and Technology 31: 121–187

Fox EA, Betrabet S, Koushik M, Lee WC (1992) Extended Boolean Models, in Frakes WB and Baeza-Yates R (Eds.) Information Retrieval: Data Structures and Algorithms, Prentice Hall

Grefenstette G (1998) Cross-Language Information Retrieval. Kluwer

Jones KS (1972) A statistical interpretation of term specificity and its application in retrieval. Journal of Documentation 28(1): 11–21

Lee WC, Fox EA (1988) Experimental compartison of schemes for interpreting Boolean queries. Technical Report TR-88-27, Computer Science, Virginia Polytechnic Institute and State University

Littman ML, Dumais ST, Landauer TK (1998) Automatic cross-language information retrieval using latent semantic indexing. In Grefenstette G (Ed.), Cross-Language Information Retrieval. Kluwer

Liu TY (2009) Learning to Rank for Information Retrieval. Foundations and Trends in Information Retrieval 3(3): 225–331

Lv Y, Moon T, Kolari P, Zheng ZH, Wang XH, Chang Y (2011) Learning to Model Relatedness for News Recommendation. In Proceedings of the International Conference on World Wide Web (WWW)

Manning CD, Raghavan P, Schütze H (2008) Introduction to Information Retrieval, Cambridge University Press

Mitra B, Craswell N (2018) An Introduction to Neural Information Retrieval. Foundations and Trends® in Information Retrieval, 13(1): 1–126, https://www.microsoft.com/en-us/research/uploads/prod/2017/06/INR-061-Mitra-neuralir-intro.pdf Accessed 6 July 2021

Nordbotten JC (2008) Multimedia Information Retrieval Systems, http://nordbotten.com/ADM/ADM_book/MIRS-frame.htm Accessed 6 July 2021

Robertson SE, Jones KS (1976) Relevance weighting of search terms. Journal of the American Society for Information Science 27: 129–146

Robertson SE, Walker S, Jones S, Hancock-Beaulieu M, Gatford M (1994) Okapi at TREC-3. In Proceedings of the Third Text REtrieval Conference (TREC 1994). Gaithersburg, USA

Robertson SE, Zaragoza H (2009) The Probabilistic Relevance Framework: BM25 and Beyond. Foundations and Trends in Information Retrieval, 3(4): 333–389

Rocchio JJ (1971) Relevance feedback in information retrieval. In Salton G (Ed.) The SMART Retrieval System – Experiments in Automatic Document Processing. Prentice Hall, pp. 313–323

Salton G, Wong A, Yang CS (1975) A vector space for automatic indexing. Communications of the ACM 18(11): 613–620

Saracevic T (1975) Relevance: A review of and a framework for the thinking on the notion of information science. Journal of American Society for Information Science, 26(6):321-343

Wang JQ, Oard DW (201) Matching meaning for cross-language information retrieval. Information Processing & Management, 48(4): 631–653

White R, Capra R, Golovchinsky G, Kules B, Smith C, Tunkelang D (2013) Introduction to Special Issue on Human-computer Information Retrieval. Journal of Information Processing and Management, 49(5): 1053–1057

Witten IH, Moffat A, Bell TC (1999) Managing Gigabytes: Compressing and Indexing Documents and Images. San Francisco, Morgan Kaufmann Publishing

Xuan JF, Monperrus M (2014) Learning to Combine Multiple Ranking Metrics for Fault Localization. In Proceedings of the IEEE International Conference on Software Maintenance and Evolution, pp. 191–200

Yu CT, Salton G (1976) Precision weighting: An effective automatic indexing method. Journal of ACM 23(1): 76–88

13 Content Analysis

In the previous two chapters we have reviewed different methods aiming at classifying and searching for documents, as well as identifying the most relevant vocabulary words with respect to a given set of documents or domain. In this chapter, we continue in a similar direction but differently from the previous ones here we pay attention to the contents of the documents from a different perspective. Rather than focusing on topics or domains, we will be focusing on the entities being mentioned, their properties and their subjective or opinionated aspects.

This chapter is organized as follows. First, in section 13.1, we present a brief experiment illustrating the ability to independently discriminate between the thematic and subjective natures of text data. Then, in section 13.2, we focus our attention on the specific problem of polarity detection and intensity estimation. In section 13.3, we pay attention to the features and qualifiers of the entities being described in texts. Finally, in sections 13.4, we present and illustrate pattern-match based methods for discovering specific terminology associated to entities, their properties, and the different types of relationships among them.

Before starting, an operational notice must be given for this chapter. As we will be mainly working with user generated content collected from the web, some of which are protected under copyright, some of datasets will not be provided. In order to be able to reproduce some of the experiments in this chapter you will be given only the mathematical models derived from such datasets. In other cases, where the dataset is required, you will need to get the dataset directly from the original source and apply the indicated preprocessing steps to it.

13.1 Dimensions of Analysis

In the previous two chapters we have analyzed documents, paragraphs, and words according to their specific thematic nature. This kind of information mainly reflects the different topics or domains that are present in the given document collection. There are, however, other dimensions within the nature of a given document or collection of documents that we might be interested in analyzing. Such other dimensions can include, for instance, the writing style or genre of the documents, their authorship, the kind of communicative functions they are addressing, their emotional and subjective characteristics, etc. In this section, and the next couple of sections, we focus our attention in one specific aspect of the subjective nature of written documents, the polarity of the opinions expressed in them.

R. E. Banchs, *Text Mining with MATLAB®*, https://doi.org/10.1007/978-3-030-87695-1_13

Here, we illustrate with a simple experiment two important issues related to the analysis of opinionated texts. First, that the previously described mathematical models for representing text are prone to capture and discriminate the main characteristics of the opinionated nature of documents; and second, that such discriminative power is somehow independent from the representation of the thematic or domain nature of the documents.

Let us start by loading the vector model for a small set of documents on service reviews gathered from a customer review website. The model is provided in the data file `datasets_ch13a.mat`:

```
>> clear                                                                      (13.1a)
>> load datasets_ch13a
```

```
>> whos                                                                       (13.1b)
   Name          Size            Bytes  Class        Attributes
   domain        140x1             396  categorical
   polarity      140x1             382  categorical
   tfidfmtx      140x819        172768  double       sparse
```

As seen from (13.1b), the provided model consists of a *140×819* TF-IDF matrix. Only *140* documents compose the collection, and the vocabulary size is also relatively small, only *819* vocabulary terms are considered (vocabulary pruning was applied to this dataset by using a standard stopword list). Two additional categorical variables are also provided along with the model, which are `domain` and `polarity`. They indicate which category in these two dimensions each document belongs to. Two different categories are available for each one.

Regarding the domain categories, reviews were collected from two major categories in the source website, namely *financial* and *transportation*. The number of documents belonging to each category can be seen as follows:

```
>> summary(domain)                                                            (13.2)
      financial            60
      transportation       80
```

As seen from (13.2), *60* documents belong to the *financial* domain and *80* documents belong to the *transportation* domain.

Similarly, regarding the polarity categories, the collected reviews belong to one of the two extreme categories in the valuation scale provided on the website: extremely negative reviews (*1* star) and extremely positive reviews (*5* stars). The number of documents within each of these two categories can be seen as follows:

```
>> summary(polarity)                                                          (13.3)
      negative            70
      positive            70
```

As seen from (13.3), half of the documents in the collection belongs to the *negative* polarity category and the other half of the documents belongs to the *positive* polarity category.

In the following example we generate a pie chart illustrating the proportions of documents in each of the categories along both, the domain and the polarity, dimensions:

```
>> % computes the number of documents per category groups          (13.4a)
>> type_TN = sum((domain=="transportation")&(polarity=="negative"));
>> type_TP = sum((domain=="transportation")&(polarity=="positive"));
>> type_FP = sum((domain=="financial")&(polarity=="positive"));
>> type_FN = sum((domain=="financial")&(polarity=="negative"));
>> doctypes = ["TRN-NEG","TRN-POS","FIN-POS","FIN-NEG"];

>> hf = figure(1); % generates the pie chart                        (13.4b)
>> figname = 'Proportion of documents across domains and polarities';
>> set(hf,'Color',[1 1 1],'Name',figname);
>> pie3([type_TN,type_TP,type_FP,type_FN],ones(1,4),doctypes);
```

The resulting chart is depicted in Figure 13.1. As seen from the figure, although there are about *14%* more documents in the *transportation* domain that in the *financial* domain, the dataset can still be considered balanced across domains and polarities. Regarding polarity, exactly one half of the documents is in the *negative* category and the other half in the *positive* category.

Next, we apply dimensionality reduction to the model of the document collection. We do this with the objective of visualizing, in a two-dimensional space, the relative distances among the different document types. For computing the projections, we use the non-linear dimensionality reduction method described in section 10.3, which is based on multidimensional scaling. We proceed as follows:

```
>> dims = 2; % defines the dimensionality of the reduced space      (13.5)
>> cdist = pdist(tfidfmtx,'cosine'); % computes cosine distances
>> dmap = mdscale(cdist,dims,'Criterion','stress'); % computes the projections
```

Now, we are ready to represent the documents in the small data collection described in Figure 13.1 into a two-dimensional map. In the map, we assign different markers to each different domain-polarity combination. The resulting map is depicted in Figure 13.2, which can be generated as follows:

```
>> hf = figure(2); % creates the figure                             (13.6a)
>> figname = 'Two-dimensional map of the dataset described in Figure 13.1';
>> set(hf,'Color',[1 1 1],'Name',figname);

>> % gets indexes for each type of documents                        (13.6b)
>> TN = (domain=="transportation")&(polarity=="negative");
>> TP = (domain=="transportation")&(polarity=="positive");
```

```
>> FN = (domain=="financial")&(polarity=="negative");
>> FP = (domain=="financial")&(polarity=="positive");
>> doctypes = ["TRN-NEG","TRN-POS","FIN-NEG","FIN-POS"];

>> % generates the plot                                                    (13.6c)
>> hp = plot(dmap(TN,1),dmap(TN,2)'.',dmap(TP,1),dmap(TP,2),'.',...
             dmap(FN,1),dmap(FN,2),'.',dmap(FP,1),dmap(FP,2),'.');
>> mc = 'rbrb'; ms = '*dvo'; % sets colors and shapes for the markers
>> for k=1:length(hp)
      set(hp(k),'Marker',ms(k),'LineStyle','none','MarkerSize',5);
      set(hp(k),'MarkerFaceColor',mc(k),'MarkerEdgeColor',mc(k));
   end

>> % adjusts axis and prints legend and labels                             (13.6d)
>> axis([-0.5,0.5,-0.6,0.5]);
>> legend(doctypes,'Location','South','Orientation','horizontal');
>> xlabel('Domain discriminating dimension')
>> ylabel('Polarity discriminating dimension');
```

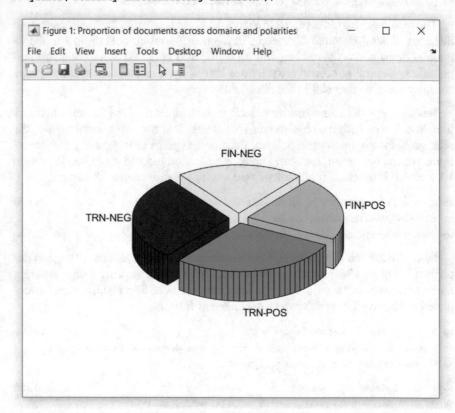

Fig. 13.1. Proportion of documents in each domain-polarity category combination

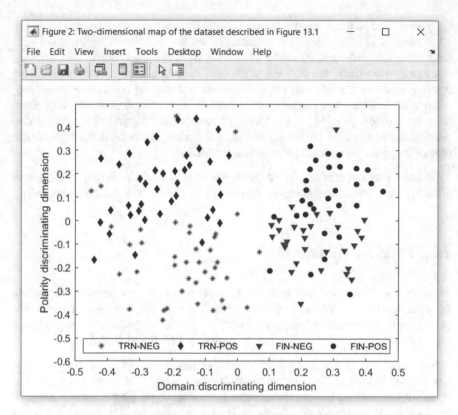

Fig. 13.2. Two-dimensional representation of the dataset described in Figure 13.1. Different markers have been assigned to each different domain-polarity category combination

As seen from Figure 13.2, each entry in the map represents a document in the data collection. As a different marker has been assigned to each different domain-polarity category combination, it is easy to visually track the distribution of each category across the two-dimensional map.

Notice the clear separation between the two domain categories along the horizontal direction. While all documents belonging to the *financial* category are located in the right-hand side of the map, all documents belonging to the *transportation* category are located in the left-hand side of the map.

A similar separation can be observed along the vertical direction for the two polarity categories under consideration. While the documents belonging to the *positive* review category tend to be located in the upper side of the map, the documents in the *negative* review category tend to appear towards the lower side of the map. However, as seen from the figure, there are quite a few exceptions, which make the separation between the two polarity categories less clear than the separation between the two domain categories.

We can derive some important conclusions from the results depicted in Figure 13.2. First, it is indeed possible to discriminate the polarity of opinionated contents by using mathematical models based on word frequencies and distributions, and more specifically, the vector space model approach used here. Moreover, the discriminating power of both, the domain-oriented and the polarity-oriented, representation dimensions generated by the models seem to be somehow independent from each other. In addition, the obtained results also suggest that polarity discrimination is a harder task than domain discrimination, at least for the specific type of text data we have consider in this section.

In the following section, we continue working in this direction but, more specifically, we discuss the problem of detecting polarity and estimating its intensity.

13.2 Polarity Estimation

Some of the most important tasks when analyzing opinionated contents include the detection of polarity, the estimation of its intensity, the detection of subjectivity and the measurement of emotions. Here, we will focus on the first two, and the other two problems will be addressed in exercises 13.6-5 and 13.6-6.

While polarity detection deals with the identification of the two polarity categories defined in the previous section, *negative* and *positive*,[1] intensity estimation has to do with measuring how much negative or how much positive a given content is. In other words, while polarity detection is a binary classification problem, intensity estimation constitutes either a discrete multi-category classification problem or a continuous regression problem.

In this section we approach these two problems from the statistical perspective. More specifically, a naïve Bayes approach, based on the likelihood ratio method described in section 11.4, will be applied. Different from what we did in section 11.4, here we use word unigram models instead of LDA models.

Regarding the dataset, we use a reference experimental polarity dataset that has been originally derived from the IMDb collection.[2] This dataset, known as *polarity dataset v2.0*, contains full texts for *1,000* positive and *1,000* negative reviews about movies.[3] You will need to get this dataset directly from the website at Cornell University in order to be able to reproduce most of the examples presented in this and the following section.

---

[1] Sometimes, the third category *neutral* is also included in the polarity detection problem.

[2] IMDb (The Internet Movie Database) https://www.imdb.com/ Accessed 6 July 2021.

[3] "Movie Review Data" at Cornell University http://www.cs.cornell.edu/people/pabo/movie-review-data/ Accessed 6 July 2021.

Once you have retrieved the dataset and uncompressed the files you will see the two directories `review_polarity/txt_sentoken/pos` and `review_polarity/txt_sentoken/neg`, each of which contains the *1,000* documents for the corresponding polarity. Next, you need to apply the following preprocesses to the files in each directory in order to create the tokenized document objects we will be working with:

```
>> % preprocesses and prepares the 1000 positive reviews              (13.7a)
>> dirname = "review_polarity/txt_sentoken/pos/";
>> filelist = dir(dirname); counter = 0;
>> for k=1:length(filelist)
       if filelist(k).isdir==0
           counter = counter+1;
           temp = string(evalc('type(dirname+filelist(k).name)'));
           tempdocs(counter) = strtrim(regexprep(temp,'\s*',' '));
       end
   end
>> docs_pos = erasePunctuation(tokenizedDocument(tempdocs));

>> % preprocesses and prepares the 1000 negative reviews              (13.7b)
>> dirname = "review_polarity/txt_sentoken/neg/";
>> filelist = dir(dirname); counter = 0;
>> for k=1:length(filelist)
       if filelist(k).isdir==0
           counter = counter+1;
           temp = string(evalc('type(dirname+filelist(k).name)'));
           tempdocs(counter) = strtrim(regexprep(temp,'\s*',' '));
       end
   end
>> docs_neg = erasePunctuation(tokenizedDocument(tempdocs));
```

Next, we compute histograms of document lengths for each of the two categories of documents in the dataset.

```
>> hf = figure(3); % creates the figure                               (13.8a)
>> set(hf,'Color',[1 1 1],'Name','Distributions of document lenghts');

>> % computes and plots the histogram for positive reviews            (13.8b)
>> lpos = doclength(docs_pos);
>> subplot(2,1,1); hist(lpos,min(lpos):max(lpos));
>> axis([0,2500,0,8]); text(1800,7,'Positive Reviews');
>> xlabel('Document length (in words)'); ylabel('Frequency');

>> % computes and plots the histogram for negative reviews            (13.8c)
>> lneg = doclength(docs_neg);
>> subplot(2,1,2); hist(lneg,min(lneg):max(lneg));
>> axis([0,2500,0,8]); text(1800,7,'Negative Reviews');
>> xlabel('Document length (in words)'); ylabel('Frequency');
```

The resulting histograms are depicted in Figure 13.3. As seen from the figure, both document length distributions seem to be quite similar, although the positive review category contains some few larger documents.

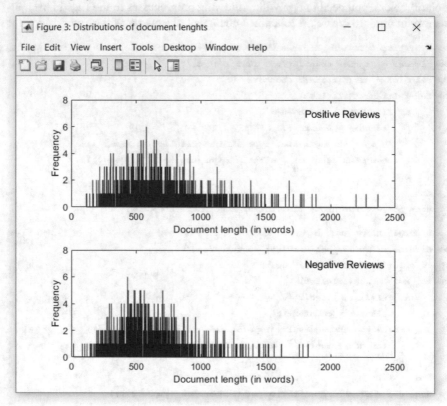

Fig. 13.3. Histograms of document lengths for the two categories of documents

The basic statistics for each category subset, as well as for the overall data collection, are summarized in Table 13.1.

Table 13.1. Basic statistics for both, positive and negative, review subsets and the overall *polarity dataset v2.0* collection

	Positive Reviews	Negative Reviews	Full Collection
Number of documents	1,000	1,000	2,000
Running words	684,354	610,760	1,295,114
Vocabulary size	34,460	32,233	46,598
Minimum document length	116	16	16
Maximum document length	2,367	1,825	2,367
Average document length	684.35	610.76	647.56

Next, we generate randomly permuted indexes for all documents in each of the two category subsets, which are used to generate the data partition for test, development, and train sets. We reset the random number simulator to its default value to ensure all experiments are reproducible.

```
>> rng('default');                                                    (13.9)
>> randselect_pos = randperm(1000);
>> randselect_neg = randperm(1000);
```

where the first *100* indexes contained in `randselect_pos` and `randselect_neg` define the test set, the next *100* indexes define the development set, and the last *800* indexes define the train set.

Word unigram models have been already computed for each data category by using the corresponding train sets. The models are available in the data file named `datasets_ch13b.mat`, which contains the two structures `unigram_pos` and `unigram_neg`. Each structure is composed of three fields: `vocab`, `prob` and `unkp`, which contain the vocabulary terms, the unigram probability estimates, and the unseen event probability estimate, respectively. The model probabilities have been discounted by using a procedure like the one described in section 8.2.

Logarithmic likelihood ratios can be computed for a given document, or set of documents, by using the function `compute_loglirat`. This function implements a unigram version of the log likelihood ratio computation procedure we implemented with LDA models in section 11.4. The syntax of the function is as follows:

```
loglirat = compute_loglirat(docs,model1,model2)                       (13.10)
```

where `model1` and `model2` are structures containing word unigram models in the same format as described above for `unigram_pos` and `unigram_neg`, and `docs` is a tokenized document array containing one or more documents like both document arrays, `docs_pos` and `docs_neg`, created in (13.7a) and (13.7b).

The output returned by function `compute_loglirat` is a numeric vector containing as many values as documents are given in the input variable `docs`. For each individual document d, the function returns the result of the following operation: $loglirat = log(p(d|model1)) - log(p(d|model2))$, where each document probability p is estimated by means of the product of unigram probabilities.

Before starting any computation, let us extract the test documents from both tokenized document arrays generated in (13.7). For this, we use the randomly permuted indexes from (13.9). We proceed as follows:

```
>> % gets the test documents for both categories                      (13.11)
>> tst_docs_pos = docs_pos(randselect_pos(1:100));
>> tst_docs_neg = docs_neg(randselect_neg(1:100));
```

Now, we are ready to compute log likelihood ratios for the documents in the test sets of each of the two categories (positive reviews and negative reviews) by using the word unigram models provided in `datasets_ch13b.mat`.

```
>> load datasets_ch13b % loads the models                           (13.12a)

>> % computes the log likelihood ratios                             (13.12b)
>> loglirat_pos = compute_loglirat(tst_docs_pos,unigram_pos,unigram_neg);
>> loglirat_neg = compute_loglirat(tst_docs_neg,unigram_pos,unigram_neg);
```

The resulting log likelihood ratios are depicted in Figure 13.4 (the generation of the plots is left to you as an exercise).

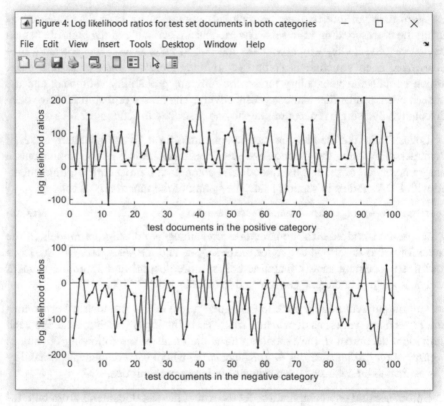

Fig. 13.4. Log likelihood ratios for test documents in both, positive and negative, categories

Following the order in which we have provided the unigram models in the two function calls in (13.12b), Figure 13.4 depicts for each test document the logarithm of the ratio between its probability estimated with the unigram model `unigram_pos` and its probability estimated with the unigram model `unigram_neg`. This means that a log likelihood ratio greater or less than zero implies the given document belongs to the positive or negative category, respectively.

As both categories have been depicted separately in Figure 13.4, it is relatively easy to visually assess the performance of the implemented method as a means for polarity detection. Notice, from the upper panel in the figure, that log likelihood ratios are greater than zero for most of the test documents in the positive category. Similarly, as can be seen from the lower panel, log likelihood ratios are less than zero for most of the test documents in the negative category. In both plots, the zero level has been indicated with segmented lines for an easier visualization of the correctly and wrongly classified documents.

In addition to the visual inspection conducted on Figure 13.4, we can compute the accuracy of the proposed technique with regards polarity detection. In the following example, we compute both the confusion matrix and the overall accuracy:

```
>> % sets both the estimated and the actual category vectors            (13.13a)
>> targets = [ones(length(loglirat_pos),1);zeros(length(loglirat_neg),1)];
>> estimated = double([loglirat_pos>0;loglirat_neg>0]);

>> % computes the confusion matrix                                      (13.13b)
>> confusion_mtx = confusionmat(estimated,targets,'ORDER',[1,0]);
>> cmtx = array2table(confusion_mtx,'VariableNames',["positive";"negative"]);
>> cmtx.('Classified as') = ["positive";"negative"];
>> disp(cmtx)
      positive      negative      Classified as
        78            18            "positive"
        22            82            "negative"

>> % computes the overall accuracy                                      (13.13c)
>> accuracy = sum(diag(confusion_mtx))/sum(sum(confusion_mtx))*100
accuracy =
    80
```

As seen from (13.13c), the achieved accuracy for polarity detection is *80%*. Moreover, as can be verified from (13.13b), such accuracy happens to be balanced across the categories. Indeed, from the *100* documents belonging to the positive category, *22* have been mistakenly classified as negative, while from the *100* negative documents, *18* has been classified as positive.

Let us consider next the problem of intensity estimation, in which we are interested in assigning a certain degree of positivity or negativity to a given document. Notice from Figure 13.4 that, although we only used the sign of the log likelihood ratios in (13.13a) to determine the polarity of the documents, the actual values of the log likelihood ratios already provide a means for intensity estimation as the computed values are distributed over a wide range of real values. Indeed, we can intuitively think about documents exhibiting larger positive log likelihood ratios as being more positive than those exhibiting smaller positive log likelihood ratios and, similarly, we can think about documents with larger negative ratios as being more negative than those with smaller negative ratios.

In the following example, we illustrate the use of the log likelihood ratio as estimator of polarity intensity. Consider, for instance, the following three sample sentences: *"the movie is bad"*, *"the actor performance is excellent"*, and *"the actor performance is excellent but the movie is bad"*.

```
>> % inputs the sample sentences in the required format        (13.14a)
>> newdata(1) = "the movie is terrible";
>> newdata(2) = "the actor performance is excellent";
>> newdata(3) = newdata(2)+" but "+newdata(1);
>> newdocs = tokenizedDocument(newdata);

>> % estimates the polarity intensity                          (13.14b)
>> loglirat = compute_loglirat(newdocs,unigram_pos,unigram_neg)'
loglirat =
   -2.6394    2.5212   -0.1476
```

As seen from (13.14b), the intensity estimates for the first two sentences, although opposite in polarity, are somehow similar in absolute value. On the other hand, the intensity estimate for the third sentence, which combines both positive and negative comments, is much smaller in magnitude. However, pure log likelihood ratio values do not seem to be a good choice for intensity scores. Indeed, in theory, such values can range from minus infinity to infinity, making it difficult to appreciate how positive or negative are the values obtained in (13.14b).

In order to have a more appropriate range of values for estimating polarity intensity, we should use a normalization function to convert pure log likelihood ratios into a more useful polarity intensity score. Here we use a sigmoid function to perform the normalization. More specifically, we use the following formula:

$$intensity\_score = 2 \: / \: (1 + e^{-loglirat}) - 1 \qquad (13.15)$$

The normalized intensity score proposed in (13.15) maps polarity intensities between the negative and positive extremes into the interval from *−1* to *+1*, respectively. We can apply this normalization to the results in (13.14b):

```
>> scores = 2*(1+exp(-loglirat)).^(-1)-1              (13.16)
scores =
   -0.8667    0.8512   -0.0737
```

The score values in (13.16) are easier to interpret than the ones in (13.14b) as we already know that *−1* and *+1* are the extremes of the scale.

Table 13.2 presents some additional examples on polarity detection and intensity estimation which follow the procedure described in this section. In the table, each text segment being analyzed is presented along with its corresponding values of log likelihood ratio, polarity, and intensity score. The twelve examples presented in the table have been manually ranked in ascending order according to their polarity intensities, from the most negative (rank *1*) to the most positive (rank *12*).

Although the provided rankings are certainly subjective, and you could not completely agree with them, they provide a reference for comparison against the computed intensity scores.

Table 13.2. Some additional examples on polarity detection and intensity estimation within the movie review domain

	Example	loglirat	polarity	score
1	It was as good as garbage.	−2.4069	negative	−0.8347
2	This actor is terrible; his performance was pathetic.	−3.1293	negative	−0.9162
3	The music was bad, and the script was boring.	−4.4053	negative	−0.9759
4	This film has some problems with the plot.	−1.2401	negative	−0.5512
5	Not as good as the previous movie in the saga.	2.1414	positive	0.7897
6	Not so bad. I was expecting something worse.	−3.8648	negative	−0.9589
7	The plot was simple, but I enjoyed the movie anyway.	−1.1864	negative	−0.5322
8	Interesting film, which is full of action and excitement.	−0.2224	negative	−0.1108
9	A very funny and pleasantly entertaining film.	1.8993	positive	0.7396
10	Wonderful script and beautiful photography. A great movie!	1.9856	positive	0.7585
11	Excellent movie, as expected from such a great director.	1.7281	positive	0.6983
12	Exceptional production I will be watching again and again.	1.9173	positive	0.7437

As seen from the table, the method is performing fairly well in general terms. However, in some of the cases, it is still far from conducting a good work. In particular, it is evident that the methodology is having problems with negated constructions. Consider for instance examples 5 and 6, in which negations are used. In the case of example 5, which has received a relatively high positive score, the comment is actually saying that the movie was "not as good" as expected. Similarly, in the case of example 6, which has received a very negative score (surely because of the occurrences of *bad* and *worse*), the comment is actually implying that the film was better than expected.

Another interesting case is example 7, which presents one negative and one positive opinion in the same comment. The assigned score, however, seems to be giving more importance to the negative part of the comment. In the case of example 8, it certainly constitutes an error as a small negative score has been given to a content which is definitively a positive one. Other interesting error is the one related to examples 11 and 12, in which extremely positive comments have been given only moderately positive scores.

In the same way we have detected polarity and computed its intensity for documents and sentences, we can also think about using the procedure described in this section to estimate the polarity of individual vocabulary terms. Notice, however, that we should expect a poorer performance in this task, as the polarity estimation for the case of an isolated word is more likely to be affected by the noise present in the models. Additionally, the polarity and intensity for a given word is generally affected by the context in which it occurs.

In the following example, we estimate polarity intensity scores for a list of ten vocabulary terms. The results are presented as a bar graph in Figure 13.5.

```
>> % computes polarity intensity scores for a given list of words        (13.17a)
>> words = ["awful","bad","excellent","good","hilarious","normal",...
            "perfect","poor","regular","terrible"];
>> loglirat_words = compute_loglirat(words,unigram_pos,unigram_neg);
>> scores_words = 2*(1+exp(-loglirat_words)).^(-1)-1;

>> % creates the bar graph                                               (13.17b)
>> hf = figure(5); set(hf,'Color',[1 1 1]); ha = axes;
>> set(hf,'Name','Intensity score estimates at the word level');
>> barh(scores_words); xlabel('Intensity of Polarity');
>> axis([-1,1,0,11]); set(ha,'YTickLabel',words,'YDir','reverse');
```

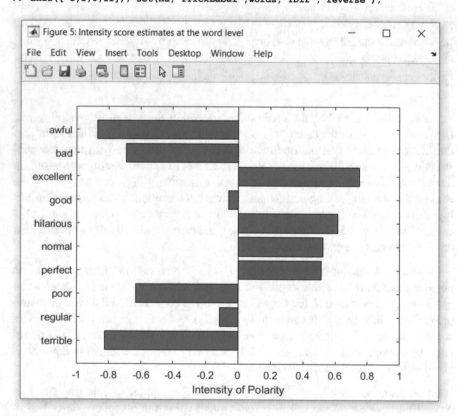

Fig. 13.5. Polarity intensity scores computed at the word level for some sample terms

As seen from Figure 13.5, with the exceptions of *good* and *normal*, most of the intensity estimates for the considered words are correct. In the case of *good*, however, it has been assigned an intensity score close to zero instead of a more appropriate high positive value. Similarly, the word *normal* received and exaggeratedly high positive score rather than a more appropriate one closer to zero, such as in the case of *regular*.

13.3 Qualifier and Aspect Identification

In the previous sections and chapters, we have presented methods that primarily exploit word frequency or probability information to assess similarities among different units of text. In this (and the following) section, we present a completely different approach to analyzing text for mining purposes, which is based on text-pattern matching. This approach, which is both simple and convenient, is one of the fundamental text mining tools for extracting information from text.

Text-pattern matching allows for extracting and discovering useful information from large volumes of text data and it has a wide variety of applications. Here, we focus our attention on the specific problem of mining aspects and qualifiers from opinionated contents. In particular, we illustrate the combined use of text-pattern matching techniques and probabilistic models for extracting information that is relevant to opinionated contents such as, for instance, the targets of the opinions, their specific aspects or features being apprised, and the different qualifiers used.

In this section we continue working with the same IMDb polarity dataset used in the previous section. As this is a relatively small dataset, most of the examples presented in this section are limited in several respects, but we use the same dataset for the sake of continuity and clarity, as well as to be able to reuse the word unigram models that are already available for this dataset.

Let us start by considering a more specific problem than the general ones of detecting polarity and estimating its intensity. Suppose we are interested in knowing what people say exactly when they speak either bad or good about a movie. What are the different aspects of movies they write about and what are the corresponding qualifiers they use for describing bad movies and good movies? A simple and convenient way of doing this is by matching some the text patterns that we observe (or might expect to occur) in movie reviews. For this purpose, we can create customized regular expressions aiming at capturing the different qualifiers and entities appearing in common text structures used in movie reviews.

Before illustrating this point, we need to convert the data collection into a format that facilitates for matching regular expression. More specifically, we concatenate all documents in each subcollection into a single string variable. Let us consider first the files within the negative polarity subset.

```
>> % copies all negative documents into a single string                    (13.18)
>> strings_neg = joinWords(docs_neg);
>> collection_neg = join(strings_neg);
```

Consider now the following basic regular expression:

```
>> pattern1 = "this movie was (\w+) ";                                      (13.19)
```

which aims at capturing all possible words used just after the sequence of words
this movie was. If we attempt to match this regular expression within the string
`collection_neg` created in (13.18), we obtain the following result:

```
>> matches = regexp(collection_neg,pattern1,'tokens');                      (13.20)
>> matches1_neg = string(matches)
matches1_neg =
  1×25 string array
  Columns 1 through 5
    "worth"     "originally"    "really"    "boring"    "pointless"
  Columns 6 through 12
    "likely"    "they"    "that"    "based"    "sheer"    "ever"    "based"
  Columns 13 through 18
    "trying"    "boring"    "marketed"    "the"    "about"    "ever"
  Columns 19 through 24
    "awful"    "all"    "obviously"    "just"    "basically"    "laughable"
  Column 25
    "that"
```

Notice from (13.20) that the proposed pattern was matched *25* times. However,
these results are very noisy! While some interesting terms such as *boring*, *point-
less* and *awful* have been found, some other irrelevant and spurious terms such as
the, *all* and *just* have been gathered too.

In order to improve the quality of the results in (13.20) we need to implement a
cleaning or filtering mechanism able to identify the terms that are actually relevant
to the objective we are pursuing, which is identifying negative qualifiers commonly use to describe movies. Implementing such a filtering mechanism happens to be
a relatively simple endeavor if we consider taking advantage of the unigram models that we have available for this dataset.

To perform this filtering, we first need to rank all vocabulary terms within the
negative polarity subset of the collection according to their relevance to the negative category. In other words, we need to discriminate those terms that are most
relevant to the negative polarity category from those that are moderately relevant,
as well as from those that are irrelevant to the negative category and relevant to
the positive one. We do this by computing the probability ratio between the negative unigram model probability (`unigram_neg`) and the positive unigram model
probability (`unigram_pos`) for each word in the negative subset:

```
>> % computes probability ratios for negative vocabulary terms          (13.21a)
>> for k=1:length(unigram_neg.vocab)
       p_neg = unigram_neg.prob(k);
       idx = find(unigram_pos.vocab==unigram_neg.vocab(k));
       if isempty(idx), p_pos = unigram_pos.unkp;
       else, p_pos = unigram_pos.prob(idx); end
       score_neg(k) = p_neg/p_pos;
   end

>> % ranks negative vocabulary according to probability ratios          (13.21b)
>> [~,all_ranks_neg] = sort(score_neg,'descend');
```

Now, we can use the resulting vocabulary rankings to assign ranks to the words extracted in (13.20). This, hopefully, should allow us to find a filtering criterion for retaining only those relevant terms we are interested in. In the following step, we assign ranks to all words obtained in (13.20):

```
>> ranks1_neg = zeros(1,length(matches1_neg));                           (13.22)
>> for k=1:length(matches1_neg)
       idx = find(unigram_neg.vocab==matches1_neg(k));
       if isempty(idx), ranks1_neg(k) = mean(all_ranks_neg);
       else, ranks1_neg(k) = find(all_ranks_neg==idx); end
   end
```

The resulting ranks are depicted, along with their corresponding words in Figure 13.6. The figure can be generated as follows:

```
>> hf = figure(6); set(hf,'Color',[1 1 1]);                              (13.23)
>> set(hf,'Name','Ranking of words matching the pattern "this movie was _"');
>> ha = axes; hp = plot(ranks1_neg,1:length(ranks1_neg),'o-');
>> set(hp,'MarkerFaceColor','b','MarkerSize',5);
>> axis([0,25000,0,length(ranks1_neg)+1]); xlabel('Rank index');
>> set(ha,'YTick',1:length(ranks1_neg),'YTickLabel',matches1_neg);
>> set(ha,'YDir','reverse','YGrid','on');
```

Notice from the figure how the most relevant words to the negative polarity category (those with small rank indexes in the left-hand side of the plot) are clearly separated from the rest of the words. As seen from the plot, any rank index in the interval from *5,000* to *11,000* would perfectly serve as a separation boundary between relevant and irrelevant words. Accordingly, we can use a rank threshold value of *10,000* to filter out the irrelevant words:

```
>> unique(matches1_neg(ranks1_neg<10000))                                (13.24)
ans =
  1×5 string array
    "awful"     "boring"     "laughable"     "marketed"     "pointless"
```

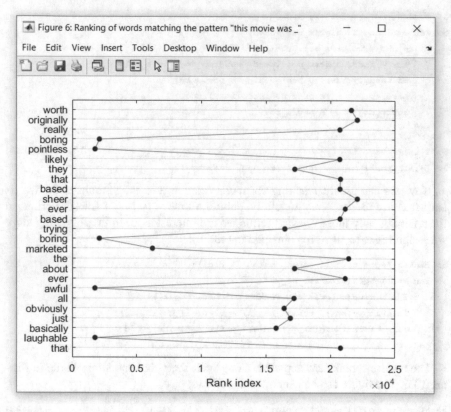

Fig. 13.6. Extracted words, and their relevance rankings with respect to the negative polarity category, that are commonly used in negative movie reviews

Alternatively, we can also retrieve the words of interest by intersecting the set of extracted words `matches1_neg` from (13.20) with the *10,000* top ranked vocabulary terms in (13.21b), without the need for the step in (13.22):

```
>> intersect(matches1_neg,unigram_neg.vocab(all_ranks_neg(1:10000)))          (13.25)
ans =

  1×5 string array

    "awful"     "boring"     "laughable"     "marketed"     "pointless"
```

Next, we apply the same extraction procedure to the positive-polarity part of the data collection:

```
>> % copies all positive documents into a single string                       (13.26a)
>> strings_pos = joinWords(docs_pos);
>> collection_pos = join(strings_pos);

>> % matches regular expression 'pattern1' from (13.19)                       (13.26b)
>> matches = regexp(collection_pos,pattern1,'tokens');
```

```
>> matches1_pos = string(matches)
matches1_pos =
  1×12 string array
  Columns 1 through 5
    "really"    "actually"    "basically"    "one"    "until"
  Columns 6 through 12
    "exceptional"    "not"    "one"    "not"    "not"    "released"    "its"
```

As seen from (13.26), with the exception of *exceptional*, most of the extracted words in this case are practically useless! But anyway, let us proceed with the filtering stage and see what happens:

```
>> % computes probability ratios for positive vocabulary terms         (13.27a)
>> for k=1:length(unigram_pos.vocab)
       p_pos = unigram_pos.prob(k);
       idx = find(unigram_neg.vocab==unigram_pos.vocab(k));
       if isempty(idx), p_neg = unigram_neg.unkp;
       else, p_neg = unigram_neg.prob(idx); end
       score_pos(k) = p_pos/p_neg;
   end

>> % ranks positive vocabulary according to probability ratios         (13.27b)
>> [~,all_ranks_pos] = sort(score_pos,'descend');

>> % filters words matched in (13.26b) with the ranked vocabulary      (13.27c)
>> intersect(matches1_pos,unigram_pos.vocab(all_ranks_pos(1:10000)))
ans =

    "exceptional"
```

As seen from (13.27c), indeed, *exceptional* is the only word that matches the pattern under consideration and is relevant to the positive polarity category.

Next, as you might be probably thinking already, we can consider using a more sophisticated regular expression than the one in (13.19), such as for instance:

```
>> word_sequence = "(?:this|that|the) (?:movie|film) (?:is|was)";    (13.28)
>> optional_word = "(?: \w*?y\>)?";
>> candidate_word = " (\w+) ";
>> pattern2 = word_sequence+optional_word+candidate_word;
```

This new regular expression is composed of three parts:

- an initial word sequence, which aims at matching patterns such as *this movie is*, *that movie was*, *this film is*, *the film was*, etc.
- an optional *y*-ending word, which aims at matching possible adverbs such as the ones retrieved in (13.20) and (13.26b): *originally*, *actually*, etc.
- the candidate word to be extracted as a possible movie qualifier.

In the following example we extract candidate words from both, the positive and the negative, subsets in the collection by using the regular expression defined in (13.28). We also assign ranks to the extracted words and generate plots for the resulting ranks, which are depicted in Figure 13.7 (the generation of the plots is left to you as an exercise).

```
>> % extracts candidate words from the positive polarity subset        (13.29a)
>> matches2_pos = string(regexp(collection_pos,pattern2,'tokens'));
```

```
>> % extracts candidate words from the negative polarity subset        (13.29b)
>> matches2_neg = string(regexp(collection_neg,pattern2,'tokens'));
```

```
>> % assigns ranks to positive polarity candidate words                (13.29c)
>> ranks2_pos = zeros(1,length(matches2_pos));
>> for k=1:length(matches2_pos)
       idx = find(unigram_pos.vocab==matches2_pos(k));
       if isempty(idx), ranks2_pos(k) = mean(all_ranks_pos);
       else, ranks2_pos(k) = find(all_ranks_pos==idx); end
   end
```

```
>> % assigns ranks to negative polarity candidate words                (13.29d)
>> ranks2_neg = zeros(1,length(matches2_neg));
>> for k=1:length(matches2_neg)
       idx = find(unigram_neg.vocab==matches2_neg(k));
       if isempty(idx), ranks2_pos(k) = mean(all_ranks_neg);
       else, ranks2_neg(k) = find(all_ranks_neg==idx); end
   end
```

As seen from the figure, a clear separation between the words that are relevant to the category under consideration (those with small ranking indexes appearing at the left-hand sides of the plots) and the rest of the words can be observed in both cases. According to the plots, the same threshold value of *10,000* used in (13.24) seems to be also appropriate here for filtering purposes.

```
>> % selects top-ranked positive candidate words                       (13.30a)
>> selected2_pos = unique(matches2_pos(ranks2_pos<10000));
>> whos matches2_pos selected2_pos
  Name                Size            Bytes  Class     Attributes
  matches2_pos        1x802          45772   string
  selected2_pos       1x102           6548   string
```

```
>> % selects top-ranked negative candidate words                       (13.30b)
>> selected2_neg = unique(matches2_neg(ranks2_neg<10000));
>> whos matches2_neg selected2_neg
  Name                Size            Bytes  Class     Attributes
  matches2_neg        1x703          40090   string
  selected2_neg       1x73            4726   string
```

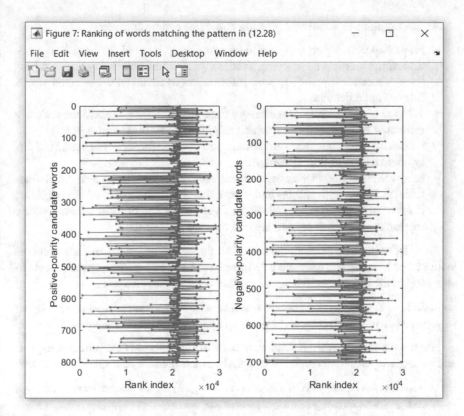

Fig. 13.7. Assigned ranks to positive and negative polarity candidate words that have been extracted with the regular expression defined in (13.28)

As seen from (13.30), from a total of *802* positive candidate words, *102* potential qualifiers were selected and, from a total of *703* negative candidate words, *73* potential qualifiers were selected. If we take a look at some of them:

```
>> disp(selected2_pos(1:10))                                    (13.31a)
   Columns 1 through 5
     "acclaimed"    "aided"    "always"    "amarcord"    "amazing"
   Columns 6 through 10
     "ambulance"    "amistad"    "balanced"    "beautiful"    "best"
```

```
>> disp(selected2_neg(1:10))                                    (13.31b)
   Columns 1 through 6
     "106"    "absurd"    "abysmal"    "annoying"    "asinine"    "atrocious"
   Columns 7 through 10
     "awful"    "bad"    "bland"    "blas"
```

we immediately discover that, regrettably, our proposed methodology seems to be still needing some further refinements. Indeed, more filtering is required!

At this point, three additional filtering strategies can be considered:

- Part-of-speech. Up to now, we have completely ignored the linguistic function of the words we are trying to identify. As qualifiers are typically adjectives, we can use part-of-speech information to filter out all selected words that are not adjectives.
- Word counts. The selection strategy used so far relies on relevance rankings estimated from probabilities, which might be noisy for the case of infrequent words. We can compute word counts and remove those words occurring only once in each of the selected sets.
- Stopwords. Although we might expect most function words to be evenly distributed across both subsets and, consequently, to get intermediate rankings; it is possible that some unbalances might push some of these words into the top ranks. Then, just in case, we can remove stopwords.

First, let us start by extracting a list of all the adjectives in the overall data collection. For this, we use `addPartOfSpeechDetails` to add part-of-speech labels to all words in the collection.[4] Then, we extract all words labeled as adjectives.

```
>> details = tokenDetails(addPartOfSpeechDetails([docs_pos,docs_neg]));  (13.32)
>> adjectives = unique(details.Token(details.PartOfSpeech=='adjective'));
```

Next, we compute word counts over each of the selected sets (`selected2_pos` and `selected2_neg`) and retain those words occurring more than once in each set.

```
>> [selected2_pos,~,idx_pos] = unique(matches2_pos(ranks2_pos<10000));  (13.33a)
>> counts_pos = hist(idx_pos,length(selected2_pos));
>> frequent_pos = selected2_pos(counts_pos>1);

>> [selected2_neg,~,idx_neg] = unique(matches2_neg(ranks2_neg<10000));  (13.33b)
>> counts_neg = hist(idx_neg,length(selected2_neg));
>> frequent_neg = selected2_neg(counts_neg>1);
```

Finally, we remove stopwords from the retained word sets and intersect them with the extracted adjectives:

```
>> frequent_pos = setdiff(frequent_pos,stopWords);                       (13.34a)
>> filtered2_pos = intersect(frequent_pos,adjectives)'
filtered2_pos =
  1×15 string array
  Columns 1 through 5
    "best"      "breathtaking"    "concerned"    "enjoyable"    "excellent"
  Columns 6 through 11
    "filmed"      "flawless"      "great"      "misleading"      "paced"      "perfect"
```

---

[4] Notice the collection we are using is provided in lowercased format, which might result in some POS labelling errors. In an ideal situation, `addPartOfSpeechDetails` should be used before lowercasing the data.

```
Columns 12 through 15
    "powerful"    "pure"    "quite"    "remarkable"
>> frequent_neg = setdiff(frequent_neg,stopWords);                    (13.34b)
>> filtered2_neg = intersect(frequent_neg,adjectives)'
filtered2_neg =
  1×12 string array
  Columns 1 through 5
    "annoying"    "awful"    "bad"    "boring"    "dull"
  Columns 6 through 9
    "imcomprehinsable"    "lacking"    "neither"    "pathetic"
  Columns 10 through 12
    "ridiculous"    "stupid"    "terrible"
```

As seen from (13.34a), after applying the proposed triple filtering strategy, *15* positive qualifiers have been extracted from the *102* that were originally selected in (13.30a). Similarly, as seen from (13.34b), *12* negative qualifiers have been extracted from the *73* that were originally selected in (13.30b). Notice that these extracted lists of positive and negative movie qualifiers, although still not completely perfect, are certainly much cleaner and more accurate.

Before moving into the problem of identifying aspects or features of movies that are commonly evaluated in reviews, let us take a quick look at Figure 13.7 again. An interesting observation, although not as evident as in the case of top ranked words, is that certain differentiation can be also appreciated for the case of words with large ranking indexes (the ones appearing at the right-hand side of the plots). Indeed, three different rank regions can be differentiated in the plots: the first and widest one, which is located towards the top ranks, corresponds to words that are more relevant to the category under consideration; the second one, located towards the central range of ranks, corresponds to words that are similarly relevant or irrelevant to both categories; and the third one, located towards the right, include words that are relevant to the opposite category.

This means that we can also extract negative qualifiers from texts belonging to the positive polarity category, as well as positive qualifiers from texts in the negative polarity category. This happens basically because of two reasons. First, movie reviews (as well as any other type of reviews, in general) are rarely either completely positive or completely negative. A positive review might include negative opinions about specific aspects of the movie under consideration, and vice-versa. Second, negative qualifiers can be used to express positive opinions by means of comparisons, negations, and other linguistic constructions. In a similar way, positive qualifiers can be validly used for expressing negative opinions.

Accordingly, if we look at the worst ranked words (the ones in the right-hand side of the plots in Figure 13.7), we should find words that are relevant to the opposite category. Consider, for instance the following sets of words:

```
>> % positive review words that are relevant to the negative category   (13.35a)
>> disp(unique(matches2_pos(ranks2_pos>29000)))
  Columns 1 through 6
    "bad"     "comprised"    "dumb"    "rubbish"    "stupid"    "supposed"
  Columns 7 through 8
    "unwatchable"    "watchable"
```

```
>> % negative review words that are relevant to the positive category   (13.35b)
>> disp(unique(matches2_neg(ranks2_neg>27000)))
  Columns 1 through 4
    "awesome"    "breathtaking"    "generating"    "littered"
  Columns 5 through 8
    "schizophrenic"    "spielbergs"    "tremendous"    "wellpaced"
```

After having devoted a good deal of attention to the problem of extracting some of the qualifiers most commonly used to describe movies, let us consider the problem of extracting the movie aspects or features that are commonly evaluated in the reviews. To do this, we leverage the qualifiers already identified in (13.34). Accordingly, we can construct the following pattern:

```
>> candidate_word = " (\w+) ";                                          (13.36)
>> qualifiers = unique([filtered2_pos,filtered2_neg]);
>> anyqualifier = " (?:"+join(qualifiers,'|')+") ";
>> pattern3 = "(?:this|that|the)"+candidate_word+"(?:is|was)"+anyqualifier;
```

This regular expression is composed of three parts:

- the candidate word to be extracted as movie aspect or feature,
- the matching pattern **anyqualifier** that will match any of the positive or negative qualifiers from (13.14),
- additional words completing the regular expression, which aims at matching patterns such as *this ___ is [qualifier]*, *that ___ was [qualifier]*, etc.

Next, we apply the constructed regular expression to the overall dataset (including both the positive and negative subsets):

```
>> collection_all = join([collection_pos,collection_neg]);              (13.37a)
>> matches3_asp = regexp(collection_all,pattern3,'tokens');
>> selected3_asp = unique(string(matches3_asp));
```

```
>> whos matches3_asp                                                    (13.37b)
  Name              Size            Bytes    Class      Attributes
  selected3_asp     1x63            3802     string
```

where we can see that a total of *63* candidate movie aspects have been selected. However, as you might be expecting the set of terms in **selected3_asp** is very noisy and still far from a desirable list of movie aspects. These can be easily confirmed by means of a quick inspection, which we are omitting here.

In order to clean up the results from (13.37), we apply the same triple filtering procedure previously used when extracting qualifiers. Different from what we did in (13.32), here we will be looking for nouns instead, as we expect most of the aspects and attributes of movies we want to identify belong to this part-of-speech category. Regarding word counts and stopwords, we follow the same approach used in (13.33) and (13.34), respectively.

```
>> nouns = unique(details.Token(details.PartOfSpeech=='noun'));        (13.38a)

>> [selected3_asp,~,idx_asp] = unique(string(matches3_asp));           (13.38b)
>> counts_asp = hist(idx_asp,length(selected3_asp));
>> frequent_asp = selected3_asp(counts_asp>1);

>> frequent_asp = setdiff(frequent_asp,stopWords);                     (13.38c)
>> filtered3_asp = intersect(frequent_asp,nouns)'
filtered3_asp =
  1×12 string array
  Columns 1 through 5
    "action"    "actors"    "animation"    "cast"    "cinematography"
  Columns 6 through 11
    "ending"    "film"    "movie"    "plot"    "script"    "show"
  Column 12
    "writing"
```

As seen from (13.38c), after applying the filters, *12* movie aspects have been extracted from the *63* candidates originally selected in (13.37b). Notice that the obtained set of movie aspects can be plugged into the regular expression defined in (13.28) with the purpose of harvesting more qualifiers, which can be subsequently plugged into the regular expression in (13.36) to collect more movie aspects (this idea is explored in more detail in exercise 13.6-8).

In the next section, we will continue exploring pattern matching techniques in the more generic context of entity and relation extraction.

13.4 Entity, Relation and Definition Extraction

Another important and interesting problem is the identification and extraction of entities, relations, and definitions. This includes a variety of problems such as, for instance, identifying categories and their corresponding instances. In this case, we would be interested in extracting evidence that, for example, the term *animal* constitutes a category, which instantiations include terms such as *dog, bird, fish*, etc. Similarly, we can also consider more generic entity relations of type *noun-verb-object*, such as *smoking-causes-cancer, birds-eat-seeds, words-contain-letters*, etc. Automatically deriving such types of knowledge from text is of vital importance for creating and populating valuable resources such as knowledge bases, semantic networks and ontologies.

Before starting, let us load the dataset to be used along this section. It consists of a Simple Wikipedia article[5] that have been already preprocessed according to the following steps (see exercise 13.6-9 for more details):

- keywords and key-phrases were extracted by leveraging *html* markups (specifically, text within the `span` and `anchor` markups),
- key-phrase words were connected with underscores (e.g. *amino_acids*),
- sequences of consecutive whitespace characters (tabs, newlines, blanks, etc.) were replaced by a single whitespace,
- sequences of words within parenthesis and brackets were removed,
- standard sentence segmentation and word tokenization was applied, and
- part-of-speech information was added to each token.

```
>> clear; load datasets_ch13c                                       (13.39a)
>> whos
   Name        Size            Bytes  Class                Attributes
   keywds      62x1             4228  string
   stpwds      230x1           12612  string
   tokdoc      46x1            63648  tokenizedDocument

>> keywds([5,25,35]) % some examples of the extracted keywords      (13.39b)
ans =
  3×1 string array
    "protein_structure"
    "amino_acids"
    "digestion"
```

As seen from (13.39a), the dataset includes three objects: `tokdoc`, a tokenized document array that contains the preprocessed article text; `keywds`, a string array with the extracted keywords and key-phrases; and `stpwds`, a string array with an extended list of stopwords. Some examples of the extracted keywords and key-phrases are displayed in (13.39b).

Next, let us extract the token details from the tokenized document. This information will be used later to extract specific subsets of words, such as all common and proper nouns occurring in the article:

```
>> details = tokenDetails(tokdoc);                                  (13.40a)
>> size(details)
ans =
    653       7
```

---

[5] https://simple.wikipedia.org/wiki/Protein Accessed 6 July 2021.

```
>> string(details.Properties.VariableNames)                          (13.40b)
ans =
  1×7 string array
  Columns 1 through 5
    "Token"      "DocumentNumber"    "SentenceNumber"    "LineNumber"    "Type"
  Columns 6 through 7
    "Language"     "PartOfSpeech"
```

As seen from (13.40a) the document contains a total of *653* tokens for each of which six different annotations are available, in addition to the token itself. The available annotations are listed in (13.40b). Notice that part-of-speech information has been included in the document representation.

Now, we are ready to start with our first entity extraction experiment. In this first experiment, we consider the problem of identifying itemized list patterns with the purpose of extracting sets of similar or related items. More specifically, we are interested in patterns of the form *"a, b, c, and d"*, *"x, y or z"*, etc. To this end, we can define the following regular expression:

```
>> pattern1 = "(\w+ , )+(\w+ )(?:and |, and |or |, or )(\w+ )";        (13.41)
```

which will capture one or more words followed by commas `"(\w+ , )+"`, followed by two additional words `"(\w+ )"` separated by either `"and"`, `", and"`, `"or"` or `", or"`.

Next, we attempt to match the regular expression defined in (13.41) against all sentences in the document:

```
>> counter = 0;                                                       (13.42a)
>> for k = 1:length(tokdoc)
      sentence = lower(joinWords(tokdoc(k)));
      matches = regexp(sentence,pattern1,'tokens');
      if isempty(matches), continue; end
      items = split(join(string(matches{1})),[" ",",","]);6
      items = items(strlength(items)>0);
      items = setdiff(items,stpwds);
      counter = counter+1;
      itemlists(counter).sentn = k;
      itemlists(counter).items = items';
   end

>> struct2table(itemlists)                                            (13.42b)
```

---

[6] Notice we are assuming here that, if the regular expression is matched, only one match per sentence occurs: `matches{1}`. If the regular expression is matched twice or more times within the same sentence, the implementation in (13.42a) ignores any additional match after the first one. A single match per sentence is a reasonable assumption in this case, but it might not be valid if larger units of analysis such as paragraphs or full documents are used.

```
ans =

  5×2 table

    sentn        items
    ─────      ──────────
     10        {1×4 string}
     16        {1×3 string}
     21        {1×4 string}
     25        {1×7 string}
     41        {1×9 string}
```

The matching procedure implemented in (13.42a) iterates over the sentences in the document and, for each sentence, it attempts to match the regular expression from (13.41). The procedure can be described as follows. First, the corresponding sentence is recovered from the tokenized document representation `tokdoc` by joining and lowercasing the words in the sentence. Next, the regular expression match is attempted and matched tokens, if any, are captured. If there are no matches, the script continues with the next sentence. On the other hand, if there are matches, the first match is processed (notice that multiple matches are ignored). The match is processed in three steps: first, the captured tokens are converted to strings, concatenated, and split again at any comma or whitespace location; second, spurious strings of zero length are discarded; and third, stopwords are removed. Finally, the resulting list of strings `items` is saved into the structure array `itemlists`, along with the sentence number in which the match occurred.

Notice from (13.42b) that a total of five itemized lists were extracted from the document under consideration, with numbers of items ranging from three to nine. Let us print the contents of the extracted lists:

```
>> for k=1:length(itemlists)                                     (13.43)
       sentn = itemlists(k).sentn;
       items = itemlists(k).items;
       fprintf('* In sentence #%d: %s\n',sentn,join(items,', '));
   end
* In sentence #10: activity, folding, function, stability
* In sentence #16: cell_division, cell_signalling, immune_responses
* In sentence #21: animals, bacteria, fungi, plants
* In sentence #25: beans, eggs, fish, meat, milk, nuts, spinach
* In sentence #41: histidine, isoleucine, leucine, lysine, methionine,
  phenylalanine, threonine, tryptophan, valine
```

As seen from (13.43), each of the extracted lists is composed of a semantically coherent set of entities, which happen to be specific instances of more general categories such as protein properties, cellular processes, living things, aliments, and amino acids. Moving forward in our entity and relation extraction exercise, let us try to automatically find some of those instance-category pairs.

The main idea for this new task is to match patterns of the form *"x such as y"*, *"x like y"*, etc., where x is typically a category and y a specific instance of it. This task is, however, much more challenging than the previous one, as these simple types of patterns are typically matched by many uninteresting word pairs. To overcome this problem, we need to limit the kind of terms x and y that should be matched in the patterns. Ideally, we should be aiming at matching nouns as well as the keywords and key-phrases already identified in the original *html* document.

```
>> % collects common and proper nouns in the document          (13.44a)
>> mask = (details.PartOfSpeech=='noun') |...
          (details.PartOfSpeech=='proper-noun');
>> nouns = unique(details.Token(mask));
>> nouns = nouns(strlength(nouns)>2); % keeps strings of length > 2
>> nouns = setdiff(lower(nouns),stpwds); % removes stop words

>> % creates a list of valid entities by combining nouns and keywds      (13.44b)
>> entity = union(nouns,keywds);
```

By using the list of entities from (13.44b), we can restrict the possible matches for the expected categories x and specific instances y to a subset of words of potential interest. The regular expression for extracting instance-category pairs can be constructed as follows:

```
>> entwrd = "("+join(entity,'|')+")";          (13.45)
>> relwrd = "(?:like|are|is|such as|including)";
>> pattern2 = entwrd+"\W+?"+relwrd+"\W+?"+entwrd;
```

The expression in (13.45) will match patterns of the form *entity₁ relation entity₂*, where the expected relation is of the type *is_an_instance_of* and going from the second entity to the first entity: *entity₂ → is_an_instance_of → entity₁*.

The instance-category pairs can be extracted by following a similar procedure to the one implemented in (13.42a).

```
>> counter = 0;          (13.46a)
>> for k = 1:length(tokdoc)
       sentence = lower(joinWords(tokdoc(k)));
       matches = regexp(sentence,pattern2,'tokens');
       if isempty(matches), continue; end
       counter = counter+1;
       catinspairs(counter).sentn = k;
       catinspairs(counter).category = matches{1}(1);
       catinspairs(counter).instance = matches{1}(2);
   end

>> struct2table(catinspairs)          (13.46b)
ans =

  5×3 table
```

sentn	category	instance
13	"proteins"	"enzymes"
25	"foods"	"milk"
33	"proteins"	"enzymes"
41	"essential_amino_acids"	"histidine"
44	"products"	"tofu"

Notice from (13.46b) that a total of four different instance-category pairs have been extracted from the document under consideration. More interestingly, there are two specific instance entities (*milk* and *histidine*) that also appear in two of the itemized lists from (13.43). Then, by combining results from (13.46b) with results from (13.43), more instances for the corresponding categories (*foods* and *essential amino acids*) can be identified.

```
>> for k=1:length(catinspairs)                                        (13.47)
       cat = catinspairs(k).category;
       ins = catinspairs(k).instance;
       for n=1:length(itemlists)
           itm = itemlists(n).items;
           if sum(itm==ins)>0
               fprintf('* %s -> %s\n',cat,join(itm,', '));
           end
       end
   end
* foods -> beans, eggs, fish, meat, milk, nuts, spinach
* essential_amino_acids -> histidine, isoleucine, leucine, lysine, methionine,
  phenylalanine, threonine, tryptophan, valine
```

A seen from (13.47), by combining the itemized list extractions from (13.42) with the instance-category pair extractions from (13.46), we have been able to start connecting the first nodes and edges of a knowledge graph. (For a much larger scale example of this procedure, see short project 13.7-2.)

Next, let us focus on a more general class of *entity₁ relation entity₂* extraction. More specifically, let us consider patterns of the form *subject-verb-object*, which are typically referred to as triples. Notice that the case we considered in (13.45) is indeed a specific case of *subject-verb-object* pattern. The regular expression for extracting triples can be constructed as follows:

```
>> relations = ["caused by","caused","causes","cause","contain",...   (13.48)
                "contains","contained in","contained","provide",...
                "provides","provided by","provided","consists of",...
                "consists","involves","involve"];
>> relwrd = "("+join(relations,'|')+")";
>> pattern3 = entwrd+"\W*?(?:\w+ ){0,1}?"+relwrd+".*?"+entwrd;
```

Notice from (13.48) that, in this example, we already start with a predefined list of probe verbs. So, we are aiming at extracting the entities (subjects and objects) for triples containing those verbs (for a more general extraction procedure involving the extractions of verbs too, see exercise 13.6-11).

We can extract triples for the probe verbs listed in (13.48) by following a similar procedure to the ones implemented in (13.42a) and (13.46a).

```
>> counter = 0;                                                    (13.49a)
>> for k = 1:length(tokdoc)
        sentence = lower(joinWords(tokdoc(k)));
        matches = regexp(sentence,pattern3,'tokens');
        if isempty(matches), continue, end
        counter = counter+1;
        entreltriples(counter).sentn = k;
        entreltriples(counter).entity1 = matches{1}(1);
        entreltriples(counter).relation = matches{1}(2);
        entreltriples(counter).entity2 = matches{1}(3);
    end

>> struct2table(entreltriples)                                     (13.49b)
ans =

  6×4 table
```

sentn	entity1	relation	entity2
17	"egg"	"contain"	"protein"
22	"muscle"	"contain"	"protein"
34	"cancers"	"caused by"	"mutations"
42	"meat"	"contain"	"essential_amino_acids"
43	"beans"	"provide"	"essential_amino_acids"
44	"tofu"	"provide"	"essential_amino_acids"

As seen from (13.49b), a total of six different *subject-verb-object* patterns were extracted from the document under consideration. Notice, however, the extraction procedure is prone to errors as some invalid or incomplete knowledge can be eventually extracted. Consider for instance sentence *43*, from which the triple *beans → provide → essential amino acids* was extracted.

```
>> disp(joinWords(tokdoc(43)))                                     (13.50)
However , eating a mixture of plants , such as both wheat and peanut_butter ,
or rice and beans , provides all the essential_amino_acids needed .
```

Notice from (13.50) that the actual statement in sentence *43* is that the combination of both, rice and beans, is what provides all the essential amino acids, rather than beans alone. In general, you might expect a significant amount of noise arising from pattern-matching-based extraction procedures like the ones described here, which typically require the implementation of filtering mechanisms similar to the ones used in the previous section. In most of the cases, frequency-based filtering methods significantly alleviate the problem, as combining evidence from multiple matches and sources improves the precision of the extraction procedures. However, as you might expect, such filtering procedures compromise recall severely, significantly reducing the number of extracted entities and relations.

More recent approaches to entity and relation extraction include the use of deep neural network architectures, which provide a more convenient way for increasing precision while keeping the impact on recall within reasonable limits. However, this advantage comes with the price of curated training sets, which are typically expensive to produce. (For a more detailed discussion about these methods, refer to the corresponding references provided on the *Further Reading* section.)

Finally, let us briefly explore the problem of definition extraction. Before proposing a regular expression to extract definitions, let us look at some sentences in the document to infer some of the properties of definitions. Let us consider for instance sentences *12* and *14*.

```
>> disp(joinWords(tokdoc(12)))                                              (13.51a)
Proteins are essential to all cells .
```

```
>> disp(joinWords(tokdoc(13)))                                              (13.51b)
Like other biological macromolecules proteins take part in virtually every
process in cells : Many proteins are enzymes that catalyze biochemical reac-
tions and are vital to metabolism .
```

As seen from (13.51), although sentence *12* starts with *"proteins are"*, it is not providing a definition. On the other hand, sentence *13* contains a segment starting with *"proteins are"*, which is providing a definition for a certain class or group of proteins. If we study in detail the differences between both candidate definitions, we will see that, in the first case, an adjective (*essential*) is used to describe proteins while, in the second case, a noun (*enzymes*) is used.

As we might expect definitions to contain adjectives too, just looking at the part-of-speech of the word following the verb does not seem to be enough. Indeed, a much more elaborated strategy is needed to extract definitions. Then, our proposed matching strategy goes as follows: we will catch all tokens between a definition hint (such as *"proteins are"*) and the next occurring stop word. If any of such tokens happens to be a noun, we assume the segment under exploration is a definition. Accordingly, the regular expression for identifying candidate definitions can be defined as follows:

```
>> stpwrd = "(?:"+join(stpwds,'|')+")";                                     (13.52)
>> relwrd = "(?:are the|are|is the|is a|is|consists of)";
>> tokwrds = "([\w-]+ )*?";
>> pattern4 = entwrd+"\W+?"+relwrd+"\W+?"+tokwrds+stpwrd+"\W+?";
```

At this point, we are ready to extract definitions from the document under consideration. We follow a similar implementation to the ones in (13.42a), (13.46a) and (13.49a). However, different from those previous cases, we need an additional step here: once a match for a candidate definition is found, we need to look at all tokens captured by `tokwrds` to decide whether the candidate is valid definition or not. Only candidate definitions, for which at least one of the captured tokens is a noun, are considered valid and finally extracted.

```
>> counter = 0;                                                              (13.53a)
>> for k = 1:length(tokdoc)
      sentence = lower(joinWords(tokdoc(k)));
      [init, matches] = regexp(sentence,pattern4,'start','tokens');
      if isempty(matches), continue; end
      entity = matches{1}(1);
      tokens = split(strip(matches{1}(2)));
      for m = 1:length(tokens)
      % extracts the definition only if at least one of the tokens is a noun
            if sum(tokens(m)==lower(nouns))>0
                counter = counter+1;
                definitions(counter).sentn = k;
                definitions(counter).entity = entity;
                definition = string(sentence{1}(init(1):end));
                definitions(counter).definition = definition;
                break
            end
      end
  end
```

```
>> whos definitions                                                          (13.53b)
    Name              Size          Bytes   Class     Attributes
    definitions       1x5           3980    struct
```

As seen from (13.53b), five definitions were extracted. The definitions are presented in Table 13.3 along with their corresponding sentence numbers and entities.

Table 13.3. Definitions extracted from the Simple Wikipedia article under consideration

Sentence	Entity	Definition
3	proteins	proteins are long-chain molecules built from small units known as amino_acids .
6	polypeptide	polypeptide is a single linear polymer chain of amino_acids .
13	proteins	proteins are enzymes that catalyze biochemical reactions and are vital to metabolism .
15	cytoskeleton	cytoskeleton is a system of scaffolding that keeps cell shape .
35	p53	p53 is a protein which regulates cell_division .

13.5 Further Reading

In this chapter, we have devoted part of our attention to the problem of analyzing opinionated content, which exhibits the properties of subjective language (Wiebe *et al.* 2004). The use of text mining and natural language processing techniques to analyze subjective contents is generally referred to as Opinion Mining and Sentiment Analysis (Pang and Lee 2008).

Different from conventional text analysis, which focus on the factual aspects of what is said, opinion mining and sentiment analysis focus on the subjective nature of what is said. In this case, the main objective of the analysis is to determine the attitudes and emotions of the person that writes. Traditional methods for identifying the emotional orientation of a given segment of text are based on the use of lexicons, i.e. a large list of terms with a predefined emotional load. Examples of such lexicons include the General Inquirer (Stone *et al.* 1996) and the Affective Norms for English Words (Bradley and Lang 1999). More recent approaches that incorporate statistical methods (Turney 2002) and machine learning techniques (Pang and Lee 2004)[7] have proven to be very effective for addressing the problem of polarity detection.

The problems of qualifier and aspect extraction (Zhang and Liu 2014) as well as the one of entity and relation extraction (Wong *et al.* 2009), introduced in sections 13.3 and 13.4, belong to a more general area of research know as information extraction (Cowie and Wilks 1996) and (Moens 2006). The main objective of information extraction, commonly referred to as IE, is to extract structured information from unstructured text data. Some important tasks within the scope of information extraction include: coreference resolution (Rahman and Ng 2009), entity and relation extraction (Song *at al.* 2015), and event detection (Sayyadi *et al.* 2009), among others.

In addition to the text-pattern matching techniques described in this chapter, other approaches to information extraction use machine learning (Freitag 2000) and statistical models, such as conditional random fields (Peng and McCallum 2006), to implement specific methods and applications. More recent approaches use deep neural network architectures to perform information extraction tasks. For more information with regards to the use of deep learning in information extraction see (Poria *et al.* 2016), (Nguyen and Verspoor 2019), and (Nguyen 2018).

The application of information extraction technologies for mining information from the web has been gaining a lot of interest in recent years (Kowalkiewicz *et al.* 2012). Other different dimensions of content analysis include problems such as genre identification (Santini 2004), authorship attribution (Juola 2006) and discourse analysis (Gee 2005).

---

[7] The *polarity dataset v2.0* we have used in sections 12.2 and 12.3 was introduced in this work.

13.6 Proposed Exercises

1. Consider the experimental work in section 13.1, in which positive and negative reviews from two different domains were projected into two dimensions.

   - Collect new samples of positive and negative reviews from different domains. You might consider some of the following websites for collecting the data (accessed 6 July 2021):
     - https://www.tripadvisor.com/
     - https://www.yelp.com/
     - https://www.wayfair.co.uk/
     - https://www.amazon.com/
     - https://www.consumerreports.org/cro/index.htm
   - Repeat the same procedure from section 13.1 for different pairs of domains and generate two-dimensional maps for each case.
   - Compare your results with those reported in section 13.1. What similarities do you observe? What differences do you observe?
   - According to the differences and similarities observed among your results, which categories are more distinguishable from each other? Which are less distinguishable?

2. Consider the preprocessing and preparation of *1,000* positive and *1,000* negative movie reviews conducted in (13.7)

   - Reproduce the procedure and, in addition to extracting the tokens, compute both vocabulary and frequency counts for each document.
   - Compute total running words and total vocabularies for each of the two categories as well as the whole collection.
   - Compute minimum, maximum and average document lengths for each of the two categories, as well as for the whole collection.
   - Reproduce Table 13.1 from the computed parameters.

3. Consider the following modified version of the polarity intensity score defined in (13.15), which incorporates the two new parameters *offset* and *factor*:

   - *intensity_score* $= 2/(1+e^{-(loglirat-offset)/factor}) - 1$
   - Use the second *100* random indexes from (13.9) to extract the development set for each of the two categories in the collection.
   - Adjust the value of *offset* such that the classification accuracy over the development set is maximized. Compute the accuracy over the test set.
   - Adjust the value of *factor* such that the polarity intensity distribution is made as much uniform as possible (consider discretizing intensity into five intervals from *-1* to *1*). Compute the accuracy over the test set.

– What differences do you observe between the computed accuracies? How do these results compare with the ones presented in (13.13)?

4. Create a vector model for the dataset used in section 13.2 by applying the standard TF-IDF weighting with *L2*-norm normalization.

– Implement a *k*-nearest neighbors classifier for polarity detection purposes by using the procedure described in section 11.3. Use the same train-test-development partition according to the indexes in (13.9).
– Search for an optimum value of *k* by maximizing the classification accuracy over the development set.
– Compute the resulting classification accuracy of polarity detection over the test set. How does this result compare to the one presented in (13.13)?
– Propose a method for using the already trained *k*-nearest neighbors classifier to estimate polarity intensity.
– Use your proposed method to estimate polarity intensities for the examples in Table 13.2. How do your results compare to those in the table?
– Study and use the MATLAB® functions `vaderSentimentScores` and `ratioSentimentScores` to compute scores for the examples in Table 13.2.
– How do these new scores compare to those computed with your *k*-nearest neighbors classifier implementation and the ones from Table 13.2?

5. Preceding the task of polarity detection described in section 13.2, there is the need for discriminating factual text from subjective text. This problem is generally referred to as subjectivity detection.

– Download file *subjectivity dataset v1.0* (Pang and Lee 2004) from the *Movie Review Data* website at Cornell University http://www.cs.cornell.edu/people/pabo/movie-review-data/ Accessed 6 July 2021.
– Prepare the data and randomly split it into train, test, and development sets, considering an *8-1-1* proportion.
– Compute unigram models (over the train set) for each of the two categories in the collection: subjective and objective. Follow the same format used in section 13.2 for the cases of `unigram_pos` and `unigram_neg`.
– Use the function `compute_loglirat` and the procedure in (13.13) to estimate subjectivity over the test set. What is the accuracy for this task? What are your main conclusions?

6. Another interesting problem related to the analysis of subjective contents is the assessment of emotions.

– Obtain a lexicon of emotional content. You might consider one of the following (accessed 6 July 2021):
- The General Inquirer: http://www.wjh.harvard.edu/~inquirer/
- ANEW: http://csea.phhp.ufl.edu/media/anewmessage.html

- Consider either the datasets you collected in exercise 13.6-1 or the dataset you used in exercise 13.6-5.
- Count the number of words from each emotional category in the lexicon occurring at each document in the collection.
- Compute the emotional load of each document in the collection by aggregating the emotional categories occurring in it.
- Generate a graph summarizing the distribution of the emotional content in the collection. What are your main conclusions?

7. Consider the following regular expression, which aims at extracting the same kind of information as the one defined in (13.28):

   - `(?:this|that|it) (?:is|was) (?:a|an)(?: \w*?y\>)? (\w+) (?:movie|film)`
   - Repeat the extraction procedures conducted in (13.29) by using the newly proposed regular expression.
   - Apply the selection procedure used in (13.30), as well as the filtering steps from (13.32), (13.33) and (13.34).
   - Compare the new sets of extracted terms with those from (13.34). What similarities and differences do you observe?
   - Think about other regular expressions that can be used to extract the same class of information (positive and negative qualifiers of movies) and use them to extract more candidate words.

8. Consider the filtered lists of positive and negative qualifiers obtained in (13.34) and the filtered list of aspects obtained in (13.38).

   - Use the lists of aspects from (13.38) to replace the `(?:movie|film)` part of the regular expression in (13.28).
   - Use this new regular expression to extract qualifiers by following the steps implemented in (13.29), (13.30), (13.32), (13.33) and (13.34).
   - How do the new lists of qualifiers compare to the lists from (13.34)? How many new positive and negative qualifiers were obtained?
   - Use the new lists of extracted qualifiers to update the `qualifiers` component of the regular expression defined in (13.36).
   - Repeat the steps in (13.37) and (13.38) to extract a new list of aspects.
   - How does the new list of aspects compare to the list from (13.38)? How many new aspects were obtained?

9. Create a processing function that receives as input the *.html* file of a Wikipedia article, and produces as output the representation in (13.39), i.e. a tokenized document containing the preprocessed article text and a string array with the extracted keywords and key-phrases. The preprocessing should include:

   - Article segmentation. Get the section of interest, which contains the actual article, from the *html* file. You might consider using specific makers such

as `id="mw-content-text"`, `id="References"`, `id="Related_pages"`, etc.
to identify the boundaries of the article contents.

- Keyword and key-phrase extraction. Use the `span` and `anchor` markups in
  the *html* codes to extract keywords and key-phrases. For the latter, replace
  whitespaces with underscores to concatenate the different words.
- Text preprocessing. Use `extractHTMLText` to extract the article's text from
  the *html* file. Replace all sequences of consecutive whitespace characters
  (tabs, newlines, blanks, etc.) with a single whitespace. Replace whitespaces with underscores for all key-phrases in the text. Remove all sequences
  of words occurring within parenthesis and brackets.
- Segmentation and tokenization. Use `splitSentences` to segment the article into sentences and `tokenizedDocument` to tokenize the sentences.
- Annotation. Use `addPartOfSpeechDetails` to add part-of-speech labels to
  each token in the tokenized document representation.
- Test your implemented function with the Simple Wikipedia article used in
  section 13.4. Compare your obtained representation with the one provided
  in `datasets_ch13c.mat`.

10. The regular expression in (13.41) matches patters of the form *"a, b, c, and d"*,
    "x, y or z", etc. It aims at collecting lists of instances that belong to a common
    category, like the ones in (13.43). However, there are other text patterns that al-
    so provide opportunities to extract the same type of information.

- Consider the sentence *"Like other biological macromolecules (polysac-
  charides and nucleic acids), proteins take part in virtually every process in
  cells"*, which has been extracted from the same article used in section 13.4.
- In the example above, an instance-category pattern of the form *"category
  ($instance_1$ and $instance_2$), $instance_3$"* is observed.
- Consider other common variations of instance-category pairs that involve
  the use of parenthesis, such as *"category (e.g. instance)"* and so on.
- Design a few regular expressions and extraction procedures that leverage
  instance-category patterns involving parenthesis.
- Test your designed solutions by extracting categories and instances from a
  collection of Wikipedia articles. Evaluate its performance.

11. Let us consider a more general triple extraction approach than the one imple-
    mented in (13.48), where we used a list of probe verbs to extract *subject-verb-
    object* patterns. In this exercise, we attempt to extract the verb too.

- Consider the pattern *$entity_1$ [other] verb [other] $entity_2$*, where each *[other]*
  block represents a sequence of zero or more words (up to $k \approx 3$) that are
  not either valid entities or verbs.
- Design a regular expression to match the pattern above and use it for extracting triples of the form *($entity_1$, verb, $entity_2$)*.

- Test your regular expression over a set of Wikipedia pages. How many of the resulting triples are valid *subject-verb-object* patterns?
- Try different values of k and evaluate the quality of the different results.
- Design and test filtering strategies to increase the quality of the results.

13.7 Short Projects

1. Consider some of the websites you already used in exercise 13.6-1.

- Select five or six different categories of services or domains and collect positive and negative reviews for each category. Attempt to collect at least *1,000* samples for each polarity-domain subset.
- If you are not able to gather *1,000* negative comments for a given domain, you can use comments with mid to low ratings for the negative category.
- Preprocess the data to create an annotated dataset with domain labels and polarity labels.
- Split the dataset into train, test, and development subsets using the proportion *8-1-1*. Make sure all categories are well represented in all subsets.
- Implement a domain classifier (select the supervised classification method of your preference). Optimize the classifier with the development data and evaluate its performance over the test data.
- Implement a polarity classifier for each available domain. Optimize and evaluate each system with its corresponding development and test data.
- Evaluate the performance of each polarity classifier over all other domain test sets. Do you observe any difference in performance? How can you explain the observed differences?
- Create a script for cascading the domain classifier and the polarity ones.
- The script should allow for inputting a new text either from a plain text file or from a dialog window.
- As output, the script should provide the most probable domain and the corresponding polarity for the input text.
- Collect new data from the same websites used to collect the original dataset and use your script to classify domains and polarities.
- Evaluate the performance of the overall system.
- Collect new data from websites different from the ones used to collect the original dataset and use your script to classify domains and polarities.
- Evaluate again the performance of the overall system.
- What are your main conclusions?

2. Read the Wikipedia Database Download page (accessed 6 July 2021):

- http://en.wikipedia.org/wiki/Wikipedia:Database_download

- Select a topic of your interest and download the subset of Wikipedia pages related to that topic.
- Define a strategy for preprocessing the whole downloaded subset or, alternatively, a part of it.
- Create some regular expressions for extracting categories and sample instances of such categories, similar to the one in (13.45).
- Implement some filtering strategies for removing undesired results from the long list of extracted matches.
- Define regular expressions for collecting additional instances for a category, given some instance examples, like the list extraction one in (13.41).
- Store all collected information in a structure array, with each element containing two fields: *category* (a string scalar) and *instances* (a string array).
- Create a script for reducing the structure array to its minimum length by combining repeated categories.
- Implement a procedure for exploiting the information within the *category-instance* structure to find possible synonyms for the different categories. You can assume, for instance, that two different categories that have a high percentage of common instances might be synonyms.
- Similarly, implement a procedure to find potential sub-instances and meta-categories for the different instances and categories, respectively. In this case, if a term x appears as category c and, also, as an instance y. The listed category for y can be considered a meta-category of c, and the listed instances of c can be considered sub-instances of y.
- Attempt to derive a taxonomy by concatenating meta-categories and sub-instances with their corresponding categories and instances.
- Conduct a subjective evaluation of your procedures and results. What are the main limitations of your methodology?
- What kind of information, resources or tools do you think would help to improve the results of this short project?

13.8 References

Bradley MM, Lang PJ (1999) Affective norms for English words (ANEW): Stimuli, instruction manual, and affective ratings. Technical Report C-1, Center for Research in Psychophysiology, University of Florida, Gainesville, Florida

Cowie J, Wilks Y (1996) Information Extraction, https://citeseerx.ist.psu.edu/viewdoc/download;jsessionid=0C54A0B37087FD3D9381BA405F22BFA7?doi=10.1.1.61.6480&rep=rep1&type=pdf Accessed 6 July 2021

Freitag D (2000) Machine Learning for Information Extraction in Informal Domains. In Cohen W (eds) Machine Learning 39:169-202. Kluwer Academic Publishers, The Netherlands.

Gee JP (2005) An Introduction to Discourse Analysis: Theory and Method. London: Routledge

Juola P (2006) Autorship attribution. Foundations and Trends in Information Retrieval 1(3): 233–234

Kowalkiewicz M, Orlowska ME, Kaczmarek T, Abramowicz W (2012) Web Information Extraction and Integration, Web Information Systems Engineering and Internet Technologies Book Series, Vol 5, Springer

Moens MF (2006) Information Extraction: Algorithms and Prospects in a Retrieval Context, The Information Retrieval Series. Springer

Nguyen DQ, and Verspoor K (2019) End-to-end neural relation extraction using deep biaffine attention. In proceedings of the 41st European Conference on Information Retrieval (ECIR)

Nguyen TH (2018) Deep Learning for Information Extraction (dissertation), New York University, ProQuest Dissertations Publishing

Pang B, Lee L, Vaithyanathan S (2002) Thumbs up? Sentiment classification using machine learning techniques. In proceedings of the Conference on Empirical Methods in Natural Language Processing, pp. 79–86

Pang B, Lee L (2008) Opinion Mining and Sentiment Analysis, Foundations and Trends in Information Retrieval. Now Publishers Inc

Peng F, McCallum A (2006) Information extraction from research papers using conditional random fields. Information Processing and Management 42(4): 963–979

Poria S, Cambria E, Gelbukh A (2016) Aspect extraction for opinion mining with a deep convolutional neural network, Knowledge-Based Systems 108: 42–49

Rahman A, Ng v (2009) Supervised Models for Coreference Resolution. In Proceedings of the Conference on Empirical Methods in Natural Language Processing, pp. 968–977

Santini M (2004) State-of-the-art on automatic genre identification. Technical Report ITRI-04-03, ITRI, University of Brighton

Sayyadi H, Hurst M, Maykov A (2009) Event Detection and Tracking in Social Streams. In proceedings of the Third International AAAI Conference on Weblogs and Social Media, 3(1)

Song M, Kim WC, Lee D, Heo GE, Kang KY (2015) PKDE4J: entity and relation extraction for public knowledge discovery. Journal of Biomedical Informatics 57: 320–332

Stone PJ, Dunphy DC, Smith MS, Ogilvie DM (1966) The General Inquirer: A Computer Approach to Content Analysis. Cambridge: The MIT Press

Turney P (2002) Thumbs up or thumbs down? Semantic orientation applied to unsupervised classification of reviews. In Proceedings of the Annual Meeting of the Association for Computational Linguistics, pp. 417–424

Wiebe J, Wilson T, Bruce R, Bell M, Martin M (2004) Learning subjective language. Computational Linguistics 30(3)

Wong W, Liu W, Bennamoun M (2009) Acquiring Semantic Relations using the Web for Constructing Lightweight Ontologies. In Proceedings of the 13th Pacific-Asia Conference on Knowledge Discovery and Data Mining (PAKDD).

Zhang L, Liu B (2014) Aspect and Entity Extraction for Opinion Mining. In: Chu W (eds) Data Mining and Knowledge Discovery for Big Data. Studies in Big Data, vol 1. pp.1–40. Springer, Berlin, Heidelberg

14 Keyword Extraction and Summarization

In the previous chapters we focused on fundamental text mining problems related to text categorization, search, and content analysis. In this chapter, as well as in the next one, we pay attention to more complex problems, whose solutions typically require the use of higher-level cognitive processes. Although these problems are well beyond the scope of text mining in the traditional sense, we have included them here because of two main reasons. First, they are related to important and practical applications that have been gaining popularity in the recent years. Second, we can leverage solutions from some of the simpler problems studied in previous chapters to partially approach some of these more complex problems. In this chapter, we focus on the problems of keyword extraction and text summarization.

This chapter is organized as follows. First, in section 14.1, we revisit the problem of keyword extraction, which was briefly introduced in chapter 11. We introduce the concept of word centrality and describe in detail how it can be used for keyword extraction purposes. Then, in section 14.2, we introduce the problem of text summarization. More specifically, we focus our attention on the extractive summarization approach, in which the idea of centrality is extended to the sentence level and used to retrieve the most relevant set of sentences in a given document or text passage.

Like in some previous chapters, in this chapter we will be working with contents that have been collected from the web. Although most of these contents are available under Creative Commons licenses, some of them are subject to copyright and will not be provided. In such cases, you will be given the mathematical models only. All models and contents made available through the companion website of the book are intended to be used for educational and research purposes only.

14.1 Keywords and Word Clouds

In this section, we start by revisiting the notions of keywords and word clouds, before introducing the concept of centrality. Previously, in chapter 11 (at the end of section 11.3), we introduced a simple procedure to identify the sets of most representative words with respect to some predefined document categories based on the TF-IDF representation of a document collection. There, we also illustrated how such sets of words can be graphically displayed by means of either a dendrogram or a word cloud, and how such graphical representations provide useful insights about the relative importance of the words within each document category.

© The Author(s), under exclusive license to Springer Nature Switzerland AG 2021
R. E. Banchs, *Text Mining with MATLAB®*, https://doi.org/10.1007/978-3-030-87695-1_14

The procedure illustrated in chapter 11 used the TF-IDF vector representations and the knowledge about the different categories present in the document collection to construct an "average document" vector for each document category. The most relevant set of words for each category was obtained by subtracting the average vectors of the other categories from the one of the category under consideration. This differential way of extracting relevant words proved to be effective in the case in which different document categories exist in the collection.

However, in a more general scenario, we might be interested in finding the most relevant or important words within a document collection in which different document categories are not necessarily present. In this sense, we pay attention here to more general approaches to keyword extraction.

Before starting with our first experiment, let us load and describe the first dataset to be used in this chapter, which is composed of all sentences from the first volume of Don Quixote.[1] The dataset is available in the data file `datasets_ch14a`.

```
>> clear                                                         (14.1a)
>> load datasets_ch14a

>> whos                                                          (14.1b)
  Name          Size              Bytes  Class      Attributes
  data        3785x2            2253966  table

>> disp(data.Properties.VariableNames)                          (14.1c)
    {'chap'}      {'text'}
```

As seen from (14.1b) and (14.1c), the dataset is provided in a table with *3,785* entries and two variables: `chap` and `text`. The table contains all sentences from the *52* chapters of the first volume of Don Quixote, one sentence per row. The `chap` variable indicates the chapter number each sentence belongs to, and the `text` variable contains the sentences themselves in plain text string format.

The simplest approach to derive important words from a document collection is by looking at the word frequencies. The function `worldcloud` uses word frequency counts to create a word cloud from a collection of strings. However, to avoid short words and function words from appearing in the word cloud, we apply some basic preprocessing to the text first.

```
>> % creates an array of tokenized documents                    (14.2)
>> tokdocs = tokenizedDocument(data.text);
>> vocab = tokdocs.Vocabulary; % extracts the vocabulary
>> % identifies short words (with less than 5 characters)
>> shortwords = vocab(strlength(vocab)<5);
>> % removes short words and stop words
>> prepdocs = removeWords(tokdocs,union(shortwords,stopWords));
```

---

[1] Available from https://www.gutenberg.org/files/996/996-0.txt Accessed 6 July 2021.

Now, we are ready to create a word cloud with the words identified by means of the remarkably simple keyword extraction mechanism illustrated in (14.2), i.e. the list of most frequent words excluding short words and stop words. The resulting word cloud is displayed in Figure 14.1.

```
>> hf = figure(1);                                              (14.3)
>> set(hf,'Color',[1 1 1],'Name','Word Cloud of frequent words');
>> wordcloud(prepdocs,'Shape','oval','HighlightColor','blue',...
             'MaxDisplayWords',150,'SizePower',0.2);
```

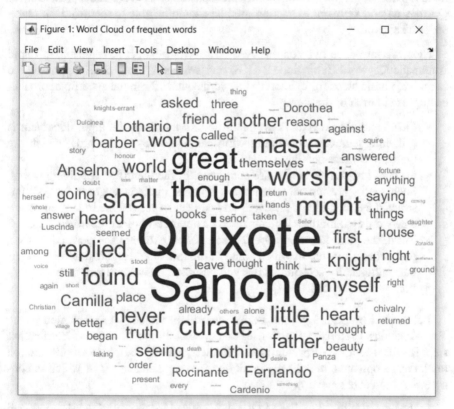

Fig. 14.1. Word cloud of the most frequent words (excluding short words and stop words) form Don Quixote

As defined in (14.3), the set of *150* most frequent words are displayed in the word cloud. As seen from the figure, the sizes of the words in the word cloud are scaled according to their frequency counts and the power parameter `SizePower`. Notice also that the largest words appear highlighted in a different color. It should be mentioned that the two largest words in the word cloud, *Quixote* and *Sancho*, are the names of the two main characters in the story. If you are familiar with the book, you will find the word cloud in Figure 14.1 is surprisingly good given the simplicity of the keyword extraction mechanism used. This puts in evidence once again the importance of word frequencies in language analysis in general.

However, just word frequencies along with the naïve filtering mechanism used in (14.2) is not enough for a robust keyword extraction. For instance, removing short words might eliminate valid keywords from the list of frequent words. Another important limitation of the described method is that the notion of keyword is reduced to a simple word or token. Ideally, the set of extracted keywords should include both single words and multiword constructs as well. Indeed, instead of *Quixote*, a good keyword extraction method should be able to identify *Don Quixote* as a keyword.

Next, we introduce two commonly used keyword extraction algorithms: Rapid Automatic Keyword Extraction (RAKE) and TextRank. Both methods follow the same three basic steps for extracting keywords but differ in the specific approaches they use in each of the steps:

- First, candidate keywords are identified. In this step, the input document is split into segments containing valid candidate keyword words, which are typically content-bearing words.
- Second, importance scores are computed for the identified candidate words. Such scores are typically computed from an undirected graph that is constructed by counting occurrences and co-occurrences of the identified candidate words within the segments or context windows of a predefined size.
- Third, the final keywords are extracted. The extraction is performed by a post-processing step in which some of the identified candidate words are either individually selected or merged based on their score and relative positions within the text.

Let us start by describing the RAKE algorithm, which is implemented in the Text Analytics Toolbox™ function `rakeKeywords`. In the candidate identification step, RAKE uses a list of pre-defined delimiters, mainly composed of punctuation marks and stopwords, to split the input text into segments, each of which is considered a candidate keyword.

In the second step, an undirected graph is created for the set of words contained in the segments. The edges of the graph are defined by word co-occurrences in the segments. Word occurrences are also included in the graph in the form of a recurrent edge for each word. The edges are weighted according to the frequency of word co-occurrences and self-occurrences. Then, the constructed graph is used to score the words. In the RAKE algorithm, the score of each word is computed as the ratio between its degree (the sum of weights of all edges associated to the word) and its frequency (the weight of the recurrent edge of the word). Finally, the score of each candidate keyword is computed by adding the scores of the individual words contained in it.

In the third step, a merging post-processing step is used before the final ranking and selection of keywords. During merging, adjacent candidate keywords that are separated by a special subset of delimiters, referred to as merging-delimiters, may be merged into a single candidate keyword. The merging is produced only if the adjacent keywords under consideration happen to appear (in the same order and separated by the same merging-delimiter) more than once in the input text. The score of a merged keyword is computed by adding the scores of the candidate keywords being merged. Then, the final list of candidate keywords is ranked according to their weights and the top-ranked ones, typically around one third of the total number of candidates, is selected as the final set of extracted keywords.

Figure 14.2 illustrates the first two steps of the RAKE algorithm for the sample sentence: *This illustrates a keyword extraction example with the RAKE algorithm for keyword extraction.*

This illustrates a keyword extraction example with the RAKE algorithm for keyword extraction .

This illustrates a keyword extraction example with the RAKE algorithm for keyword extraction .

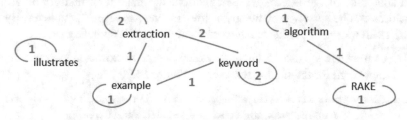

Candidate Keyword	Individual Word Scores	Total Score
illustrates	1/1	1
keyword extraction example	5/2 + 5/2 +3/1	8
RAKE algorithm	2/1 + 2/1	4
keyword extraction	5/2 + 5/2	5

Fig. 14.2. Illustrative example of keyword extraction and scoring with RAKE

As seen from the figure, in the first step, candidate keywords are identified by splitting the input sentence according to the following delimiters: *this, a, with, the,* and *for*. Then, in the second step, the graph is constructed and the scores are computed. Notice that the graph includes recurrent edges for all the words, and cross-word edges between those pairs of words that co-occur at least in one candidate keyword. As both words, *keyword* and *extraction*, occur twice in the set of candidate keywords, the weights of their recurrent edges are equal to *2*. Similarly, as both words also co-occur twice in the set of candidate keywords, the weight of the edge between them is also equal to *2*. For the rest of the words, on the other hand, their recurrent edges have a weight of *1*, as they only occur once in the set of candidate keywords. In a similar way, cross-word edges between word pairs *RAKE-algorithm*, *extraction-example*, and *example-keyword* have a weight of *1*, as those word pairs only occur once. For the rest of the word pairs, which do not co-occur in any of the candidate keywords, the corresponding edge weight is *0* and, accordingly, the links are omitted in the graph. Finally, the scores for each individual word and candidate keyword are computed as illustrated in the table in the bottom of the figure. Each word score is computed as the total weight of the edges associated to it divided by the weight of its recurrent edge, which are then added to obtain candidate keyword scores.

The third step of the process cannot be illustrated in Figure 14.2, as for such a small text sample no merging is possible. You can imagine, however, the situation in which a larger document contains a second instance of the segment *RAKE algorithm ~~for~~ keyword extraction*. In such a case, if *for* is contained in the set of merging-delimiters, the two candidate keywords *RAKE algorithm* and *keyword extraction* should be merged. The final ranking and selection process is trivial and, in the particular case of the `rakeKeywords` function, the maximum number of desired keywords can be specified by means of the `'MaxNumKeywords'` parameter. By default, it will return all candidate keywords with their corresponding scores.

Let us illustrate the use of function `rakeKeywords` to extract keywords with the RAKE algorithm for the sample sentence in Figure 14.2.

```
>> sentence = "This illustrates a keyword extraction example "+...          (14.4)
             "with the RAKE algorithm for keyword extraction.";
>> keywords = rakeKeywords(tokenizedDocument(sentence))
keywords =
  4×3 table
```

	Keyword		DocumentNumber	Score
"keyword"	"extraction"	"example"	1	8
"keyword"	"extraction"	""	1	5
"RAKE"	"algorithm"	""	1	4
"illustrates"	""	""	1	1

As seen from (14.4), the input of the function must be a tokenized document array and the output is a table containing the keywords, the document number, and the score. However, notice each keyword result is given in the form of a string array containing a single word per string. We can easily reformat the output to contain each full keyword in a single string as follows:

```
>> keywords.Keyword = strip(join(keywords.Keyword))                          (14.5)
keywords =
  4×3 table
```

Keyword	DocumentNumber	Score
"keyword extraction example"	1	8
"keyword extraction"	1	5
"RAKE algorithm"	1	4
"illustrates"	1	1

Let us now use the RAKE algorithm to extract keywords from the Don Quixote data collection, which we already tokenized in (14.2).

```
>> keywords = rakeKeywords(tokdocs);                                       (14.6)
>> keywords.Keyword = strip(join(keywords.Keyword));
>> height(keywords) % number of keywords extracted
ans =
      59499
```

As seen from (14.6), as we did not specify the desired maximum number of keywords per document, all candidate keywords found in each document have been returned. Let us see some of those candidate keywords:

```
>> % Show keywords extracted from first sentence                          (14.7)
>> disp(keywords(keywords.DocumentNumber==1,:))
```

Keyword	DocumentNumber	Score
"La Mancha"	1	4
"lean hack"	1	4
"old buckler"	1	4
"call"	1	1
"coursing"	1	1
"desire"	1	1
"gentlemen"	1	1
"greyhound"	1	1
"keep"	1	1
"lance"	1	1
"lance-rack"	1	1
"lived"	1	1
"long"	1	1
"mind"	1	1
"name"	1	1
"village"	1	1

As seen from (14.7), a significant number of candidate keywords were extracted from the first sentence in the book. Most of them are single words but there are also a few multiword keywords. Although some of them seem to be good candidates, others definitely do not.

As important keywords are expected to be repeated across multiple sentences in the book, we can compute frequency counts for the candidate keywords over the entire dataset and rank them accordingly.

```
>> [uniquekeywords,~,uidx] = unique(keywords.Keyword);                (14.8)
>> keywordfreqs = hist(uidx,length(uniquekeywords));
>> [rankedfreqs,ridx] = sort(keywordfreqs,'descend');
>> rankedkeywords = uniquekeywords(ridx);
```

Then, we can use the top-ranked candidates and their frequencies to generate the word cloud.

```
>> hf = figure(3);                                                    (14.9)
>> set(hf,'Color',[1 1 1],'Name','Word Cloud of RAKE Keywords');
>> wordcloud(rankedkeywords,rankedfreqs,'HighlightColor','red',...
             'MaxDisplayWords',150,'SizePower',0.2);
```

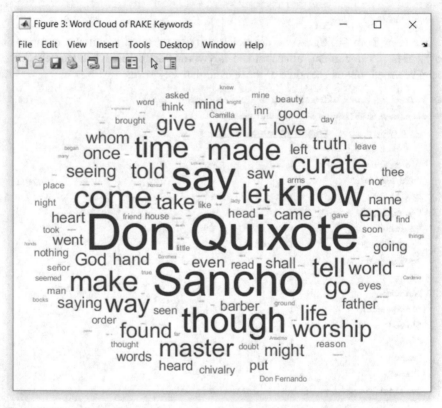

Fig. 14.3. Word cloud of the most frequent candidate keywords extracted with RAKE

Notice from Figure 14.3 that, different from the previous example (illustrated in Figure 14.1), the RAKE algorithm was able to identify multiword keywords such as *Don Quixote* and *Don Fernando*, as well as short ones such as *man* and *inn*. Like in the previous case, *Don Quixote* and *Sancho* continue to be the most important keywords. However, in this case, we can see a strong predominance of verbs, some of which do not necessarily constitute good keywords.

The TextRank keyword extraction algorithm, which is implemented in the Text Analytics Toolbox™ function `textrankKeywords`, follows the same three basic steps as RAKE. However, it uses different approaches at each step. In the first step, TextRank uses part-of-speech information to identify candidate keywords. By default, the available implementation considers only those words labeled as noun, proper-noun or adjective as valid candidate keywords.

In the second step, an undirected graph is also constructed with the identified candidate keywords. However, in this case, the graph is unweighted (i.e. edges in the graph are assigned a unitary weight) and recurrent edges are not included. Different from RAKE, rather than using the identified segments to look for keyword co-occurrences, TextRank uses a context window of predefined fixed size, which can be set by means of the `'Window'` parameter. If no window size is specified, a default value of 2 is used. Finally, a scaled version of PageRank centrality is used to score the keywords in the graph.

Figure 14.4 illustrates the corresponding TextRank graph, when using a window of size 3, for the sample sentence: *This illustrates a keyword extraction example with the TextRank algorithm for keyword extraction.* The corresponding candidate keyword scores are also depicted.

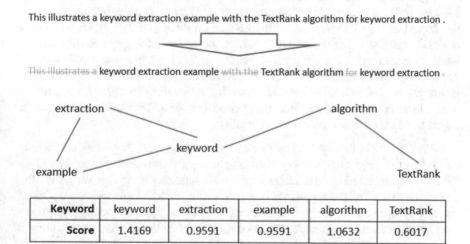

Keyword	keyword	extraction	example	algorithm	TextRank
Score	1.4169	0.9591	0.9591	1.0632	0.6017

Fig. 14.4. Illustrative example of keyword extraction and scoring with TextRank

As seen from Figure 14.4, different from the example presented in Figure 14.2, the verb *illustrates* has not been selected as a candidate keyword by TextRank. Also notice that, because a context window of size 3 has been used, edges are assigned to both word pairs that appear consecutively (*keyword-extraction*, *extraction-example*, and *TextRank-algorithm*) and word pairs that are separated by a single word (*keyword-example* and *algorithm-keyword*).

In the last step, candidate keywords are ranked by their scores and the top third is selected. Next, keywords appearing consecutively in the input text are merged into a single one and their scores are added. Finally, the number of desired keywords, as specified by `'MaxNumKeywords'`, is returned.

We can compute keywords with `textrankKeywords` for the sample sentence in Figure 14.4 as follows:

```
>> sentence = "This illustrates a keyword extraction example "+...          (14.10)
             "with the TextRank algorithm for keyword extraction.";
>> keywords = textrankKeywords(tokenizedDocument(sentence),'Window',3)
keywords =

  2×3 table

      Keyword       DocumentNumber     Score
    "keyword"             1            1.417
    "algorithm"           1            1.063
```

As seen from (14.10), the two keywords with highest score have been returned, which roughly approximates one third of the selected candidate keywords ($5/3 \approx 2$). As they do not appear consecutively in the input text, they were not merged.

In general, the scores computed for nodes in graphs like the one from Figure 14.4 are referred to as node centrality scores. This type of scores attempt to assess the importance of the nodes in a graph based on the edges they share with other nodes. In the context of word graphs, where the edges in the graph are defined by word co-occurrences, node centrality relates to the importance and relevance of keywords with respect to the document collection under consideration. Node centrality can be defined and computed according to a variety of criteria, such as degree, closeness, betweenness, etc. In the specific case of TextRank, the PageRank algorithm[2] is used to compute node centrality.

Although it was originally conceived for directed graphs, PageRank also works on undirected graphs like the keyword graphs we are considering here. The algorithm is implemented in a recursive way which simulates a random walk over the graph. The result of the algorithm is a probability distribution over the nodes of the graph, representing the probabilities of visiting the different nodes. The recursion assumes that, at each node, either edges will be followed with a probability p or a new node will be selected at random with probability $(1-p)$. If edges are followed, the specific edge to follow from a given node j into a new node i is selected with probability $t(i,j)$. This transition probability is defined as the ratio between the number of edges going from node j to node i and the total number of edges departing from node j (notice that in the case of undirected graphs it is just the num-

---

[2] PageRank is the original algorithm used by Google to rank websites based on the links among them. It works under the assumption that important pages will have more incoming links than unimportant pages, as well as that the importance of an incoming link depends on the importance of the page from which it comes.

ber of edges between j and i divided by the total number of edges associated to j). Accordingly, the PageRank algorithm is described by the following recursion:

$$\begin{bmatrix} s_1 \\ s_2 \\ \vdots \\ s_N \end{bmatrix} = (1-p) \begin{bmatrix} 1/N \\ 1/N \\ \vdots \\ 1/N \end{bmatrix} + p \begin{bmatrix} t(1,1) & t(1,2) & \cdots & t(1,N) \\ t(2,1) & t(2,2) & \cdots & t(2,N) \\ \vdots & \vdots & \ddots & \vdots \\ t(N,1) & t(N,2) & \cdots & t(N,N) \end{bmatrix} \begin{bmatrix} s_1 \\ s_2 \\ \vdots \\ s_N \end{bmatrix} \qquad (14.11a)$$

$$\sum_{i=1}^{N} t(i,j) = 1 \ \forall j \qquad (14.11b)$$

where s_1, s_2, s_3, ... s_N are the corresponding PageRank node centrality scores for the N nodes in the graph.

As seen from (14.11b), the matrix in (14.11a) is a stochastic matrix, in which the elements in each column add up to 1, as each $t(i,j)$ describes the transition probability from node j into node i. If no edge exists between a pair of nodes m and k, then $t(m,k)=0$. However, if a given node k does not have edges to any other node in the graph, the new node is selected at random with uniform probability, i.e. $t(i,k)=1/N$ for all i.

Now, let us compute PageRank node centrality scores for the graph in Figure 14.4 by using the `centrality` function. First, we define the edges by means of the pair of nodes they connect. If we enumerate the nodes as *(1) keyword, (2) extraction, (3) example, (4) algorithm* and *(5) TextRank*; the five edges in Figure 14.4 are then given by the node pairs *(1,2), (1,3), (1,4), (2,3)* and *(4,5)*.[3]

```
>> kwds = {'keyword','extraction','example','algorithm','TextRank'};          (14.12)
>> nodes1 = [1 1 1 2 4];
>> nodes2 = [2 3 4 3 5];
```

Next, we create the graph and compute the node centrality scores:

```
>> kwdsgraph = graph(nodes1,nodes2,[],kwds);                                   (14.13)
>> pageranks = centrality(kwdsgraph,'pagerank')'
pageranks =
    0.2834    0.1918    0.1918    0.2126    0.1203
```

from which the TextRank scores can be obtained by scaling the PageRank scores according to the number of nodes (keywords) in the graph:

```
>> textranks = pageranks * length(kwds)                                        (14.14)
>> textranks =
    1.4169    0.9591    0.9591    1.0632    0.6017
```

---

[3] Notice that the order of the two nodes in each node pair is irrelevant here as we are creating an undirected graph.

To explore in more detail the idea of centrality, let us consider now the names of the characters in Don Quixote's plot and compute its co-occurrence matrix over the different sentences in the dataset. Such a co-occurrence matrix defines an undirected graph of characters, for which node centrality scores can be computed. First, let us build and plot the graph of characters, which is displayed in Figure 14.5. We proceed as follows:

```
>> % identifies words with entity labels different from 'person'        (14.15a)
>> newtokdocs = addEntityDetails(tokdocs);
>> details = tokenDetails(newtokdocs);
>> nonpersons = details.Token(details.Entity~='person');

>> % creates a bag of words model                                       (14.15b)
>> bow = bagOfWords(newtokdocs);
>> bow = removeWords(bow,nonpersons); % removes non-persons
>> bow = removeInfrequentWords(bow,5); % removes infrequent persons

>> % computes the co-occurrence matrix and builds the graph             (14.15c)
>> counts = bow.Counts;
>> cooccurrencemtx = counts.'*counts;
>> pgraph = graph(cooccurrencemtx,bow.Vocabulary,'omitselfloops');

>> hf = figure(5); % plots the graph                                    (14.15d)
>> figname = 'Graph of most frequent characters in Don Quixote';
>> set(hf,'Color',[1 1 1],'Name',figname);
>> plot(pgraph); axis off
```

As seen from Figure 14.5, the graph contains *17* nodes, which correspond to the most frequent characters in the book. Pairs of nodes are connected only when the corresponding character names co-occur at least in one sentence of the book. The weights of edges (no displayed in the figure) depend on the number of sentences in which the corresponding two names co-occur. The graph topology provides good visual insights about the relative importance of the characters in the book. The importance of a node is directly related to the number of edges it has. Indeed, notice how the two main characters, *Don Quixote* and *Sancho*, which are heavily connected to most of the other nodes, appear in the center of the graph.

There are, however, a few problems that can be addressed to improve the quality of the graph in Figure 14.5 as a visual representation of character importance. First, character names that have not been properly tokenized and appear truncated, such as *Dulcinea del* (instead of *Dulcinea del Toboso*) and *Gines de* (instead of *Gines de Pasamonte*). Second, character names that are duplicated due to different mentions to the same person in the text, such as *Sancho* and *Sancho Panza*, which actually refers to the same character. And third, important characters that have not been identified as person entities, such as the *barber* and the *curate*. For a more detailed discussion about remediating these problems and improving the quality of the character graph in Figure 14.5, see exercise 14.4-3.

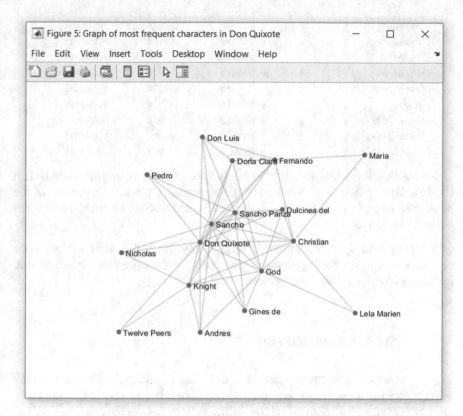

Fig. 14.5. Undirected graph of the most frequent characters in the plot of Don Quixote

Next, we compute different types of node centrality scores for the character graph in Figure 14.5. More specifically, we compute degree, closeness, betweenness and pagerank.

```
>> pgraph.Nodes.Degree = centrality(pgraph,'degree');          (14.16)
>> pgraph.Nodes.Closeness = centrality(pgraph,'closeness');
>> pgraph.Nodes.Betweenness = centrality(pgraph,'betweenness');
>> pgraph.Nodes.PageRank = centrality(pgraph,'pagerank');
>> disp(pgraph.Nodes)
```

Name	Degree	Closeness	Betweenness	PageRank
{'Nicholas' }	3	0.032258	0	0.0318
{'Knight' }	8	0.041667	6.3667	0.074737
{'Don Quixote' }	14	0.055556	27.536	0.12448
{'Dulcinea del'}	5	0.037037	0	0.047011
{'God' }	10	0.045455	11.71	0.090425
{'Andres' }	4	0.034483	0	0.039488
{'Twelve Peers'}	2	0.03125	0	0.024328
{'Sancho Panza'}	10	0.045455	7.019	0.088685

{'Sancho' }	13	0.052632	17.369	0.11424
{'Christian' }	9	0.043478	14.905	0.083947
{'Pedro' }	3	0.032258	0	0.031398
{'Gines de' }	4	0.034483	0	0.039085
{'Fernando' }	7	0.04	6.3202	0.066023
{'Maria' }	2	0.028571	0	0.024765
{'Lela Marien' }	2	0.029412	0	0.024437
{'Doña Clara' }	5	0.037037	0.575	0.047636
{'Don Luis' }	5	0.035714	0.2	0.04751

Notice that according to all different flavors of node centrality computed in (14.16), *Don Quixote* and *Sancho* are the two most important characters in the book. The rest of the characters follow with scores that reflect their relative importance in the plot.

When applied to sentences, rather than to keywords, this same idea of node centrality as a proxy for importance is the basis for the extractive summarization technique described in the next section.

14.2 Text Summarization

The same idea of centrality introduced in the previous section can be used to rank the sentences in a document according to their importance. This is the basis of a class of text summarization techniques know as extractive summarization. As implied by its name, extractive summarization aims at constructing summaries for a given document by extracting the most important set of sentences in it.

Instead of counting co-occurrences to generate a keyword graph (as it is done for keyword extraction), in this case, pairwise sentence similarities are used to generate a sentence graph. Accordingly, node centrality scores can be computed and used to rank the sentences in a document according to their importance. In this section, we introduce two specific scoring methods, LexRank and TextRank, for which Text Analytics Toolbox™ implementations are available. While LexRank computes pairwise cosine similarity scores, TextRank uses the BM25 algorithm. For computing the node centrality scores, on the other hand, both methods use the PageRank algorithm.

Let us start by computing LexRank and TextRank scores for all sentences in the first chapter of Don Quixote. We use the functions `lexrankScores` and `textrankScores`, accordingly.

```
>> ch1sentences = tokenizedDocument(data(data.chap==1,:).text);          (14.17)
>> lexrnkscores = lexrankScores(ch1sentences);
>> txtrnkscores = textrankScores(ch1sentences);
```

The resulting scores are compared in the bar plot presented in Figure 14.6 (the generation of the plot is left to you as an exercise).

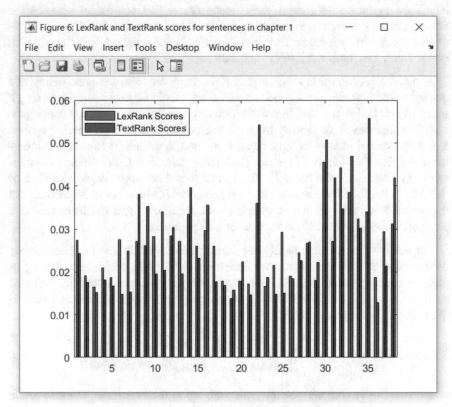

Fig. 14.6. LexRank and TextRank scores for the sentences in chapter 1 of Don Quixote

Notice from the figure that although both sets of scores exhibit significant differences at the individual sentence level, somehow, they follow similar trends. For instance, the region from sentence *30* to sentence *35* seems to concentrate the highest importance scores, followed by the region of sentences from *8* to *16*. Similarly, both methods seem to agree in giving individual sentences *22* and *38* relatively high scores.

Indeed, although the top-ranked sentences are different in both cases, if we consider some subsets of top-ranks, we will see a high proportion of intersections. Let us compute the size of the intersection set for the subsets of top-*10* ranks:

```
>> [~,idxlexrnk] = sort(lexrnkscores,'descend');                    (14.18)
>> [~,idxtxtrnk] = sort(txtrnkscores,'descend');
>> disp(length(intersect(idxlexrnk(1:10),idxtxtrnk(1:10))))
    7
```

As seen from (14.18), *7* out of *10* top-ranked sentences coincide in both cases. Overall, the sentence ranking scores produced by both methods are more similar than they appear to be. Indeed, the Pearson correlation coefficient between the rankings score depicted in Figure 14.6 is *0.7228*, and the corresponding mean values are *0.0263* in both cases. They mainly differ in the standard deviations, which is *0.0076* for LexRank and *0.0122* for TextRank.

A natural question that arises at this point is, apart from the consistency and quality of the rankings, how good is a summary that is constructed by selecting a set of top-ranked sentences? In practice, quality assessment of automatically generated summaries is conducted by comparing the automatic summaries against human-generated ones. The most popular method to evaluate automatic summaries is the ROUGE family of metrics (ROUGE stands for Recall-Oriented Understudy for Gisting Evaluation). The Text Analytics Toolbox™ implementation of ROUGE is the function `rougeEvaluationScore`, which receives as input variables the candidate summary and a set of references. Additionally, it admits name-value pair arguments to overwrite some of its default parameters.

Let us now describe this family of metrics in more detail. The most commonly used version of the metric is the one known as ROUGE-N, which accounts for *n*-gram overlaps between the automatic summary being evaluated and the provided set of references. The ROUGE-N score for a single reference is computed according to the following expression:

$$ROUGE\text{-}N(sum, ref) = \frac{\sum_{sent \in ref} \sum_{n\text{-}gram \in sent} Count(n\text{-}gram, sum)}{\sum_{sent \in ref} \sum_{n\text{-}gram \in sent} Count(n\text{-}gram)} \qquad (14.19)$$

where *sum* is the summary being evaluated, *ref* is the reference, *Count(n-gram, sum)* is the number of *n*-grams co-occurring in the summary and the reference, and *Count(n-gram)* is the number of *n*-grams in the reference. As seen from the formula, counts are first computed at the sentence level and then aggregated at the reference level. Notice that ROUGE-N is a recall-oriented metric as the counts in the denominator are computed over the reference.

Other variants of ROUGE include ROUGE-L, which considers Longest Common Subsequences (LCS) rather that fixed size *n*-grams; ROUGE-W, which considers weighted LCS statistics; ROUGE-S, which considers skip-bigrams[4] instead of *n*-grams, and ROUGE-SU, which considers both skip-bigrams and unigram co-occurrence statistics. In all metric variants, when multiple references are available, multi-reference ROUGE scores are computed by taking the maximum value of the corresponding set of single-reference scores. In the rest of this section, we use ROUGE-N (for a more complete exploration of other variants of the metric, refer to exercise 14.4-5).

---

[4] A skip-bigram is an ordered pair of words w_i and w_j within the same sentence and with an arbitrary number of words in between, i.e. $j - i \geq 1$.

Let us now proceed to evaluate some of the automatic summaries we can extract for Don Quixote. For this, we will need human-generated summaries as references. The data file `datasets_ch14b.mat` contains a set of summaries for different episodes in the book, which are publicly available from Wikipedia.[5]

```
>> load datasets_ch14b                                          (14.20a)
>> summaries.Properties.VariableNames
ans =
  1×4 cell array
    {'episode'}     {'from'}     {'to'}     {'text'}

>> height(summaries)                                            (14.20b)
ans =
     9
```

As seen from (14.20) there are nine summaries in total. The corresponding variables in the table of summaries are the name of the `episode` being summarized, the chapter spans of the episodes (`from`, `to`), and the `text` of the reference summaries. The nine available summaries are further described in Table 14.1.

Table 14.1. Human-generated summaries for nine episodes in Don Quixote (summary sizes are given in number of sentences)

Episode #	Episode Name	Chapter Span	Summary Size
1	The First Sally	1 to 5	12
2	Destruction of Don Quixote's library	6 to 7	6
3	The Second Sally	8 to 10	10
4	The Pastoral Peregrinations	11 to 15	9
5	The inn	16 to 17	10
6	The galley slaves and Cardenio	19 to 24	6
7	The priest, the barber, and Dorotea	25 to 31	6
8	Return to the inn	32 to 42	6
9	The ending	45 to 52	5

Now, we are ready to evaluate automatic summaries for the corresponding episodes in Don Quixote. Let us start by generating summaries for the first episode by using both scoring methods described at the beginning of this section. First, we create a tokenized document for the first episode summary, which will serve as the refence to evaluate the extracted summaries.

```
>> reference = summaries.text(1);                               (14.21)
>> referencetok = tokenizedDocument(reference);
```

---

[5] The summaries are available at https://en.wikipedia.org/wiki/Don_Quixote#Summary Accessed 6 July 2021.

Next, we select all sentences within the first episode chapter span, already to-kenized in the tokenized document array `tokdocs` used in previous exercises.

```
>> chapterspan = (data.chap>=summaries.from(1))&...                    (14.22)
                 (data.chap<=summaries.to(1));
>> episodetok = tokdocs(chapterspan);
```

Then, we extract top-ranked sentences with both LexRank and TextRank. For this, we use the function `extractSummary`.

```
>> referencesize = length(splitSentences(reference));                  (14.23a)

>> % extracts summary with LexRank scores                              (14.23b)
>> extractedlex = extractSummary(episodetok,'ScoringMethod','lexrank',...
                                 'SummarySize',referencesize);
>> summarylextok = tokenizedDocument(join(joinWords(extractedlex)));

>> % extracts summary with TextRank scores                            (14.23c)
>> extractedtxt = extractSummary(episodetok,'ScoringMethod','textrank',...
                                 'SummarySize',referencesize);
>> summarytxttok = tokenizedDocument(join(joinWords(extractedtxt)));
```

Notice from (14.23) that we extracted summaries of the same length, in number of sentences, as the reference.[6] Also, we converted the resulting tokenized docu-ment arrays into tokenized document scalars for evaluation purposes.

Finally, we can compute ROUGE scores for both generated summaries:

```
>> rougelex = rougeEvaluationScore(summarylextok,referencetok)         (14.24a)
>> rougelex =
       0.6235

>> rougetxt = rougeEvaluationScore(summarytxttok,referencetok)         (14.24b)
>> rougetxt =
       0.6306
```

As seen from (14.24) both ROUGE scores are quite similar, being the score of the summary generated with TextRank slightly higher than the score of the sum-mary generated with LexRank.

We can still ask how good these summaries are indeed, as we do not have any reference to judge what a ROUGE score of 0.6 means. We only know that if the automatic summary fully covers all n-grams in the reference, the resulting score would be 1.0, while if it completely misses all of them, the score would be 0.0. To get a better feeling about how good the summaries computed in (14.23) are, let us

---

[6] As ROUGE is a recall-based metric, it can be typically pushed up by just extracting larger summaries. However, one of the main characteristics of good summaries is brevity. Accordingly, summary optimality is typically assessed under predefined length constraints. For a more de-tailed exploration about this problem, see exercise 14.4-6.

compare them to summaries generated by randomly picking sentences from the episode under consideration. This will provide us with a lower-bound for ROUGE scores in the dataset under consideration. Any automatically generated summary with scores equal or lower than such a random baseline should be deemed as a bad summary. Similarly, an upper-bound for ROUGE scores can be computed by comparing human-generated summaries among themselves. Such a human performance level is typically a good reference to what can be expected as best achievable performance (regrettably, as we only have one summary per episode available, we will not be able to illustrate the computation of this upper-bound).

Next, let us compute a lower-bound for ROUGE scores in Don Quixote's first episode with a random baseline system. The baseline extraction method is quite simple, we just extract a random sample set of sentences from the sentences in the first episode. We extract as many sentences as the reference has, so the random summary is of the same length. Then, we compute the corresponding ROUGE score. We repeat this exercise multiple times (*500* times actually) and look at the resulting distribution of scores, their mean and standard deviation. The results are presented in Figure 14.7.

```
>> rng('default');                                              (14.25a)
>> randsummscore =[];

>> for k=1:500                                                  (14.25b)
        % generates a random summary
        randindx = randperm(length(episodetok));
        rndsumm = episodetok(sort(randindx(1:referencesize)));
        rndsumm = tokenizedDocument(join(joinWords(rndsumm)));
        % computes its ROUGE score
        randsummscore(k) = rougeEvaluationScore(rndsumm,referencetok);
    end

>> hf = figure(7);                                              (14.25c)
>> figname = 'Distribution of ROUGE scores for random summaries';
>> set(hf,'Color',[1 1 1],'Name',figname);
>> histogram(randsummscore,'Normalization','probability');
>> text(0.42,0.12,sprintf('mean = %6.4f',mean(randsummscore)));
>> text(0.42,0.11,sprintf('std = %6.4f',std(randsummscore)));
```

As seen from the figure, the distribution of ROUGE scores for the summaries generated with the random baseline system resembles a normal distribution, with a mean value of ~0.5 and a standard deviation of ~0.03. Recall from (14.24) that ROUGE scores obtained for summaries generated with LexRank and TextRank were *0.6235* and *0.6306*, respectively. These values are well above ~0.53, which is one standard deviation over the mean of the random baseline. Accordingly, we can say both methods perform significantly better than the lower-bound defined by the random baseline.

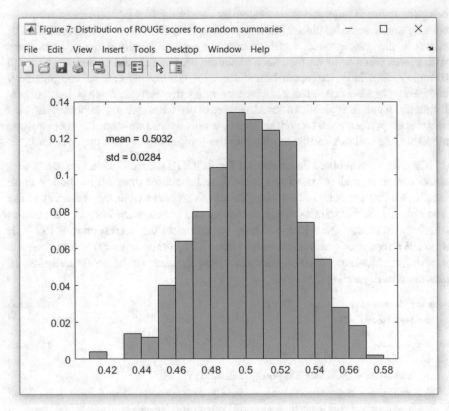

Fig. 14.7. Distribution of ROUGE scores for randomly generated summaries of the first episode (chapters 1 to 5) in Don Quixote

Next, let us explore in more detail the effects of summary lengths over the scores. Notice that all scores computed in (14.25) correspond to random summaries of the same length (in number of sentences) as the reference.

In the next exercise, we repeat the step in (14.25b) multiple times for different summary lengths (ranging from *1* single sentence up to *30* sentences). We collect all means and standard deviation values of ROUGE scores computed for each of the considered summary lengths. We also compute TextRank summaries of different lengths along with their corresponding ROUGE scores. We plot the results to visually compare the performance of random summaries and TextRank summaries of different lengths. The results are illustrated in Figure 14.8 (the generation of the plot is left to you as an exercise).

```
>> summarysizes = 1:30;                                                    (14.26a)
>> meanvalues = [];
>> stdvalues = [];
>> txtrnkscore = [];
```

```
>> for k=1:length(summarysizes)                                    (14.26b)
      randsummscore = [];
      for n=1:100
          % generates and evaluates random summaries
          randindx = randperm(length(episodetok));
          rndsumm = episodetok(sort(randindx(1:summarysizes(k))));
          rndsumm = tokenizedDocument(join(joinWords(rndsumm)));
          randsummscore(n) = rougeEvaluationScore(rndsumm,referencetok);
      end
      meanvalues(k) = mean(randsummscore);
      stdvalues(k) = std(randsummscore);
      % generates and evaluates TextRank summaries
      extracted = extractSummary(episodetok,'SummarySize',summarysizes(k));
      summarytok = tokenizedDocument(join(joinWords(extracted)));
      txtrnkscore(k) = rougeEvaluationScore(summarytok,referencetok);
  end
```

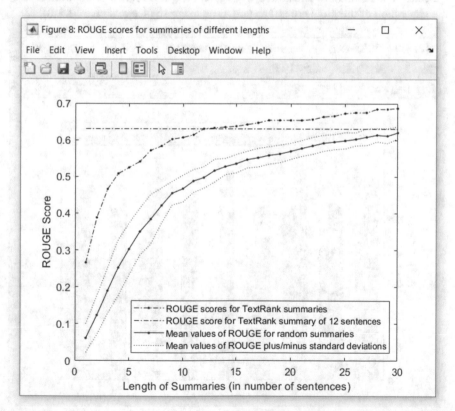

Fig. 14.8. Comparison of ROUGE scores for summaries of different lengths generated with TextRank and a random baseline for the first episode (chapters 1 to 5) in Don Quixote

Notice from Figure 14.8, how ROUGE scores monotonically increase as summary sizes are increased for both, TextRank and the random baseline, systems. In all cases, TextRank performs significantly better than the baseline, being the former's performance well above the mean plus one standard deviation of the latter's. As the summary sizes continue to increase beyond *30* sentences, it is expected for both curves to get closer and closer, until they reach the same maximum score value when all the sentences (*38* in this case) are contained in the summaries.

Interestingly, we can also see that the score of a randomly generated summary of *30* sentences is equivalent, on average, to the score of a TextRank generated summary of *12* sentences (segmented horizontal line in Figure 14.8), which is the one previously computed in (14.24b). Notice this last observation, while valid for the episode under consideration, might vary significantly across different datasets.

Finally, we extract and evaluate summaries for all nine episodes in Table 14.1, for which we compare the performances of LexRank, TextRank and the random baseline. The results are illustrated in Figure 14.9. All related computations and the generation of the plots are left to you as an exercise (see exercise 14.4-8).

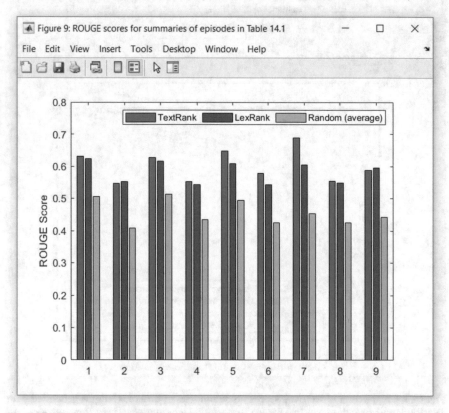

Fig. 14.9. Comparative evaluation of LexRank, TextRank, and random baseline generated summaries for all nine episodes of Don Quixote described in Table 14.1

As seen from Figure 14.9, with the exception of two or three cases, the summaries extracted with LexRank and TextRank achieve very similar performances, which are always consistently above the average performances of the randomly extracted summaries.

It is important to mention that all summaries involved in the evaluations presented in Figure 14.9 contain the same number of sentences as their corresponding reference summaries. Accordingly, score comparisons among the different extraction methods for a given episode are fair to all methods. Overall, observed average ROUGE scores for LexRank and TextRank are *0.5817* and *0.6015*, respectively, and the average ROUGE score for the random baseline system is *0.4560*.

14.3 Further Reading

We started this chapter by presenting two specific keyword extraction methods: Rapid Automatic Keyword Extraction, also known by its acronym RAKE, (Rose *et al.* 2010) and TextRank (Mihalcea and Tarau 2004). These methods are based on the concept of node centrality, which aims at assessing the relative importance of the nodes in a graph (Newman 2006) and (Benzi and Klymko 2013). In the specific case of TextRank, the PageRank node centrality score is used (Page 1997), which was the first ranking algorithm used by Google to rank webpages retrieved during search (Brin and Page 1998).

The type of text summarization techniques described in section 14.2 are commonly referred to as extraction-based or extractive summarization (Hirohata et al. 2005). The LexRank approach to extractive summarization was introduced by (Erkan and Radev 2004), and the TextRank approach was introduced by (Mihalcea and Tarau 2004). Another important paradigm to text summarization, which was not covered here, is the abstraction-based or abstractive summarization approach (Cheung and Penn 2013). Different from the extractive approach, in the abstractive approach, summaries are produced by using techniques such as sentence fusion (Barzilay and McKeown 2005) and Natural Language Generation (Zhang *et al.* 2019). More recent implementations of both extractive and abstractive summarization that are based on deep learning architectures can be found in (See *et al.* 2017), (Zhou *et al.* 2018), (Paulus *et al.* 2018), (Bhargava and Sharma 2020) and (Syed *at al.* 2021).

The Recall-Oriented Understudy for Gisting Evaluation (ROUGE) metric for automatic summarization evaluation is due to (Lin 2004). Another metric used in summarization evaluation, although originally conceived for machine translation evaluation (Papineni *at al.* 2002), is the Bilingual Evaluation Understudy (BLEU). Different from ROUGE, which is a recall-oriented score, BLEU is a precision-oriented score.

14.4 Proposed Exercises

1. Consider the word cloud of Don Quixote presented in Figure 14.3 for keywords extracted with the RAKE algorithm.

   – Repeat the steps in (14.6), (14.8) and (14.9) to generate a new word cloud but use the function `textrankKeywords` instead (consider the default window size value of 2).
   – Compare the resulting word cloud with those from Figures 14.1 and 14.3. What similarities and differences do you observe?
   – Recompute the word cloud again by using different window sizes: 1 and 3. What are your main observations?

2. Consider the sample sentence from Figure 14.4.

   – Follow the steps in (14.12) and (14.13) to create the TextRank graphs for windows of sizes 1, 2 and 3. Plot the graphs and follow the step in (14.14) to compute PageRank node centrality scores for the three graphs.
   – Create a script to implement the recursive procedure described in (14.12). The script must iterate m times over a vector s of node centrality scores for a given node transition probability matrix T, total number of nodes n and follow probability p. The recursion should start with the uniform probability vector $s_i = 1/n$ for $i = 1, 2, ... n$.
   – Use the function to compute PageRank node centrality scores for the three graphs. Compare the new results with the ones previously computed.
   – Multiply the obtained scores by the number of nodes to get the TextRank scores. Extract keywords with `textrankKeywords` and compare the resulting keywords scores with the ones previously derived with the script.
   – Use the script to generate plots of centrality scores vs. number of iterations for the five nodes in the three graphs. What are your main observations?

3. Create an improved version of the character graph in Figure 14.5 by preprocessing the data as follows:

   – Consider the following normalized list of characters: *Don-Quixote, Sancho-Panza, Twelve-Peers, Don-Luis, Doña-Clara, Dulcinea-del-Toboso, Gines-de-Pasamonte, Maria, Fernando, Christian, Nicholas, Lela-Marien, Pedro, Andres, God, curate, barber.*
   – Join multiword character names into single tokens by using dashes such as, for instance, *Don Quixote → Don-Quixote, Don Luis → Don-Luis.*
   – Map different mentions of the same entity into a single normalized one as, for instance, *Sancho → Sancho-Panza, Dulcinea → Dulcinea-del-Toboso.*
   – Repeat the procedure in (14.15) to create the new character graph. (Suggestion: use `setdiff` to obtain the vocabulary subset to be removed).

- Plot the graph and compute node centrality scores. How these new results compare to those from Figure 14.5 and (14.16)?

4. Consider the LexRank and TextRank scores depicted in Figure 14.6.

    - Compute basic statistics (min, max, mean and std) for each case.
    - Generate a cross-plot and compute the correlation coefficient between the two sets of scores. Is the resulting correlation statistically significant?
    - Follow a procedure like the one in (14.18) to compute overlap proportions for all ranks and generate a plot of overlaps vs ranks.
    - Repeat the previous steps for a random selection of *10* different chapters. What are your main observations?

5. Reproduce the plots in Figure 14.8 by using LexRank generated summaries. How these new results compare to the ones in the figure?

    - Generate similar plots for both LexRank and TextRank summaries of the other eight episodes described in Table 14.1.
    - What differences and/or similarities do you observe between LexRank and TextRank summaries among the different episodes?
    - Consider the different variants of the ROUGE score. Select a few of the plots generated in the previous step and regenerate them by using different variants of ROUGE. What are your main observations?

6. Consider again the plots in Figure 14.8. As observed in the figure, the scores increase as the length of the summaries increases.

    - Propose and implement a strategy to compensate the observed effect by penalizing summary lengths. Notice your proposed strategy must penalize for summary length only when the candidate summary being evaluated is larger than the reference summary.
    - Reproduce the plots in Figure 14.8 by using your newly implemented length-penalization strategy.
    - Can you identify an optimal "region" of summaries by looking at both the ROUGE scores and the summary lengths?

7. Study the Text Analytics Toolbox™ implementation of the BLEU metric given in `bleuevaluationScore`.

    - Reproduce the plots in Figure 14.8 by using BLEU instead of ROUGE. What are your main observations?
    - Implement an F1-score oriented metric for text summarization by combining BLEU (precision-oriented) and ROUGE (recall-oriented).
    - Reproduce the plots in Figure 14.8 by using it. How these results compare with the ones obtained with BLEU and ROUGE alone?

8. Generate the bar plots in Figure 14.9 by implementing the following procedure:

   – Gather the corresponding set of sentences for each of the nine episodes de-
     scribed in Table 14.1. You can use the same approach used in (14.22).
   – Generate summaries for each of the nine episodes with both the LexRank
     and Text Rank extraction methods. Each extracted summary must be of the
     same length (in number of sentences) as the corresponding reference sum-
     mary. You might want to follow the procedure used in (14.23) for this.
   – Generate a set of *100* random summaries for each of the nine episodes by
     following the approach used in (14.25).
   – Compute ROUGE scores for all generated summaries and compute the av-
     erages of random summary scores corresponding to the same episode.
   – Use these results to generate the bar plots presented in Figure 14.9.

9. Study the Text Analytics Toolbox™ function `mmrScores`, which scores documents with respect to their relevance to a given query by means of the Maximal
   Marginal Relevance algorithm (Carbonell and Goldstein 1998).

   – The idea of this exercise is to retrieve the sets of most relevant sentences
     for each of the episodes described in Table 14.1.
   – Accordingly, use function `mmrScores` to score all sentences in `data.text`
     from (14.1) using the `summaries` from (14.20) as queries.
   – Notice the parameter `lambda` in `mmrScores` is used to control the tradeoff
     between relevance and redundancy. In this exercise, we only care about
     relevance and, accordingly, you should set the value of `lambda` to *1*.
   – Generate plots of relevance vs sentence number for the overall data collec-
     tion. (Hint: to make sense of the resulting plots, you might want to smooth
     the score curves by computing their average values over a sliding window
     of a fixed number of sentences). What are your main observations?
   – Implement a method to select sentences for each episode. Use the actual
     episode spans defined in Table 14.1 to evaluate your selection method (i.e.
     compute accuracy, precision and recall for each set of selected sentences).

14.5 Short Projects

1. Consider the novel Pride and Prejudice by Jane Austen.

   – Download all the text of the novel from https://www.gutenberg.org/files/
     1342/1342-0.txt Accessed 6 July 2021.
   – Preprocess the content by removing all preliminary and trailing text that is
     not part of the novel. Segment the content into sentences and create a table,
     like the one in (14.1), containing all sentences and their chapter numbers.

- Extract keywords and create word clouds following the different procedures described in section 14.1. How do they compare to each other?
- Consider the character list provided in https://en.wikipedia.org/wiki/Pride_and_Prejudice#Characters (Accessed 6 July 2021) and create a character graph similar to the one in Figure 14.5.
- Compute node centrality scores for the different characters. What conclusions can you derive from the scores with regards to the characters?
- Consider the summary provided in https://en.wikipedia.org/wiki/Pride_and_Prejudice#Plot_summary (Accessed 6 July 2021) and use it as a reference to evaluate automatic summaries generated with both the TextRank and LexRank approaches. What ROUGE scores do you obtain? How do they compare to scores obtained for randomly generated summaries?
- Generate individual chapter summaries and create a new summary for the full novel by combining the individual summaries. Evaluate the new summary. How does it compare to the ones generated in the previous step?
- Select your preferred methods for keyword extraction and text summarization. Create a script that, given the text of a novel, automatically generates a word cloud and a summary of it. Use the script to process other classical novels that are available from the Project Gutenberg website.

2. Look for an online library of conference proceedings with an open access policy (make sure the papers include an abstract section and a list of keywords).

- Download a small collection of papers (~50) in *.pdf* format.
- Use the function `extractFileText` to read the text from the *.pdf* files and create a script to process the documents to extract the abstracts, keyword lists and main technical contents.
- Make sure you remove, as much as you can, the different text artifacts resulting from the *.pdf* to text conversion, as well as section titles, subtitles, citations, figure and table captions, and other non-relevant contents.
- Use the procedures described in section 14.1 to extract keywords from the technical contents. Evaluate the extracted set of keywords against the corresponding keyword lists provided in the papers.
- What keyword coverage do you get on average? What is the resulting precision and recall of your keyword extraction procedure?
- Repeat the keyword extraction but use the abstracts instead of the technical content sections. Evaluate the results. How do these new results compare to the previous ones?
- Use the procedures described in section 14.2 to conduct text summarization for the technical content sections. Evaluate the obtained summaries against the corresponding paper abstracts.
- What are the resulting ROUGE scores of the summaries? How do they compare to scores obtained for randomly generated summaries?

14.6 References

Barzilay R, McKeown KR (2005) Sentence fusion for multidocument news summarization, Computational Linguistics, 31(3): 297–328

Benzi M, Klymko C (2013) A matrix analysis of different centrality measures. SIAM Journal on Matrix Analysis and Applications, 36(2): 686–706

Bhargava R, Sharma Y (2020) Deep Extractive Text Summarization, Procedia Computer Science, 167: 138–146

Brin S, Page L (1998) The anatomy of a large-scale hypertextual Web search engine, Computer Networks and ISDN Systems, 30(1–7): 107–117

Carbonell JG, Goldstein J (1998) The use of MMR, diversity-based reranking for reordering documents and producing summaries. In proceedings of *SIGIR* 98: 335–336

Cheung JCK, Penn G (2013) Towards robust abstractive multi-document summarization: A case-frame analysis of centrality and domain. In proceedings of the 51st Annual Meeting of the Association for Computational Linguistics (ACL), pp. 1233–1242

Erkan G, Radev DR (2004) Lexrank: Graph-based Lexical Centrality as Salience in Text Summarization, Journal of Artificial Intelligence Research, 22: 457–479

Hirohata M, Shinnaka Y, Iwano K, Furui S (2005) Sentence extraction-based presentation summarization techniques and evaluation metrics. In proceedings of IEEE International Conference on Acoustics, Speech, and Signal Processing (ICASSP), pp. 1065–1068

Lin CY (2004) ROUGE: a Package for Automatic Evaluation of Summaries. In proceedings of the Workshop on Text Summarization Branches Out (ACL), pp: 74–81

Milhacea R, Tarau P (2004) TextRank: Bringing Order to Texts. In proceedings of the Conference on Empirical Methods in Natural Language Processing, pp. 404–411

Newman MEJ (2006) The mathematics of networks, http://www-personal.umich.edu/~mejn/papers/palgrave.pdf Accessed 6 July 2021

Page L (1997) PageRank: Bringing Order to the Web, Stanford Digital Library Project, http://infolab.stanford.edu/~page/papers/pagerank/index.htm Accessed 6 July 2021

Papineni K, Roukos S, Ward T, Zhu WJ (2002) BLEU: a method for automatic evaluation of machine translation. In proceedings of the 40th Annual Meeting of the Association for Computational Linguistics (ACL), pp: 311–318

Paulus R, Xiong C, Socher (2018) A deep reinforced model for abstractive summarization, https://arxiv.org/abs/1705.04304 Accessed 6 July 2021

Rose S, Engel D, Cramer N, Cowley W (2010) Automated keyword extraction from individual documents. In Berry MW, Kogan J (Eds) Text Mining: Applications and Theory, pp. 1–20 John Wiley & Sons, Ltd

See A, Liu PJ, Manning CD (2017) Get to the point: Summarization with pointergenerator networks. In proceedings of the 55th Annual Meeting of the Association for Computational Linguistics (ACL), pp: 1073–1083

Syed AA, Gaol FL, Matsuo T (2021) A Survey of the State-of-the-Art Models in Neural Abstractive Text Summarization, IEEE Access, 9(2021): 13248–13265

Zhang H, Cai J, Xu J, Wang J (2019) Pretraining-Based Natural Language Generation for Text Summarization. In proceedings of the 23rd Conference on Computational Natural Language Learning (ACL), pp: 789–797

Zhou Q, Yang N, Wei F, Huang S, Zhou M, Zhao T (2018) Neural document summarization by jointly learning to score and select sentences. In proceedings of the 56th Annual Meeting of the Association for Computational Linguistics (ACL), pp: 654–663

15 Question Answering and Dialogue

In this final chapter, we focus our attention on two problems related to conversational systems: question answering and dialogue. In recent years, conversational systems have started to play a very important role in human-computer interaction. Like in the case of information retrieval, in question answering the objective is to address user specific information needs by searching for relevant information in a given data collection. However, different from information retrieval, in question answering the user query and the system response are typically formulated in a more concise and human-like manner.

In the case of dialogue systems, although used to address informational needs too, they typically focus on addressing tasks of a more transactional nature, such as making reservations, completing purchase orders, supporting learning and training activities, etc. Different from question answering, in which user-computer interactions tend to be atomic (i.e. just one or a very small number of turns), in the case of dialogue, the system is required to handle multiple rounds of communication with the user. To properly resolve and make sense out of the current interaction, dialogue systems need to be aware of the overall context of the conversation by tracking the results of the previous interactions.

This chapter is organized as follows. First, in section 15.1, we introduce and discuss in more detail the problem of question answering. We illustrate some basic approaches to question answering with practical examples in the restaurant domain. Then, in section 15.2, we focus our attention on the problem of dialogue. More specifically, we introduce the concepts of state tracking and dialogue management. Similarly, we illustrate some basic approaches with practical examples in the restaurant domain.

Like in some previous chapters, in this chapter we will be working with contents that have been collected from the web. All models and contents made available through the companion website of the book are intended to be used for educational and research purposes only.

15.1 Question Answering

In this section we focus our attention on the problem of question answering, which deals with the extraction and generation of answers to questions formulated in the form of natural language. Like information retrieval, question answering involves a search step in which text segments or other type of information elements are

© The Author(s), under exclusive license to Springer Nature Switzerland AG 2021
R. E. Banchs, *Text Mining with MATLAB®*, https://doi.org/10.1007/978-3-030-87695-1_15

ranked according to their relevance to the question. Like text summarization, question answering needs to process sources of information to produce a precise and concise output, by either extractive or abstractive methods, satisfying the information need conveyed in the question being asked. Different from information retrieval and text summarization, which mainly deal with text as their main source of information, question answering can also leverage other structured information sources such as databases, knowledge graphs, etc.

Question answering constitutes a very active research field due to its growing demand and importance in commercial applications. Many different approaches to question answering have been proposed and studied during the last decade or two. In this section, we will cover just a little part of what is available, focusing our attention on some fundamental subproblems related to question understanding and answer generation. We also focus on use cases in which the sources of information needed to generate the answers are available either in structured form or in curated collections of question-answer pairs.

As we mainly focus on fundamental concepts of question answering here, some of the most recent work in this area is out of the scope of this book and, consequently, not covered. For a more comprehensive overview of the field of question answering, you should refer to the references provided in the *Further Reading* section at the end of the chapter.

Let us start by describing the generic architecture of the type of question answering systems discussed here. Figure 15.1 illustrates such a generic system. As seen from the figure, the architecture is composed of four different subsystems (question understanding, search, ranking and answer generation) and an information source or repository, which can be of different nature (document collection, index, database, knowledge base, etc.) depending on the specific system.

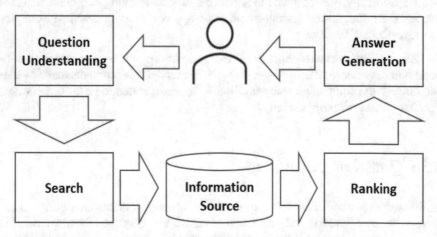

Fig. 15.1. Generic architecture of an automatic question answering system

The question understanding subsystem is responsible for converting the natural language question entered by the user into a machine understandable representation that can be used to trigger the search for a valid answer. Two important pieces of information must be determined at this stage: what to search for? and, how to search for it? The first issue is related to the nature or intention of the input question, and it is typically addressed by a classification component known as intent detection classifier or question classifier. The second issue is related to the identification of the specific parameters (typically a keyword or entity detection subcomponent) that are needed to conduct the search, and which are highly dependent on the type of search component and information repository used by the system.

The search and ranking components are responsible for retrieving and selecting the pieces of information that are needed to answer the input question. Depending on the nature of the information repository available to the system, these components can be of different types. For instance, if the information repository is a database, the search component should build queries according to the corresponding query language needed and the ranking or selection component should be able to apply the required filters to organize the search results accordingly. If the information repository is an index of text passages, the search component should construct keyword-based queries to search within the index and the ranking component should rank the retrieved passages accordingly. If the information source is a document, the search component should be able to find relevant segments within it and the ranking component must select or rank the most appropriate segments and parse the required information from them. Similarly, any other type of information repository will require specific search and ranking components.

Finally, the answer generation module is responsible for assembling the final answer to be provided to the user as system output. This component can be as simple as passing the answer through, when the information pieces retrieved from the information repository are complete answers already, or as complex as a full natural language generation system that generates complete answers based on atomic pieces of information provided to it. Other intermediate solutions that are commonly used involve the use of templatized answers that are filled in with the pieces of information derived from the information repository.

More advanced question answering systems include additional components, which perform other peripheral tasks that augment the capabilities of the system. For instance, in some cases, some clarification might be needed to resolve possible ambiguities of the original question intent. In other cases, the system might need to memorize some of the previous question context to be able to properly answer follow up questions. Similarly, in more complex architectures, an additional preprocessing module might be used to identify valid questions and redirect other types of user inputs to other modules or systems.

Other important distinctions among automatic question answering systems are related to the nature of the user's information need and the types of questions they are designed to support. With regards to the user's information need, we can distinguish between open domain and close domain systems. In an open domain system, it is expected that the user information need is general in nature and topic. This type of systems typically relies in resources like Wikipedia to generate the answers. On the other hand, close domain systems assume the user information need is specific and related to a particular topic. This type of systems typically uses domain specific ontologies or curated databases to search for the answers.

With regards to the types of questions, we can distinguish between factoid questions, confirmation questions, causal questions, and procedural questions, among others. Factoid questions are fact-based questions that commonly involve named entities as answers. This type of questions is typically formulated with question particles such as *what*, *which*, *where*, *when* and *who*. Confirmation questions, on the other hand, are questions that require either *yes* or *no* as an answer. Causal questions and procedural questions aim at explanations and descriptions as answers, respectively. Causal questions are typically formulated as *why* questions, and procedural questions are typically formulated as *how* questions.

In the experimental part of this section, we concentrate our attention on factoid questions over a close domain. More specifically, we use a manually curated dataset of restaurants in the surroundings of the San Francisco Bay Area,[1] which contains information about *104* restaurants including: restaurant name, type of cuisine, price range, location (city), address, telephone number, website and a text-based description.

The dataset, which is available in the file `datasets_ch15.mat`, is provided in the form of a structure array containing the information mentioned above. Let us load and look at the dataset:

```
>> clear; load datasets_ch15; whos                                          (15.1a)
   Name           Size              Bytes  Class      Attributes
   data           104x1            357424  struct

>> fieldnames(data)'                                                        (15.1b)
ans =
  1×8 cell array
  Columns 1 through 5
    {'restaurant'}    {'cuisine'}    {'price'}    {'location'}    {'address'}
  Columns 6 through 8
    {'telephone'}    {'website'}    {'description'}
```

As seen from (15.1), the dataset contains *104* records and *8* fields. All fields, except for *cuisine*, are single valued. In the case of *cuisine*, depending on the restaurant, multiple values might occur. Let us take a look at the first record:

---

[1] The data was directly collected from the restaurant websites between April and May 2021. The corresponding URLs are provided as part of the dataset.

```
>> data(1)                                                             (15.2)
ans =
  struct with fields:
     restaurant: "Evvia_Estiatorio"
        cuisine: ["Greek"      ""      ""      ""]
          price: "expensive"
       location: "Palo_Alto"
        address: "420 Emerson St, Palo Alto, CA 94301"
      telephone: "(650) 326-0983"
        website: "https://evvia.net/"
    description: "Evvia Estiatorio brings the warmth and charm of Greece ..."²
```

As seen from (15.2), all values are provided in the form of string scalars, except for *cuisine*, which is a string array of up to four elements. Next, let us convert the data into a table and extract unique values for the first four fields:

```
>> datatbl = struct2table(data);                                       (15.3a)

>> restaurants = unique(datatbl.restaurant);                           (15.3b)
>> cuisines = setdiff(unique(datatbl.cuisine),"");
>> prices = unique(datatbl.price);
>> locations = unique(datatbl.location);

>> whos restaurants cuisines prices locations                          (15.3c)
  Name           Size          Bytes  Class     Attributes
  cuisines       64x1           4032  string
  locations      7x1             570  string
  prices         3x1             290  string
  restaurants    104x1          7984  string
```

As seen from (15.3c), the restaurant dataset includes information about *104* restaurants, *64* types of cuisines, *7* locations and *3* price ranges. The four lists of unique values created in (15.3b) will play a fundamental role in the question processing strategy presented next. The fields they represent are typically referred to as informable fields, as users can provide values for them when formulating their questions. Those values are used by the question answering system as constrains during the search step. For instance, users can ask questions about restaurants in a specific *location* or offering certain type of *cuisine*, but they are not expected to ask questions about restaurants with a given *website* or *phone* number. On the other hand, requestable fields are those fields for which users can request their value. Accordingly, all eight fields in the dataset are requestable fields.

Next, let us consider the sample question: *what kind of food is served at Sandrino Pizza?* As discussed before, the two main tasks of the question understanding module in Figure 15.1 are: first, to detect the intent of the question and, sec-

---

² The actual description field has been truncated for display purposes.

ond, to identify the parameters needed for the search. Accordingly, the intent of the sample question under consideration can be defined as *ask_for_cuisine*, and the only search parameter required to find the answer is the name of the restaurant *Sandrino Pizza*. Notice that with these two pieces of information, the answer can be derived from the table in (15.3a) as restaurant name and intent uniquely define the row and column of the table where the answer is to be found.

Let us start by illustrating the intent detection step. For this, we use the function `get_intent`, which is available in the companion site of the book along with the other functions and datasets used in this chapter. The function uses a pretrained classifier to assign an intent, out of seven different intent categories, to the input question. The training set for the intent detection classifier includes both factoid and confirmation questions. Table 15.1 presents some examples of these questions along with their corresponding intents (for more details about the classifier and the implementation of the intent detection function, see exercise 15.4-1).

Table 15.1. Sample questions from the intent detection classifier training set

Question	Intent	Type
What is the address of The Lighthouse?	ask_for_address	factoid
What style of cuisine is offered at Rich Table?	ask_for_cuisine	factoid
Is Gypsy's Trattoria a Northern Italian food restaurant?	ask_for_cuisine	confirm
Where is Peddler Restaurant located?	ask_for_location	factoid
Is Dosa's Restaurant in San Mateo?	ask_for_location	confirm
What prices can be expected at Sartaj Cafe?	ask_for_price	factoid
Is Copita Tequileria unexpensive?	ask_for_price	confirm
Is there any Mexican restaurant in San Francisco?	ask_for_restaurant	factoid
What is the contact number for Ippuku?	ask_for_telephone	factoid
What is the website of Sandrino Pizza & Vino?	ask_for_website	factoid

As seen from Table 15.1 and (15.1b), the last token in each intent category label matches one of the field names in the dataset. This was purposely done in such manner by design, as the intent of the question uniquely determines the field in the dataset from which the answer must be retrieved. Notice also there is no intent category associated to the *description* field.

Let us illustrate the intent detection step for the sample question under consideration. For this, we use the function `get_intent`:

```
>> question = "What kind of food is served at Sandrino Pizza?";          (15.4)
>> intent = get_intent(question)
intent =
    "ask_for_cuisine"
```

Notice the intent completely defines the search strategy to be used by the system. It determines the field from which the answer must be fetched, *cuisine* in this case, as well as the entities needed to conduct the search, which in this case is the name of the restaurant.

In the second step of the question understanding submodule, which aims at extracting the entities or parameters needed for the search, we need to determine the name of the restaurant the user wants to know about. For this, we use the function `property_match`, which is also available in the companion site of the book. This function implements a string-matching algorithm that scores potential matches of properties/entities in a list against the input question (for more details about the implementation of the provided matching function, see exercise 15.4-3).

To complete the question understanding step for the sample question under consideration, we attempt to identify candidate restaurant names for which the user is asking about by using the list of unique restaurant names from (15.3b) and the function `property_match`. We proceed as follows:

```
>> matches = property_match(restaurants,question)                        (15.5)
matches =

  5×3 table
```

match	value	score
"Sandrino Pizza"	"Sandrino_Pizza_&_Vino"	0.94281
"food"	"Brenda's_French_Soul_Food"	0.25
"food"	"Imm_Thai_Street_Food"	0.25
"Pizza"	"Patxi's_Pizza"	0.2357
"Pizza"	"The_Cheeseboard_Pizza"	0.2357

As seen from (15.5), the matching function returns a table containing the string portions of the question that represent potential matches, their corresponding nominal match, and the assigned score. In this case, five candidate restaurants were matched and scored according to the quality of the match, and the top-ranked match happens to be the actual restaurant name mentioned in the question.

The steps illustrated in (15.4) and (15.5) implement the question understanding module from Figure 15.1 so far. At this point, we are ready to search for the information needed to compose the answer to the question. For the example under consideration, the search is relatively simple and can be implemented by means of simple table operations as follows:

```
>> restaurantname = matches.value(1);                                    (15.6)
>> idx = datatbl.restaurant==restaurantname;
>> result = datatbl.cuisine(idx,:);
>> result = result(strlength(result)>0)
result =

  1×2 string array

    "Pizza"     "Northern_Italian"
```

The search in (15.6) uses the top-ranked restaurant from (15.5) to get the row index of interest. Then, the field corresponding to the intent detected in (15.4) is used to retrieve the required information. Notice that, in this case, no ranking is needed as the retrieved information is already the answer to the question.[3]

Next, the final answer is produced, which is the function of the answer generation module in Figure 15.1. We illustrate this component, for the example under consideration, by using a template-based generation approach. In this type of approach, a predefined answer template is filled in with the information retrieved by the search and ranking modules.

```
>> template = "%s specializes in %s food.";                                  (15.7)
>> if length(result)>1
        result = join(result(1:end-1),', ')+" and "+result(end);
    end
>> answer = compose(template,replace([restaurantname,result],'_',' '))
answer =
    "Sandrino Pizza & Vino specializes in Pizza and Northern Italian food."
```

The answer generation mechanism implemented in (15.7) starts by defining an answer template of the form *"___ specializes in ___ food."*, which is to be filled in with the restaurant name and cuisine type information from (15.6). Notice the template is strongly coupled to the type of question being asked and the answer result. Typically, different templates should be available and used according to the specific type of question asked, its intent and the obtained search result. For instance, if the search in (15.6) results in an empty string because the cuisine information is not available for that particular restaurant name, a different answer template would be certainly needed.

Then, continuing with the procedure implemented in (15.7), if the search result happens to be a string array of more than one element, the different elements are concatenated into a single string scalar. Finally, the answer is generated by filling in the template with the restaurant name and the search result. Also, in this last step, all underscores occurring in the dataset entities are removed.

The overall procedure implemented in (15.4), (15.5), (15.6) and (15.7) illustrates the question answering architecture in Figure 15.1 applied to the sample question under consideration: *what kind of food is served at Sandrino Pizza?*

The same steps can be followed to generate answers to similar questions. Table 15.2 presents answers generated for a small set of different sample questions (the generation of the table is left to you as an exercise).

---

[3] Actually, a ranking was already conducted during the entity matching step in (15.5). In more complex scenarios, or different question answering implementations, where multiple candidate answers are retrieved during the search step, a post-search ranking or selection mechanism is typically needed.

Table 15.2. Sample questions and generated answers by following the question answering implementation steps in (15.4), (15.5), (15.6) and (15.7)

Question	Answer
What type of food is served at Barrel House?	Barrel House Tavern specializes in Californian and New American food.
What is Agave's cuisine?	Agave Uptown specializes in Oxacan and Mexican food.
Is Galeto Steakhouse a Japanese restaurant?	Galeto Brazilian Steakhouse specializes in Steaks and Brazilian food.
Which type of cuisine is there at Sartaj Cafe?	Sartaj India Cafe specializes in Vegetarian options and Indian food.
What cuisine does Zazie restaurant specializes in?	Zazie specializes in French food.

Next, we illustrate how the procedure described before, with just a few little changes, can be used to generate answers to questions from different intent categories. Indeed, questions belonging to six out of the seven available intent categories can be answered by following the same procedure described above. As we will discuss later, the *ask_for_restaurant* intent requires a different search strategy and, accordingly, a different implementation of the question understanding and answer generation components.

First, let us define one template for each question intent to be supported. We do this by means of a table containing two fields: the intent type and the template.

```
>> templatetbl = table();                                              (15.8)
>> templatetbl.type = ["cuisine"; "location"; "price"; "website"; ...
                       "address"; "telephone"];
>> templatetbl.template = ["%s specializes in %s food."; ...
                           "%s is located in %s."; ...
                           "%s price range is %s."; ...
                           "The website of %s is %s"; ...
                           "The address of %s is %s."; ...
                           "%s can be reached at %s"];
```

Now, we are ready to implement a more general version of the question answering system presented above. We illustrate the procedure with the new sample question: *where is Sandrino's Pizza located?*

```
>> question = "Where is Sandrino's Pizza located?";                    (15.9a)
>> intent = get_intent(question);
>> matches = property_match(restaurants,question);

>> restaurantname = matches.value(1);                                  (15.9b)
>> idx = datatbl.restaurant==restaurantname;
>> searchfield = split(intent,'_');
>> searchfield = searchfield(end);
```

```
>> result = datatbl.(searchfield)(idx,:);
>> result = result(strlength(result)>0);

>> templateidx = templatetbl.type==searchfield;                          (15.9c)
>> template = templatetbl.template(templateidx);
>> if length(result)>1
        result = join(result(1:end-1),', ')+" and "+result(end);
   end
>> answer = compose(template,replace([restaurantname,result],'_',' '))
answer =

    "Sandrino Pizza & Vino is located in Sausalito."
```

As seen from (15.9), two small changes have been introduced into the search (15.9b) and answer generation (15.9c) submodules. In (15.9b), the `searchfield` variable was introduced to capture the name of the table field associated to the detected intent. In this manner the search field, rather than fixed as in (15.6), is now controlled by the question intent.

In (15.9c), rather than using a fixed template like in (15.7), the template to be used is selected from the set of templates created in (15.8) in accordance with the detected question intent. In this case, the `searchfield` variable is used again to find the index of the corresponding template in the table.

For the rest of it, the procedure followed in (15.9) reproduces the one previously implemented in (15.4), (15.5), (15.6) and (15.7). The steps in (15.9) can be used to generate answers to questions in any of the six intent categories for which templates were defined in (15.8). Table 15.3 presents answers generated for a small set of different sample questions (again, the generation of the table is left to you as an exercise).

Table 15.3. Sample questions and generated answers by following the question answering implementation steps in (15.9)

Question	Answer
What city is Marufuku located at?	Marufuku Ramen is located in Redwood City.
How are the prices at Scott Grill?	Scott's Seafood Grill & Bar price range is expensive.
Is there a contact number for Nopa?	Nopa can be reached at (415) 864-8643
What is the address of waterbar?	The address of Waterbar Restaurant is 399 The Embarcadero, San Francisco, CA 94105.
Is there a website for catch seafood?	The website of Catch is http://catchsf.com/

As mentioned before, the *ask_for_restaurant* intent, which is not covered by the question answering strategy implemented in (15.9), requires special attention. Notice this type of question typically requires more than one constraint to be applied during the search step. Consider for instance the sample question: *is there any japanese restaurant in Oakland?* There are two fundamental differences on how the answer to this question is computed as compared to the previous types of questions. First, more than one constraint can be imposed to the search. For instance, in the sample question under consideration, two constraints are imposed to the restaurant search by means of the informable fields: *cuisine* and *location*. Second, the answer might involve either multiple records (e.g. multiple Japanese restaurants are retrieved), a single record (e.g. only one restaurant is retrieved) or no records at all (e.g. no Japanese restaurants are found in Oakland).

Next, let us illustrate how the new sample question can be processed. First, we search for all table indexes matching the informable field *location*:

```
>> question = "Is there any japanese restaurant in Oakland?";        (15.10a)
>> locationvalue = property_match(locations,question)
locationvalue =
  1×3 table
    match        value       score
   "Oakland"    "Oakland"      1
>> locationidx = datatbl.location==locationvalue.value(1);           (15.10b)
>> disp(sum(locationidx))
   15
```

In (15.10a), we used the `property_match` function to find all candidate location matches within the question. Then, in (15.10b), we retrieved the indexes of all restaurant records containing the top-ranked match (*Oakland*) as location. A total of *15* records in the table have their *location* field value set to *Oakland*.

Second, we conduct a similar index search for the informable field *cuisine*. However, the implementation in this case is slightly different as the *cuisine* field is a multivalued one. We proceed as follows:

```
>> cuisinetype = property_match(cuisines,question)                   (15.11a)
cuisinetype =
  1×3 table
    match         value       score
   "japanese"   "Japanese"      1
>> cuisineidx = sum(datatbl.cuisine==cuisinetype.value(1),2);        (15.11b)
>> disp(sum(cuisineidx))
   11
```

In (15.11a), we used again the `property_match` function to find all candidate cuisine matches within the question. Then, in (15.11b), we retrieved the indexes of all records containing the top-ranked match (*Japanese*) as cuisine. Notice that, in this case, as the string comparison returns a logical-valued matrix instead of a vector, we need to add up all the matrix columns to generate the index vector. A total of *11* entries were found in the table.

As third and final step, we search for the information needed to answer the input question by using both constraints derived in (15.10) and (15.11). As determined by the question intent, the information needed to answer the question must be retrieved from the *restaurant* field of the table. Accordingly, we retrieve the values of the *restaurant* field for all records in the intersection (logical *AND* operation) between the indexes derived in (15.10) and (15.11):

```
>> intent = get_intent(question)                                    (15.12a)
intent =

    "ask_for_restaurant"

>> datatbl.restaurant(locationidx&cuisineidx)'                      (15.12b)
ans =

  1×2 string array

    "Delage"      "Cafe_Umami"
```

Notice that, in general, we do not have any a priori knowledge about what informable fields are involved in the question. Accordingly, the question answering mechanism should probe for all informable fields and use those that produce matches to constraint the search. For instance, when the *ask_for_restaurant* intent is detected, we can test for all informable fields that can possibly occur in the question: *cuisine*, *location* and *price*.

For the example under consideration, attempting to match price values results in an empty table:

```
>> pricerange = property_match(prices,question)                     (15.13)
pricerange =

  0×0 empty table
```

A more convenient implementation for handling the search constraints is provided with the function `find_restaurant`, which is also available in the companion website. The main advantage of this function is that it allows for testing informable fields that are not present in the input question, which simplifies the logic of the overall system implementation.

The function `find_restaurant` receives as inputs the dataset and a structure array with the informable field names and their corresponding matches (regardless of whether they are empty tables or not). It returns the composed index that must be used to retrieve the answers (for more details about the `find_restaurant` function, see exercise 15.4-4).

```
>> constraints(1).name = "cuisine";                                 (15.14a)
>> constraints(1).values = table2struct(cuisinetype);
>> constraints(2).name = "location";
>> constraints(2).values = table2struct(locationvalue);
>> constraints(3).name = "price";
>> constraints(3).values = table2struct(pricerange);
```

```
>> searchindex = find_restaurant(datatbl,constraints);              (15.14b)
>> datatbl.restaurant(searchindex)'
ans =
  1×2 string array
    "Delage"    "Cafe_Umami"
```

Additionally, `find_restaurant` also can be used with the other six intent categories, where the informable field is the *restaurant* field. Consider the following example:

```
>> question = "What is the location of La Ciccia Restaurant?";      (15.15a)
>> intent = get_intent(question);
>> matches = property_match(restaurants,question);

>> newconstraint(1).name = "restaurant";                           (15.15b)
>> newconstraint(1).values = table2struct(matches);
>> searchindex = find_restaurant(datatbl,newconstraint);
>> responsefield = split(intent,'_');
>> datatbl.(responsefield(end))(searchindex,:)
ans =
    "San_Francisco"
```

As a final exercise in this section, we illustrate the complete implementation of a question answering system for the case of questions in the *ask_for_restaurant* intent category. We illustrate the overall procedure with the sample question: *is there any spanish restaurant with a moderate price range?* Let us start with the question understanding subsystem:

```
>> question = "Is there any spanish restaurant with a " + ...      (15.16a)
            "moderate price range?";

>> intent = get_intent(question); % detects the intent            (15.16b)

>> % looks for all possible informable field mentions             (15.16c)
>> variables = ["cuisine","price","location"];
>> valuelists = {cuisines,prices,locations};
>> constraints = struct();
>> for k=1:length(variables)
       matches = property_match(valuelists{k},question);
       constraints(k).name = variables(k);
       constraints(k).values = table2struct(matches);
   end
```

We continue with the implementation of the search and ranking module. Here, we combine strategies from (15.9) and (15.14). Additionally, we create the new variable `nresponses` to keep track of the number of results retrieved. This variable will be used to inform the answer generation subsystem later, which will select a different answer template depending on the number of restaurants retrieved.

```
>> searchindex = find_restaurant(datatbl,constraints);                    (15.17)
>> responsefield = split(intent,'_');
>> responsefield = responsefield(end);
>> result = datatbl.(responsefield)(searchindex,:);
>> result = result(strlength(result)>0);
>> if length(result)>1
       result = join(result(1:end-1),', ')+" and "+result(end);
   end
>> nresponses = sum(searchindex);
```

Finally, we implement the answer generation subsystem. We want this component to generate four different types of answers depending on the total number of results retrieved. If the number of results is greater than four, the systems must just report about the number of restaurants it found. On the other hand, if two, three or four restaurants are found, the system must reply by reporting the number of restaurants it found and listing the names of the restaurants. If only one restaurant is found, the system must provide the name of the restaurant. In the last case, if no restaurant is found, the system must reply accordingly. In all answer types, we also want the system to restate the informable fields mentioned in the question.

This answer generation module is quite more complex than the one in (15.9), which used the set of intent-specific templates defined in (15.8). In this implementation, we break down the answer generation module into three components: template assembly, template filling in, and post-edition.

In the template assembly step, a template is generated based on the required type of answer:

```
>> commontemplate = "_price_ _cuisine_ restaurants in _location_";      (15.18)
>> if nresponses < 1
       template = "I didn't find any "+commontemplate+".";
   elseif nresponses < 2
       template = "_results_ is a "+commontemplate+".";
   elseif nresponses < 5
       template = "I found _number_ "+commontemplate+": _results_.";
   else
       template = "I found _number_ "+commontemplate+".";
   end
```

As seen from (15.18), templates are generated by combining a common template with answer specific template segments. The common template accounts for restating the informable field values mentioned in the question. The answer specific template segments, on the other hand, help to formulate the answer according to the specific answer type requirements by including the additional variables required in each case. Also notice that different from the templates in (15.8) the implementation in (15.18) uses variable-specific insertion patterns.

In the template filling in step, the insertion patterns in the template to be used are replaced with the corresponding information, if available. Two different types of replacements are needed: the search results, which might involve the number of retrieved results, the results, or both; and the informable field values, which complete the common template part of the template.

```
>> % replaces insertion patterns corresponding to results          (15.19a)
>> template = replace(template,'_number_',string(nresponses));
>> if not(isempty(result))
       template = replace(template,'_results_',result);
   end

>> % replaces insertion patterns corresponding to informable fields   (15.19b)
>> for k=1:length(variables)
       pattern = "_"+variables(k)+"_";
       if isempty(constraints(k).values), value = "";
       else, value = constraints(k).values(1).value; end
       template = replace(template,pattern,value);
   end

>> answer = template                                               (15.19c)
answer =
    "I found 2 moderate_price Spanish restaurants in : La_Marcha_Tapas_Bar and
Coqueta."
```

Notice from (15.19c) that after filling in the template, apart from the obvious removal of the underscores, some additional editions are needed.

Indeed, as seen from (15.16a), there was no mention to any location value in the input question under consideration. Consequently, after filling in the template, a meaningless *in* was left hanging in the middle of the answer, which must be corrected in a post-edition step. Other important post-editions include removals of extra spaces, replacing plural forms with singular forms, replacing *a* with *an*, etc.

Next, we implement some of these strategies in the final answer post-edition step and display the final answer being produced:

```
>> answer = regexprep(answer,'_',' ');                             (15.20a)
>> answer = regexprep(answer,' +',' ');
>> if nresponses==1, answer = replace(answer,'restaurants','restaurant'); end
>> answer = replace(answer," in .",".");
>> answer = replace(answer," in :",":");
>> answer = replace(answer," a expensive "," an expensive ");
>> answer = replace(answer," moderate price "," moderate-price ");

>> disp(answer)                                                    (15.20b)
I found 2 moderate-price Spanish restaurants: La Marcha Tapas Bar and Coqueta.
```

Sample questions triggering the different answer types from (15.8) are presented in Table 15.4 along with their corresponding answers. All answers were generated by following the steps in (15.16), (15.17), (15.18), (15.19) and (15.20) (the generation of the table is left to you as an exercise).

Table 15.4. Samples of generated answers and their corresponding questions in the *ask_for_restaurant* intent category

Question	Answer
Are there any restaurants offering new american cuisine?	I found 8 New American restaurants.
Are there any pizza places in Sausalito?	I found 2 Pizza restaurants in Sausalito: Sandrino Pizza & Vino and Bar Bocce.
Is there any restaurant in Palo Alto serving steaks?	Sundance The Steakhouse is a Steaks restaurant in Palo Alto.
Is there any greek restaurant in Oakland?	I didn't find any Greek restaurants in Oakland.

15.2 Dialogue Systems

Different from question answering, where user-computer interactions tend to be atomic in nature (i.e. a single turn: a question followed by an answer), in the case of dialogue, multiple turns of interaction between the user and the system are needed to accomplish the task at hand. In this scenario, the history of the previous communication becomes of critical importance for both, the user and the system, to make sense of the current state of the communication. In this section, we will focus our attention on the specific dialogue system component that makes possible the selection of a response while taking the previous history of the dialogue into consideration: the dialogue manager.

Figure 15.2 illustrates a high-level architecture of a generic dialogue system. As seen from the figure, this architecture is similar to the one of the generic question answering system previously presented in Figure 15.1. Notice the dialogue system in Figure 15.2 includes a natural language understanding (NLU) module and a natural language generation (NLG) module. These components are analogous to the question understanding and answer generation modules of the question answering system from Figure 15.1. Accordingly, the NLU module is responsible for converting the natural language input entered by the user into a machine understandable representation that can be consumed by the system, and the NLG module is responsible for assembling the final output to be provided to the user. On the other hand, and differently from a question answering system, in a dialogue system a dialogue manager component replaces the search and ranking. As seen from Figure 15.2, the dialogue manager is composed of two subcomponents: the dialogue state tracker and the dialogue response selection.

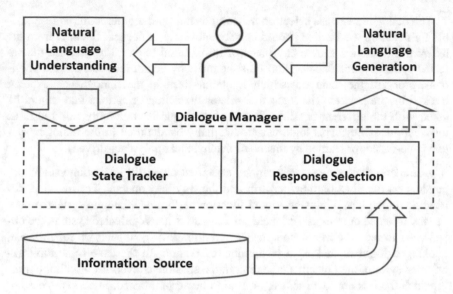

Fig. 15.2. Generic architecture of an automatic dialogue system

The function of the dialogue manager is to decide what response is to be given to the user at each dialogue turn. This decision is made based on two pieces of information: an external source of information (index, database, knowledge graph, etc.) and the previous history of the dialogue. Notice the external source of information, plays the same role it plays in a question answering system. It provides the knowledge needed by the system to fulfill the user information needs. On the other hand, the history of the dialogue is an internal source of information that allows the system to put the current interaction into context to make sense of it.

As implied by its name, the dialogue state tracker subcomponent is in charge of keeping track of the internal knowledge that is incrementally acquired as the interaction evolves turn after turn. Dialogue interactions can be seen as sequences of state transitions from an initial state at the beginning of the dialogue to a final state that determines the end of the dialogue. The role of the dialogue state tracker is to determine at what state the system is after each turn in the dialogue. Depending on the type of dialogue, different goals or objectives must be accomplished for the dialogue to be considered as successfully completed.

The state transitions in a dialogue are jointly determined by the actions taken by both the user and the system at each turn. In the context of dialogue interactions, such communicative actions are referred to as *dialogue acts*. The dialogue act of a given communicated message is composed of the different pieces of information it provides (the values) and the specific communicative function it plays (the speech act). Examples of speech acts include asking, answering, clarifying, promising, complaining, greeting, apologizing, requesting, etc.

The dialogue response selection subcomponent, on the other hand, is responsible for deciding what action should be taken next and, if needed, fetching from the external information source any piece of information required to perform the selected action. Notice this decision must be made by taking into account the previous history of the dialogue, which is summarized in the dialogue state being tracked by the tracker. The logic that relates the action or actions that should be taken with the different possible states in a dialogue is commonly referred to as the *dialogue policy*. This logic can be manually handcrafted in the dialogue system (rule-based) or learned by means of machine learning (data-driven).

Automatic dialogue systems can be classified along different dimensions depending on the function they perform and the way they operate. For instance, depending on the type of conversational function performed, dialogue systems might be task-oriented or non-task-oriented. In the case of a task-oriented system, the objective is to help the user accomplish a certain task or goal such as, for instance, booking a flight or making a restaurant reservation. In the case of a non-task-oriented system, the objective is to maintain a sensible conversation with the user, in which there is not specific task or goal to be accomplished. Such is the case of chit-chat dialogue systems, which are often referred to as chatbots.

Additionally, task-oriented dialogue systems can be transactional, informational or hybrid. In the case of a transactional dialogue system, a transaction is to be completed, such as a bank transference or a restaurant reservation. In the case of an informational dialogue system, specific information is to be provided to the user in a similar way it is done in a question answering system. Examples of informational dialogue systems include flight departure/arrivals information, bus routes and schedules, and navigational systems. Hybrid task-oriented dialogue systems combine both transactional and informational modalities.

Regarding the way they operate, dialogue systems can be also classified in different ways. For instance, depending on who leads the communication, dialogue systems can be system-initiative, user-initiative or mixed-initiative. Depending on how many users can simultaneously participate in the dialogue, dialogue systems might be single-party or multi-party. Depending on the input/output interaction mechanism, dialogue systems might be speech-based or text-based.

In the experimental part of this section, we concentrate our attention on a system-initiative informational dialogue system,[4] which uses the same manually curated dataset of restaurants from the previous section. The proposed system goal is to ask questions to determine the user preferences and retrieve the set of restaurants that satisfy such preference constraint. In the exercise, we illustrate the use of a data-driven approach to learn a dialogue policy for the system. More specifi-

---

[4] Other examples of dialogue systems are addressed in the *Proposed Exercises* section. More specifically, a user-initiative informational system is discussed in exercise 15.4-7, a system-initiative transactional system is discussed in exercise 15.4-8, and a non-task-oriented retrieval-based system is discussed in exercise 15.4-9.

cally, we use the Q-learning algorithm, which is a specific type of reinforcement learning algorithm. Let us start with a brief introduction to reinforcement learning. For a more comprehensive treatment of this topic, you must look at the references provided in the *Further Reading* section of this chapter.

Given the nature of dialogue, in which actions are taken by a system based on the states of its observed environment, reinforcement learning constitutes the most suitable machine learning framework for it. Different from unsupervised learning, in which ground truth information about data categories is not available, and supervised learning, in which ground truth is available for a subset of samples, in the reinforcement learning paradigm there is not an absolute notion of ground truth.

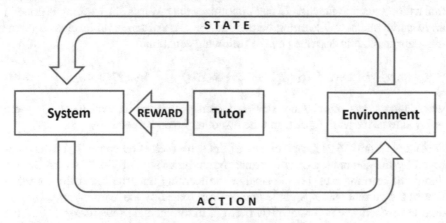

Fig. 15.3. Illustration of the reinforcement learning framework

In the type of problem addressed by reinforcement learning, which is illustrated in Figure 15.3, a system that observes its environment needs to take actions in response to the different states the environment goes through. During the learning process, instead of an error signal that compares the action taken with a ground truth action, the system receives either a reward or a penalization from an interpreter or tutor that is able to judge the actions of the system. Accordingly, in this paradigm, the system is required to "experiment" with the actions to discover what are the desirable actions to take at each stage. That discovery process is guided by the target of maximizing the reward.

Another important aspect of the reinforcement learning framework is that the individual actions taken at each step are not what matters the most but, instead, the overall sequence of actions taken to complete the task at hand. Accordingly, reinforcement learning aims at maximizing the expected cumulative reward, rather than the immediate rewards.

When the states of the environment are directly observable by the system, the problem scenario depicted in Figure 15.3 can be modeled by means of a Markov

Decision Process (MDP), which is a mathematical framework for decision making.[5] An MDP is defined by a tuple of four elements: a state space S, an action space A, a transition probability space $P(s_+|s,a)$, and an immediate reward function $R(s,a)$. The state and action spaces are defined by the set of possible states s the environment can go through and the set of actions a the system can take, respectively. The transition probability space is defined by the conditional probability of the environment transitioning into a new state s_+ given the previous state s and the action a taken by the system. The immediate reward function determines the reward given by the tutor based on the action a taken by the system in state s.

For a given finite MDP, the Q-learning algorithm allows for finding and optimal action selection function Q that maximizes the reward. The learning is done in an iterative manner. Q-learning belongs to the class of temporal-difference learning algorithms. It is described by the following equation:

$$Q_+(s,a) = (1 - \alpha)\, Q(s,a) \;+\; \alpha \left(R(s,a) + \beta \max_{a_+} Q(s_+, a_+) \right) \quad \text{(15.21)}$$

where parameters α and β are scaling factors within the interval $(0,1]$. They are referred to as the learning rate and the discount factor, respectively.

As seen from (15.21), each update of the action selection value $Q_+(s,a)$ is computed by the summation of two terms: the current value of $Q(s,a)$ scaled by one minus the learning rate (i.e. the smaller the learning rate, the larger the contribution of the current value), and the update term scaled by the learning rate. The update term itself is composed of two terms: the immediate reward $R(s,a)$ and the discounted maximum possible value of the future action selection $Q(s_+,a_+)$.

Next, we proceed to illustrate the use of Q-learning to find an optimal action selection function Q for a dialogue system in the restaurant domain. First, let us describe the intended system in more detail. As previously mentioned, the proposed system should ask questions to the user (who represents the environment in the diagram illustrated in Figure 15.3) with the objective of determining the user preferences and, accordingly, it retrieves the set of restaurants that satisfy such preference constraint. Given the characteristics of the restaurant information that is available to the system, there are four different actions the system can perform at each turn in the dialogue: ask for the preferred type of food (a_1), ask for the desired location (a_2), ask for the desired price-range (a_3), and report on the restaurant search result (a_4).

The different states the intended dialogue can eventually transition through are presented in Figure 15.4 along with the system actions governing those transitions. Notice that in the state diagram depicted in the figure, it is assumed that the user is

---

[5] When the states of the environment are not directly observable by the system due to the presence of noise or other sources of uncertainty, the problem is typically modeled by means of a Partially Observable Markov Decision Process (POMDP). For more information on the POMDP framework, see the corresponding references in the *Further Reading* section of this chapter.

well behaved in the sense that the user always provides the information requested by the system. Consequently, for a given current state, the next state is uniquely determined by the action taken by the system. This means that, in the example under consideration, the dialogue state transitions are deterministic, i.e. all transition probabilities $P(s_+|s,a)$ are either one or zero (for an example of a similar dialogue system with probabilistic state transitions see exercise 15.4-10).

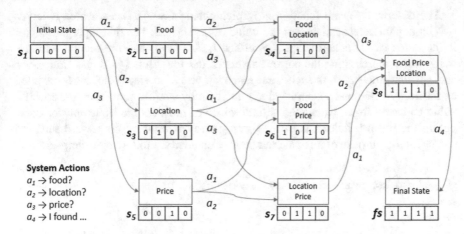

Fig. 15.4. State diagram with system actions and associated state transitions

Notice also from the figure, how the different dialogue states can be tracked based on the information collected by the system. Indeed, as the dialogue proceeds, the system and the user must jointly complete a form with four slots: *food*, *location*, *price*, and *acknowledgement*. The dialogue starts at an initial state s_1, in which all four slots are empty, and ends when the final state *fs* is reached, in which all four slots have been completed. At each turn in the dialogue the state is defined by the specific slots that have been completed.

Before proceeding to train the dialogue system illustrated in Figure 15.4 by using Q-learning, we need to define the reward function to be used by the tutor during the training process. The logic of the reward for this example is based on two simple rules: first, the system should not provide the result of the restaurant search unless all user information has been collected (i.e. the system cannot take action a_4 until the first three slots are completed); second, the system must not ask about a preference already collected (i.e. if the slot *food* has been already completed, the system should not ask for food preferences, and so on).

The reward function we implement here returns a negative reward, or penalty, whenever any of these two rules are violated. More specifically, if the first rule is violated, the returned reward is *-20*; and if the second rule is violated, the returned reward is *-10*. Notice the first rule is more critical than the second one as reporting a search result before collecting all user preferences might lead to wrong information being provided. On the other hand, asking for a preference that have been already informed, although not ideal for a good dialogue experience, does not lead to wrong information being provided to the user. In all other cases, the returned reward is zero. This means the overall reward value for a good dialogue execution will be equal to zero and, accordingly, any negative overall reward is an indication of a rule violation during the dialogue.

The described reward function is implemented in `get_reward`, which receives two input parameters and returns the value of the reward. The first input parameter is the index of the action taken (i.e. either *1, 2, 3* or *4*) and the second one is a string array representing the current status of the four slots (*food, location, prize* and *acknowledgement*). In such a `slots` string array, an empty slot is represented by an empty string and a completed slot is given by a string containing the specific value collected for it. Notice the actual values of the slots are irrelevant for computing the reward. Indeed, the `slots` string array is used by the reward function solely for the purpose of inferring the state. Consider the following examples:

```
>> % violates the first rule                                    (15.22a)
>> get_reward(4,["Chinese","","Expensive",""])
ans =
   -20
```

```
>> % violates the second rule                                   (15.22b)
>> get_reward(2,["","Palo_Alto","",""])
ans =
   -10
```

```
>> % does not violate any of the two rules                      (15.22c)
>> get_reward(3,["Mediterranean","","",""])
ans =
   0
```

Alongside `get_reward`, there are two additional functions used to support the reinforcement learning implementation shown in this section: `system_response` and `user_response`. The first one is responsible for producing the system response based on the selected action and the current status of the slots. The action determines the type of response to be produced by the system and the slots are used to parametrize the restaurant search. The result of the search is reported only if the selected action is a_4. Consider the following examples:

```
>> system_response(1,["","","Expensive",""])                    (15.23a)
ans =
   "What type of food would you like?"
```

```
>> system_response(4,["","San_Francisco","",""])                (15.23b)
ans =
   "I found 24 restaurants!"
```

Notice from (15.23) that the output of `system_response` is in the final form to be delivered to the user. This means the natural language generation (NLG) component of the system (as depicted in Figure 15.2) is built in into this function.

In the case of the `user_response` function, it implements a simulated user (i.e. the environment of the framework depicted in Figure 15.3). As explained before, in the reinforcement learning paradigm, the system is meant to learn by dynamically interacting with the environment while exploring the solution space to maximize the reward. In the scenario under consideration, this means the system needs to conduct multiple dialogue sessions with a user. When it is feasible, a simulated user is a convenient resource to automate and speed up the learning process. In the exercise illustrated here we implement a well-behaved user, who always replies to the system in the expected manner (for the case of a bad-behaved user, who might provide random responses some of the times, see exercise 15.4-10).

The function `user_response` receives two inputs and returns three outputs. The first input is the system provided response, like the ones from (15.23). The second input is a set of goals that determines the user preferences for each dialogue session. The main purpose of these second variable is to provide consistency along the dialogue with regards to user preferences. In the first call to `user_response`, an empty goal structure should be passed, and the function will return it populated with a set of preferences. In all subsequent calls during the same dialogue session, the populated structure should be passed back to `user_response` to ensure user preferences are kept consistent along the session.

The three outputs of `user_response` include: the value of the informed slot, the user response, and the goal structure. Notice that the returned value actually constitutes the output of the natural language understanding (NLU) component of the system, which means the NLU component is built in into this function. The user response, on the other hand, corresponds to the raw natural language response given by the user. Finally, as already explained, the goal structure provides the user preferences for the given dialogue session. This output is only meaningful in the first call to the function. Consider the following examples:

```
>> rng('default');                                                        (15.24a)
>> systemquery = "What city are you looking for?";
>> [value,response,goals] = user_response(systemquery,struct())
value =
    "San_Francisco"
response =
    "San Francisco"
goals =
  struct with fields:
     myfood: "Mexican"
     mycity: "San_Francisco"
    myprice: "cheap"
```

```
>> systemquery = "I found 1 restaurant!";                          (15.24b)
>> [value,response,goals] = user_response(systemquery,goals)
value =
    "Thanks!"
response =
    "Thanks!"
goals =
  struct with fields:
     myfood: "Mexican"
     mycity: "San_Francisco"
     myprice: "cheap"
```

At this point, we are finally ready to implement the reinforcement learning procedure and train the proposed dialogue system. In the implemented procedure, a total of *50* dialogues (episodes, in reinforcement learning terminology) are simulated. At each dialogue session, a maximum number of *20* turns is allowed. The overall procedure is implemented as follows:

```
>> % initializes the training session                              (15.25a)
>> maxepisodes = 50; maxturns = 20;
>> Qmtx = zeros(16,4); finalstate = 16;

>> % reinforcement learning training session                       (15.25b)
>> for episode = 1:maxepisodes
        % initializes a new dialogue session
        slots = ["","","",""]; goals = struct();
        cumreward = 0; dialogue = "";
        for k=1:maxturns % simulates a new dialogue session
            state = (strlength(slots)>0)*[1;2;4;8]+1;
            if state==finalstate, break; end
            % selects an action randomly from max Qmtx values
            possibleactions = find(Qmtx(state,:)==max(Qmtx(state,:)));
            idx = randperm(length(possibleactions));
            action = possibleactions(idx(1));
            % computes the system response and updates the dialogue log
            systemresponse = system_response(action,slots);
            dialogue = dialogue + "System: " + systemresponse + newline;
            % computes the immediate reward and updates the cumulative one
            reward = get_reward(action,slots);
            cumreward = cumreward + reward;
            % computes the simulated user response and updates the log
            [value,userresponse,goals] = user_response(systemresponse,goals);
            dialogue = dialogue + "  User: " + userresponse + newline;
            % fills in the corresponding slot and computes the new state
            slots(action) = value;
```

```
        newstate = (strlength(slots)>0)*[1;2;4;8]+1;
        % implements the Q-learning step
        alpha = 0.7; beta = 0.9;
        Qmtx(state,action) = (1-alpha)*Qmtx(state,action)+...
                          alpha*(reward+beta*max(Qmtx(newstate,:)));
    end
    % saves relevant info about the concluded dialogue session
    rewards(episode) = cumreward;
    turns(episode) = k-1;
    Qmtxs{episode} = Qmtx;
    dialogues{episode} = dialogue;
end
```

Let us explain in more details the different steps in (15.25). In the first step, in (15.25a), the overall training session is initialized. In addition to the overall number of episodes and the maximum number of turns allowed for each individual dialogue session, the action selection function Q is also initialized, and the final state is defined according to the used enumeration convention. Notice Q is represented by the *16x4* matrix Qmtx with all its values initialized to zero. The second dimension of Qmtx refers to the different actions the system can take, and its first dimension refers to the total number of states the system can potentially visit. Notice from Figure 15.4 that there are only *9* valid states for the proposed dialogue system. However, depending on how the different slots are filled in, the system can certainly visit a total of *16* different states.[6] Regarding the final state, the mapping convention used to enumerate the sates given the slots that have been completed is just a binary count shifted by one (see Figure 15.4). Accordingly, the final sate value is *16*.

Next in (15.25b) two nested loops are used to conduct the reinforcement learning training. The external loop, which iterates over dialogue sessions (episodes), is composed of three main steps: initialization, in which each new dialogue session is initialized; simulation, in which a dialogue session is simulated; and logging, in which relevant information about each concluded dialogue simulation is saved. In the initialization step, four different variables are initialized: the slots string array, the user's goals structure, the cumulative reward cumreward, and the dialogue session log, which will collect all system and user responses. In the logging step, four arrays are updated with information of the concluded dialogue: rewards, which stores the overall reward accumulated during the dialogue session; turns, which keeps track of the total number of turns taken to complete the dialogue; Qmtxs, which stores the state of the action selection matrix Qmtx at the end of the dialogue session; and dialogues, which stores the dialogue session log.

---

[6] Notice that although *16* states can be potentially visited, a well-trained dialogue system will only visit the *9* states represented in the diagram in Figure 15.4. This is basically because any of the other *7* states not represented in the figure imply a violation of the first rule (i.e. reporting the search result before collecting all user preferences), which is penalized by the reward function.

The simulation step corresponds to the inner loop in (15.25b), which iterates over the turns of the dialogue session being simulated. Each simulation iteration goes through seven different steps, which can be described as follows:

- State computation. In this step, the current state index is computed based on the completion status of slots. The mapping is implemented by means of a shifted binary count, in which the length of each string in slots is used as a binary indicator of completion. If the resulting state happens to be the final state, the dialogue simulation is concluded.
- Action selection. In this step, an action is selected according to the action selection matrix Qmtx and the current state. The implemented selection mechanism is as follows. For the current state state index, the maximum value of the Qmtx(state,:) vector is identified and all corresponding action indexes exhibiting such a maximum value are retrieved as possible actions. The action to be taken is then randomly selected from the set of possible actions.
- System response. In this step, the system response is computed by means of function system_response, which was previously introduced and illustrated in (15.23). Additionally, the dialogue session log is updated with the generated system response.
- Reward computation. In this step, the immediate reward for the given state and action taken by the system is computed, and the cumulative reward is updated accordingly. The immediate reward is computed by means of function get_reward, which was previously introduced and illustrated in (15.22).
- User response. In this step, the response of the simulated user is computed. This is done by means of function user_response, which was previously introduced and illustrated in (15.24). Additionally, the dialogue session log is also updated with the generated user response.
- Slot filling and state update. In this step, the slot corresponding to the action taken by the system is updated with the value returned along with the user response by user_response.[7] The new state index, resulting from the updated slots string array, is also computed in this step.
- Q-learning. In this step, the incremental update of the Q-learning algorithm, as described in (15.21), is implemented. Notice that a single entry of the action selection matrix Qmtx is updated at each iteration. Consequently, different entries are sequentially updated over the turns of a dialogue session as the dialogue visits different states and the system takes different actions. Several dialogue sessions are required for Qmtx to converge to a final action selection policy.

Let us now explore in more detail what happened during the training process conducted in (15.25). Let us start by looking at the overall number of turns and cumulative rewards for each of the 50 training episodes (simulated dialogues). For

---

[7] Be reminded this simplistic approach is possible because the simulated user is well-behaved, which means the user will always provide the information requested by the system.

this, we plot arrays `turns` and `rewards` versus the number of episodes. The resulting plots are depicted in Figure 15.5.

```
>> hf = figure(5);                                                    (15.26)
>> set(hf,'Color',[1 1 1],'Name','Reinforcement Learning Training Session');
>> subplot(2,1,1); plot(turns,'.-');
>> ylabel('Number of dialogue turns')
>> xlabel('Number of simulated dialogue sessions (episodes)')
>> subplot(2,1,2); plot(rewards,'.-');
>> ylabel('Overall accumulated reward')
>> xlabel('Number of simulated dialogue sessions (episodes)')
```

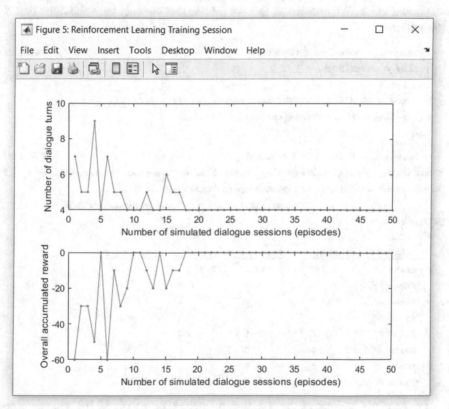

Fig. 15.5. Dialogue turns and cumulative rewards versus training episodes

As seen from the figure, during the first *17* simulated dialogue sessions, both
the number of turns per dialogue and the overall accumulated reward tend to
change abruptly from one session to another. On the other hand, at simulated session *18* and onwards, the total number of turns per dialogue gets stabilized at *4*
turns per session and the overall reward also gets stabilized at a value of zero. Ac-

cordingly, it can be concluded that the action selection matrix Qmtx has converged to its final value at episode *18*. Indeed, it can be confirmed that from episode *18* onwards, the values of Qmtx do not change anymore (see exercise 15.4-11). To better understand how the final action selection function looks like, let us represent the final Qmtx into the form of a stochastic matrix, which is commonly referred to as the dialogue policy.

To convert the action selection matrix Qmtx into a stochastic matrix we follow a procedure similar to the action selection step from (15.25b). First, we look at the maximum value of Qmtx(state,:) at each different state and, then, we uniformly distribute a unitary probability mass over the different actions exhibiting such a maximum value. We proceed as follows (notice that we only care about the first *8* rows in Qmtx):

```
>> finalQmtx = Qmtxs{maxepisodes};                                    (15.27a)
>> policy = zeros(8,4);
>> for k=1:8
        idx = find(finalQmtx(k,:)==max(finalQmtx(k,:)));
        policy(k,idx) = 1/length(idx);
    end

>> % writes the learned dialogue policy into a table                  (15.27b)
>> policytbl = array2table(policy,'VariableNames',compose('Action %d',1:4));
>> policytbl = addvars(policytbl,compose("State %d",1:8)',...
                    'Before','Action 1','NewVariableNames','State')

policytbl =

  8×5 table
```

State	Action 1	Action 2	Action 3	Action 4
"State 1"	0.33333	0.33333	0.33333	0
"State 2"	0	0.5	0.5	0
"State 3"	0.5	0	0.5	0
"State 4"	0	0	1	0
"State 5"	0.5	0.5	0	0
"State 6"	0	1	0	0
"State 7"	1	0	0	0
"State 8"	0	0	0	1

By carefully looking at the resulting dialogue policy in (15.27b), we can confirm that the system has actually learned the state transition diagram illustrated in Figure 15.4. Notice that, for instance, starting at state *1* the system can take any of the first three actions *1*, *2* or *3* with equal probability. At state *2*, the system is allowed to take either action *2* or action *3*; and so on. Finally, action *4* is only possible at state *8*, which is the state at which all user preferences have been collected.

To continue our exploration of the learning exercise in (15.25), let us look in more detail at some of the dialogues produced at different stages of the learning process. First, let us look at the dialogue session generated during the first training episode, which took seven turns to be completed and received an overall reward value of *-60* (see Figure 15.5).

```
>> disp(dialogues{1})                                              (15.28)
System: What type of food would you like?
  User: Any type
System: What price range do you have in mind?
  User: Moderate price
System: I found 52 restaurants!
  User: Thanks!
System: What type of food would you like?
  User: Any type
System: I found 52 restaurants!
  User: Thanks!
System: What price range do you have in mind?
  User: Moderate price
System: What city are you looking for?
  User: Palo Alto
```

As seen from (15.28), the first dialogue session is very chaotic in nature, as the dialogue manager is completely unaware of how the dialogue is to be conducted. Indeed, within this dialogue session the system reports twice the result of the search without having collected all user preferences. Similarly, it asks for both food and price preferences after it had collected such information already.

Next, let us consider the simulated dialogue at training episode *15*, which was completed in six turns and received an overall reward of *-20*.

```
>> disp(dialogues{15})                                             (15.29)
System: What city are you looking for?
  User: Berkeley
System: What price range do you have in mind?
  User: Cheap
System: What price range do you have in mind?
  User: Cheap
System: What city are you looking for?
  User: Berkeley
System: What type of food would you like?
  User: Thai
System: I found 1 restaurant!
  User: Thanks!
```

Notice from (15.29) that, at episode *15*, the system is still asking for preferences that have been previously collected. However, it seems to have learned already that the search result must be reported only after all user preferences have been collected. Indeed, as seen in Figure 15.5, this is the last training episode in which the overall reward received is *-20*. After this episode, the system only received overall rewards of *-10* and *0*, which suggests that it already learned to report the restaurant search result only at the very end.

Next, let us look at some of the simulated dialogue sessions that occurred after episode *18*, when convergence was achieved and, consequently, the dialogue policy remained unchanged. As seen from Figure 15.5, all such dialogues were completed in four turns and received an overall reward of zero.

```
>> disp(dialogues{20})                                          (15.30a)
System: What city are you looking for?
  User: Oakland
System: What price range do you have in mind?
  User: Moderate price
System: What type of food would you like?
  User: Italian
System: I found 1 restaurant!
  User: Thanks!

>> disp(dialogues{30})                                          (15.30b)
System: What type of food would you like?
  User: Any type
System: What city are you looking for?
  User: San Francisco
System: What price range do you have in mind?
  User: Cheap
System: I found 8 restaurants!
  User: Thanks!

>> disp(dialogues{45})                                          (15.30c)
System: What city are you looking for?
  User: San Francisco
System: What type of food would you like?
  User: Pizza
System: What price range do you have in mind?
  User: Cheap
System: I found no restaurants!
  User: Thanks!
```

Notice from the simulated dialogue samples in (15.30) that the system has been able to learn effectively the two basic rules governing this type of dialogue interaction: not asking for information already known and waiting until all user preferences have been collected to report the result of the search. Although these rules could have been manually handcrafted into a rule-based system for the specific problem under consideration, in more complex scenarios, as the number of possible states increases and tasks become more complex, handcrafting a rule-based system becomes more challenging and, eventually, infeasible. In such more complex scenarios, reinforcement learning provides an excellent framework to learn dialogue policies automatically, as far as a meaningful reward function is available to properly guide such a learning process.

Another important aspect of the learned dialogue policy is that the rules enforced by the implemented reward function are completely independent of the user behavior. Accordingly, the dialogue policy learned from (15.25), although trained with a well-behaved simulated user is somehow independent of the actual user behavior and can be used in more general contexts (see exercise 15.4-10).

Finally, a last observation with regards to Figure 15.5 must be made. It is about the erratic behavior exhibited by both the turn and reward curves, which results from both values being computed over individual simulated dialogue sessions. To better assess how much the dialogue policy improves after each training episode, average values over a large set of simulated dialogues must be computed instead. This is illustrated in Figure 15.6, which is left to you as an exercise (see 15.4-12).

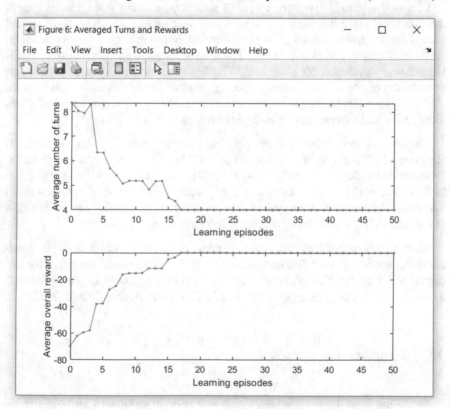

Fig. 15.6. Average number of turns and average rewards versus training episodes

15.3 Further Reading

Question Answering is a very active and broad field. A pair of brief but useful reviews about question answering systems and question types is found in (Pundge *et al.* 2016) and (Reddy and Madhavi 2017). For a more comprehensive introductions to both restricted domain and open domain question answering, see (Mollá and Vicedo 2007) and (Strzalkowski and Harabagiu 2008), respectively. A more recent survey can be found in (Bouziane *et al.* 2015).

Some commonly used approaches to question answering include systems that leverage structured knowledge sources such as knowledge graphs and ontologies (Frank *et al.* 2007), (Athira *et al.* 2013) and (Zou *et al.* 2014), retrieval-based systems (Moldovan and Surdeanu 2003), (Kolomiyets and Moens 2011) and (Bao *et al.* 2018), as well as interactive question answering (Quarteroni and Manandhar 2009), visual question answering (Teney *et al.* 2017), and more recent deep learning based approaches, (Yang *et al.* 2019) and (Pîrtoacă *et al.* 2019).

Dialogue systems have gained a lot of attention in recent years, mainly due to their practical feasibility resulting from recent advances in speech recognition and natural language understanding. A good introduction to this topic is given in chapter 24 of (Jurafsky and Martin 2009). The widely accepted speech act theory of dialogue was developed by (Austin 1962). Some early works on statistical approaches to dialogue as a decision-making process can be found in (Levin *et al.* 1998), (Young 2002) and (Williams and Young 2007), and a good general overview on statistical dialogue modelling is provided in (Gašić 2011).

The foundations of the Markov Decision Process framework can be found in (Bellman 1957), and further discussion on Partially Observable Markov Decision Processes is available in (Kaelbling *et al.* 1998). For a comprehensive introduction to Reinforcement Learning, see (Sutton and Barto 1998). Good discussions about Temporal Difference Learning and, more specifically, *Q*-learning can be found in (Tesauro 1995) and (Watkins and Dayan 1992), respectively.

More recent data-driven approaches (Serban et al. 2018), which typically model dialogue systems as end-to-end architectures, use deep neural networks for implementing task-oriented systems (Eric *et al.* 2017) and (Liu *et al.* 2018), as well as non-task-oriented systems (Vinyals and Le 2015) and (Humeau *at al.* 2020).

15.4 Proposed Exercises

1. Study the function `get_intent`, which was used in section 15.1 to classify input questions according to their intent, and the dataset `qatable.csv` used to

train the classification model `intent_classifier.mat` in `get_intent` (all function, data and model files are available in the companion website of the book).

- Column `testset` in `qatable.csv` consists of a binary flag, which indicates whether the given sample belongs to the train (`0`) or the test (`1`) partition. Use the questions in the test partition to evaluate the classifier.
- Use the same data partition to train and evaluate a new classifier for detecting the type of input questions according to the information in column `type` (*request* vs *confirm*). Create the function `get_type` accordingly.
- Look at *request* questions belonging to the *ask_for_restaurant* intent category in `qatable.csv`. Notice all these questions are of the general form: *is/are there any ... restaurant(s) ... ?* Test `get_intent` with questions like *can you recommend a Mexican restaurant in San Francisco?* What is the predicted intent? Can you find other examples with wrong predictions?
- Create a script to generate ~*100* new *request* questions with intent category *ask_for_restaurant* by using different wordings such as *can you find ... ?*, *please recommend ... ?*, etc. (you can just modify existing questions). Add these questions to the train set, and train and evaluate a new model.

2. Consider questions of the form: *what are the address and contact number of [restaurant]?, what type of food is served at [restaurant] and where is it located?*, etc. in which more than one intent are combined.

   - Consider the function `fitcecoc`, which is used to train multiclass classifiers, and study how to use its `'FitPosterior'` parameter to train a model that returns prediction probabilities for the different categories.
   - Use the `qatable.csv` dataset to train a new model that returns prediction probabilities for the seven available intent categories. How would you use this new model to handle multi-intent questions?
   - Notice your new model still implements a multiclass classifier, while this particular problem must be actually tackled with a multilabel classifier!
   - Train a multilabel classifier by following the steps in https://www.math works.com/help/deeplearning/ug/multilabel-text-classification-using-deep-learning.html Accessed 6 July 2021.

3. Study the function `property_match`, which was used in section 15.1 to identify potential informable field values mentioned in the input question (the function is available in the companion website of the book).

   - As observed from the function, it only accounts for exact string matching between the individual tokens in the value list and the ones in the question. Consequently, similar tokens and misspellings cannot be identified!
   - Consider methods to perform fuzzy or partial string matches, such as computing the edit distance between tokens or the cosine similarity over character frequency vectors, etc.

- Select the method of your preference and implement a new version of `property_match` that accounts for partial string matches. Modify the existing scoring mechanism to include the partial matching score too.
- How does the new implementation improve the performance of the question answering system? Can you illustrate it with some examples?

4. Consider questions of the type: *are there any spanish or seafood restaurants in San Francisco?*, *are there japanese restaurants in Oakland and Berkeley?*, etc. in which multiple values of the same informable field are provided.

   - Implement a new version of function `find_restaurant`, previously used in (15.14b), such that it can handle questions including multiple values of the same informable fields.
   - Notice the function must first apply logical *OR* operations between index vectors corresponding to different values of the same variable, followed by logical *AND* operations among the indexes from different variables.
   - Also notice that as multiple matches are commonly returned by function `property_match`, you need to define and calibrate a score threshold to ensure, as much as possible, that invalid value matches are filtered out.
   - Create a script that provides answers to this type of questions and test its performance with a few different sample questions.

5. Study the Text Analytics Toolbox™ function `mmrScores`, which scores documents with respect to their relevance to a given query by means of the Maximal Marginal Relevance algorithm (Carbonell and Goldstein 1998).

   - Consider the `description` column of table `datatbl` from (15.3a), which contains textual descriptions of the different restaurants in the dataset.
   - Split all descriptions into sentences and create a document collection with them. Apply standard preprocessing steps such a tokenization, punctuation removal, stopword removal, etc.
   - Propose a set of less structured questions such as *who offers great food and wine pairings?*, *what is the best and freshest seafood in town?*, etc.
   - Use function `mmrScores` to compute scores and rank the sentences according to their relevance to the proposed questions. Look at the top-ranked questions and assess up to what degree they answer the questions.
   - Repeat the previous step by using function `cosineSimilarity` instead of `mmrScores`. What differences and similarities do you observe?
   - Design a methodology for combining the scores generated in the two previous steps. Use the combined scores to select the top-ranked sentences for each question under consideration.
   - How the new results compare to the previous ones? Can you come up with some score threshold values to determine whether the retrieved sentences are actually answering the questions?

6. Create a function `qamodule` that implements the complete question answering system for the restaurant domain described by pieces in section 15.1.

   – The function should receive as inputs a restaurant `data` structure, like the one in from (15.1a), and the asked `question` in string format.
   – The function should return as outputs, the `answer` in string format and the structure `outputs` that contains the intermediate outputs of the system (i.e. `intent`, `constraints`, `nresponses`, `result`, `template`, and `answer`).
   – Breakdown the overall implementation into three specific submodules for question understanding, search and ranking, and answer generation.
   – Design the workflow in such a way that the constraints are defined depending on whether the question intent is *ask_for_restaurant* or not, and the template is selected depending on the intent and the retrieved response.
   – Add some post-edition rules to account for additional ungrammatical constructions that might result during the template filling in step. Consider, for instance, fixing segments of text such as *pizza food*, *seafood food*, *price range moderate-price*, etc.
   – Test your implemented function with the sample sentences presented in the different examples and tables in section 15.1

7. Consider a user-initiative informational dialogue system, in which the user asks questions to the system about the requestable fields in the dataset. Such a type of dialogue system, which is also commonly referred to as interactive question answering, must be able to handle follow up questions.

   – Design and implement additional logic to handle follow up questions. Notice this requires maintaining a set of context variables from the previous question-answer pair to make it possible to understand the new question.
   – Consider, for example, questions like *and where is it located?*, which follows up on the previously mentioned restaurant, or *what about [restaurant]?*, which asks about the same feature previously asked about, but on a different restaurant.
   – Also, the system should be able to understand indirect references to restaurants in a previously provided list of results. Consider, for instance, questions such as *what is the contact number of the second one?*, or *what about the last one?*
   – Create a dataset of question pairs, where each pair contains an initial question and a follow up question. Use the dataset to evaluate your implemented follow up question handling strategy.

8. Let us consider the case of a system-initiative transactional dialogue system for making a restaurant reservation.

- Assume that, at the dialogue initial state, the user intent for making a reservation has been already confirmed and the specific restaurant for the intended reservation is already known.
- The system must collect three different slot values from the user: the desired date, the time of the day and the number of people attending.
- The system must use a registry to confirm for table availability given the
desired date and time, depending on which the system should confirm the
reservation or request the user for a different date and time.
- The user should be able to cancel the reservation process if the reservation
cannot be made for the desired date and time.
- Describe the state diagram for the proposed reservation system, define a
reward function and implement both user and system response functions.
- Implement a reinforcement learning procedure, like the one in (15.25), to
learn a dialogue policy. Is the learned policy working as expected?

9. Let us consider the case of a non-goal-oriented dialogue system, in which there
is not a specific task to be completed. In this case, the system is just expected to
maintain a meaningful conversation with the user.

- Consider the retrieval-based dialogue system described in (Banchs and Li,
2012), and the Cornell Movie Dialogue Corpus (Danescu-Niculescu-Mizil
and Lee 2011) available from https://www.cs.cornell.edu/~cristian/Cornell
_Movie-Dialogs_Corpus.html Accessed 6 July 2021.
- Process the corpus to extract the collection of dialogues (i.e. the sequences
of consecutive interactions between two characters) and build two TF-IDF
models: one for each individual utterances in the collection and one for the
corresponding previous dialogue histories.
- Use the constructed models to implement a chat-oriented retrieval-based
system. For a given user input, the system must find the best match in the
collection of dialogues and use its response to reply to the user.
- The best match is selected according to the combination of two cosine similarity scores: one computed at the utterance level and the other computed
at the previous dialogue history level.
- Interact with the implemented chatbot by using different sets of weights to
combine the two cosine similarity scores. Conduct a manual assessment of
the system performance. What are your main conclusions?

10. The use of a well-behaved simulated user in section 15.2 produced the state
transitions of the simulated dialogues to be deterministic in nature. This means
that the next state is always uniquely determined by the current state and the
action taken by the system. Let us consider the case of a bad-behaved user.

- Study the function `user_response`, which was introduced in (15.24), and
create a new function `rnd_user_response` such that it simulates a probabilistic user behavior regarding preferences as described next.

- When prompted for a preference (e.g. *food*) it replies correctly with probability *0.5*, it replies with any of the other two possible preferences (e.g. *location* and *price*) with probability *0.2* each, or it replies with a meaningless answer with probability *0.1*.
- Notice this bad-behaved user implies that the state transitions are now probabilistic. Can you draw the corresponding state diagram (similar to the one illustrated in Figure 15.4) for dialogues with this user?
- Notice also that, because of the probabilistic nature of the state transitions, the slot filling step in the inner loop of (15.25b) must be modified as the action taken by the system is not a valid slot index anymore.
- To compute the index of the slot to be updated and its corresponding update value for a given user response, you need to either implement a separated NLU logic or, alternatively, built it into `rnd_user_response`. In the latter case, the function must return the slot index as an additional output.
- Use your new simulated user and NLU logic to simulate a few dialogues with the final `Qmtx` learned in (15.25). Look at the simulated dialogues an assess the performance of the dialogue system. What do you observe?
- Use your new simulated user and NLU logic to learn a new dialogue policy by using Q-learning, like it was done in (15.25). Generate plots, like the ones in Figure 15.5, to display the number of turns and cumulative rewards versus the training episodes. What are your main conclusions?
- Compare the newly learned dialogue policy with the one from (15.27b). How do they compare to each other? Can you explain the result?

11. Consider the `Qmtxs` cell array from (15.25b), which stores the status of the action selection matrix `Qmtx` at the end of each training episode.

- Compute the norms of the difference between consecutive pairs of `Qmtx` for all training episodes and generate a plot of the norm versus the training episodes. Can you confirm the observation, previously derived from Figure 15.5, that `Qmtx` converged at training episode *18*?
- Repeat the reinforcement learning training exercise conducted in (15.25) multiple times by using different initialization of the random number generator at each time. Consider, for instance, calling function `rng(n)` before each training session with different values of $n=1,2\ 3,...50$
- Compute the number of training episodes at which convergence occurs in each case. What are the minimum and maximum number of training episodes observed? What are the mean and standard deviation values?

12. Create a script to generate Figure 15.6, in which the average values of turns and rewards, computed over a large number of simulated dialogues, are plotted against the number of training episodes.

- Consider the initial action selection matrix `Qmtx = zeros(16,4)`, defined in (15.25a), as well as all subsequent updates stored in cell array `Qmtxs` at the end of each training episode in (15.25b).
- Simulate a large number of dialogues for each `Qmtx` and compute, also for each `Qmtx`, the average number of turns needed to complete each dialogue (you might want to increase the value of `maxturns` to avoid truncating some of the simulated dialogues), and the average overall reward.
- To speed up the simulations you might omit the calls to `system_response` and `user_response`, as only the overall reward value and the total number of turns are needed for each of the simulated dialogues.
- Reproduce the plots depicted in Figure 15.6. To obtain the exact same results, you must: reset the random generator `rng('default')`, set `maxturns` to *100*, and simulate *10,000* dialogues for each `Qmtx`.

13. In the reinforcement learning implementation in (12.25), the action selection function $Q(s,a)$ is updated at each turn in each dialogue. Accordingly, the dialogue policy constantly changes during each episode. In an alternative modality of reinforcement learning, known as episodic learning, $Q(s,a)$ is kept constant during each episode and updated only after each episode is concluded.

- Study Algorithm 2 in (Gašić 2011), which describes an episodic learning approach known as the Monte Carlo Control Algorithm.
- Implement a reinforcement learning script, similar to the one in (15.25), and learn a new dialogue policy by means of episodic learning.
- Generate plots of turns and rewards versus episodes, similar to the ones in Figures 15.5 and 15.6, for the new training session. How do the new plots compare to the ones from the Q-learning exercise in (15.25)?
- Explore some of the dialogues produced at different episodes during the learning process. What are your main observations?

15.5 Short Projects

1. Consider a knowledge graph representation of the restaurant dataset used in section 15.1. In this project, you are required to build a triplestore for the dataset and implement a question answering system that operates with it.

- To construct the triplestore, you need to represent each value in the table `datatbl` from (15.3a) in the form of a triple *entity–property–value*, where *entity* is a restaurant name, *property* is any of the variables other than restaurant (*cuisine, location, price*, etc.) and *value* is the corresponding entry in the table (*Mediterranean, San Francisco, expensive*, etc.)

- Notice that, in `datatbl`, the *cuisine* variable is multivalued. However, triples are not multivalued. Then, for those restaurants with two or more *cuisine* values you will need to generate as many triples of the form *[restaurant]–cuisine–[value]* as *cuisine* values are available.
- While both question understanding and answer generation components remain similar to the ones described in section 15.1, the search mechanism is different in this case.
- Consider the sample question from (15.4): *what kind of food is served at Sandrino Pizza?* To retrieve the informational elements needed to answer this question, the search mechanism needs to find all triples in the triplestore that match the triple template *Sandrino_Pizza–cuisine–[?]*
- Similar search strategies should be implemented for all other question intents except for the case of the *ask_for_restaurant* intent. Consider, for instance, the sample question in (15.10a): *is there any Japanese restaurant in Oakland?* In this case, first, the search mechanism must find triples of the forms: *[?]–cuisine–Japanese* and *[?]–location–Oakland*. Then, the restaurant intersection set should be computed to derive the final answer.
- Implement a full question answering system for the restaurant domain by using the triplestore search approach. Test the system with the sample sentences presented in the different examples and tables in section 15.1.

2. Consider the dialogue system described in section 15.2 along with those from exercises 15.4-7, 15.4-8 and 15.4-9. In this project you are required to build a complete restaurant reservation system by combining those four systems. The reservation system should work as follows:

- First, the system should collect the user preferences and search for the set of restaurants that satisfy such constraints in a similar way the system described in section 15.2 does it.
- Different from the system in 15.2, in this case, the system must provide the names of the restaurants if the number of restaurants retrieved in the search is equal or greater than one but equal or less than four.
- If the number of restaurants is larger than four, the system should inform about the total number found and provide the names of four of them.
- If the system does not find any restaurant it might either try to find some by relaxing the constraints or invite the user to start the dialogue again.
- Second, once the first dialogue is concluded the interactive question answering system form exercise 15.4-7 should take over and allow the user to ask questions about the restaurant, or restaurants, previously found.
- During the interactive question answering session, the system must invite the user to inform when she/he is ready to make a reservation.
- Third, when the user requests to make a reservation, the system from exercise 15.4-8 must take over and assist the user to complete the reservation. The dialogue ends when the reservation is completed.

- If the user decides to cancel the reservation the system must invite the user to start the process again. Based on the user response the system should decide whether to end the dialogue or start it over again.
- At any point in the dialogue, the user should be able to make irrelevant or out-of-topic comments (you will need an intent classifier to decide whether the user input is related to the reservation task at hand or not).
- When an irrelevant or out-of-topic comment is detected, the chatbot from exercise 15.4-9 must suggest a response to handle the turn and return the control back to the active sub-dialogue system, which should reply to the user with a composite answer that combines the chatbot response with a nudge inviting the user to go back into the main topic of the dialogue.

15.6 References

Athira PM, Sreeja MP, Reghuraj PC (2013) Architecture of an Ontology-Based Domain-Specific Natural Language Question Answering System, International Journal of Web & Semantic Technology, 4(4): 31–39

Austin JL (1962) How to Do Things with Words, Harvard University Press

Banchs R, Li H (2012) IRIS: a chat-oriented dialogue system based on the vector space model, in proceedings of the Annual Conference of the Association for Computational Linguistics (ACL), system demonstrations, pp: 37–42

Bao J, Duan N, Zhou M, Zhao T (2018) An Information Retrieval-Based Approach to Table-Based Question Answering, Huang *et al.* (eds) Natural Language Processing and Chinese Computing, Lecture Notes in Computer Science, 10619: 601–611

Bellman R (1957) A Markovian Decision Process, Journal of Mathematics and Mechanics 6(5): 679–684

Bouziane A, Bouchiha D, Doumi N, Malki M (2015) Question Answering Systems: Survey and Trends, Procedia Computer Science, 73: 366–375

Carbonell JG, Goldstein J (1998) The use of MMR, diversity-based reranking for reordering documents and producing summaries. In proceedings of *SIGIR* 98: 335–336

Danescu-Niculescu-Mizil C, Lee L (2011) Chameleons in imagined conversations: a new approach to understanding coordination of linguistic style in dialogs. In proceedings of the Workshop on Cognitive Modeling and Computational Linguistics, pp: 76–87

Eric M, Krishnan L, Charette F, Manning CD (2017) Key-Value Retrieval Networks for Task-Oriented Dialogue, in Proceedings of the 18th Annual Meeting of the Special Interest Group on Discourse and Dialogue (SIGDIAL), pp: 37–49

Frank A, Krieger HU, Xu F, Uszkoreit H, Crysmann B, Jörg B, Schäfer U (2007) Question answering from structured knowledge sources, Journal of Applied Logic 5(1): 20–48

Gašić M (2011) Statistical dialogue modelling (Dissertation), University of Cambridge

Humeau S, Shuster K, Lachaux MA, Weston J (2020) Poly-encoders: Transformer architectures and pretraining strategies for fast and accurate multi-sentence scoring, in Proceedings of International Conference on Learning Representations (ICLR)

Jurafsky D, Martin JH (2009) Speech and Language Processing: An Introduction to Natural Language Processing, Computational Linguistics, and Speech Recognition (2nd edition), Pearson Education, Inc. Upper Saddle River, New Jersey

Kaelbling LP, Littman ML, Cassandra AR (1998) Planning and acting in partially observable stochastic domains, Artificial Intelligence, 101: 99–134

Kolomiyets O, Moens MF (2011) A survey on question answering technology from an information retrieval perspective, Information Sciences, 181(24): 5412–5434

Levin E, Pieraccini R, Eckert W (1998) Using Markov decision process for learning dialogue strategies, in Proceedings of the International Conference on Acoustics, Speech and Signal Processing (ICASSP), pp: 201–204

Liu B, Tur G, Hakkani-Tur D, Shah P, Heck L (2018) Dialogue Learning with Human Teaching and Feedback in End-to-End Trainable Task-Oriented Dialogue Systems, in Proceedings of the Conference of the North American Chapter of the Association for Computational Linguistics (NAACL), pp: 2060–2069

Moldovan D, Surdeanu M (2003) On the Role of Information Retrieval and Information Extraction in Question Answering Systems, Pazienza M.T. (eds) Information Extraction in the Web Era, Lecture Notes in Computer Science, 2700: 129–147

Mollá D, Vicedo JL (2007) Question Answering in Restricted Domains: An Overview, Computational Linguistics 33(1): 41–61

Pîrtoacă GS, Rebedea T, Ruşeţi Ş (2019) Improving Retrieval-Based Question Answering with Deep Inference Models. In Proceedings of International Joint Conference on Neural Networks (IJCNN), pp: 1–8

Pundge AM, Khillare SA, Mahender CN (2016) Question Answering System, Approaches and Techniques: A Review, International Journal of Computer Applications, 141(3): 34–39

Quarteroni S, Manandhar S (2009) Designing an interactive open-domain question answering system, Natural Language Engineering, 15(1): 73–95

Reddy ACO, Madhavi K (2017) A Survey on Types of Question Answering System, IOSR Journal of Computer Engineering 19(6): 19–23

Serban IV, Lowe R, Henderson P, Charlin L, Pineau J (2018) A survey of available corpora for building data-driven dialogue systems, Dialogue & Discourse, 9(1): 1–49

Strzalkowski T, Harabagiu S (Eds.) (2008) Advances in Open Domain Question Answering, Text, Speech and Language Technology 32: 568, Springer Netherlands

Sutton RS, Barto AG (1998) Reinforcement Learning: An Introduction, Adaptive Computation and Machine Learning, MIT Press, Cambridge, Massachusetts

Teney D, Wu Q, van den Hengel A (2017) Visual Question Answering: A Tutorial, in *IEEE Signal Processing Magazine*, 34(6): 63–75

Tesauro G (1995) Temporal Difference Learning and TD-Gammon, Communications of the ACM, 38(3): 58–68

Vinyals O, Le QV (2015) A neural conversational model, in Proceedings of Deep Learning Workshop at International Conference on Machine Learning (ICML)

Watkins CJCH, Dayan P (1992) Q-learning, Machine Learning 8: 279–292

Williams JD, Young SJ (2007) Partially observable markov decision processes for spoken dialog systems, Computer Speech and Language 21(1): 393–422

Yang W, Xie Y, Lin A, Li X, Tan L, Xiong K, Li M, Lin J (2019) End-to-end open-domain question answering with bertserini. In Proceedings of the North American Association for Computational Linguistics (NAACL), pp: 72–77

Young SJ (2002) Talking to machines (statistically speaking), in Proceedings of the International Conference on Spoken Language Processing (ICSLP), pp: 9–16

Zou L, Huang R, Wang H, Yu JX, He W, Zhao D (2014) Natural language question answering over RDF: a graph data driven approach. In Proceedings of the ACM SIGMOD International Conference on Management of Data, pp: 313–324

Printed in the United States
by Baker & Taylor Publisher Services